ENZYBIOTICS

ENZYBIOTICS
Antibiotic Enzymes as Drugs and Therapeutics

Edited by

TOMAS G. VILLA
School of Biotechnology
University of Santiago de Compostela

and

PATRICIA VEIGA-CRESPO
Department of Microbiology
Faculty of Pharmacy
University of Santiago de Compostela

A JOHN WILEY & SONS, INC., PUBLICATION

Published by John Wiley & Sons, Inc., Hoboken, New Jersey
Published simultaneously in Canada

For general information on our other products and services or for technical support, please contact our Customer Care Department within the United States at (800) 762-2974, outside the United States at (317) 572-3993 or fax (317) 572-4002.

Wiley also publishes its books in a variety of electronic formats. Some content that appears in print may not be available in electronic formats. For more information about Wiley products, visit our web site at www.wiley.com.

Library of Congress Cataloging-in-Publication Data
Enzybiotics : antibiotic enzymes as drugs and therapeutics / [edited by] Tomas G. Villa, Patricia Veiga Crespo.
    p. ; cm.
  Includes index.
  ISBN 978-0-470-37655-3 (cloth)
  1. Enzymes–Therapeutic use. 2. Antibacterial agents. I. Gonzalez Villa, Tomas. II. Veiga Crespo, Patricia.
  [DNLM: 1. Anti-Bacterial Agents–pharmacology. 2. Anti-Bacterial Agents–therapeutic use. 3. Bacteriophages–enzymology. 4. Bacteriophages–genetics. 5. Drug Resistance–drug effects. QV 350 E612 2010]
  RM666.E55E575 2010
  615′.35–dc22
                                                                                    2009025908

Printed in the United States of America

10 9 8 7 6 5 4 3 2 1

CONTENTS

PREFACE

The writing of a preface on an old-yet-new subject, such as the one in this book, is always a difficult task. It is common knowledge that the discovery of penicillin, and subsequently the rest of the antibiotics, has probably been one of the most important scientific contributions to civilization. By saving millions of lives, antibiotics automatically increased the half-life of mankind, thus allowing scientists to give their best to society for 20 to 30 additional years.

It is also common knowledge that resistance to antibiotics is a constant possibility and unfortunately something to be considered every time a new antibiotic goes on the market. Because of this, and because the discovery and design of new antibiotics becomes more and more difficult every year, society, through the work of several worldwide research groups, is looking into the use of what one of us (Dr. Vincent Fischetti) has termed "enzybiotics" (the result of blending the words "enzymes" and "antibiotics"), for treating bacterial and fungal diseases, either alone or in combination with antibiotics.

The book starts with four chapters in which the potential, advantages, and phylogeny of enzybiotics are reviewed. Then, the new ways of controlling infections by Gram-negative bacteria and an updated view of bacteriophage holins are presented. After a review of anti-staphylococcal lytic enzymes, the book goes on to discuss membrane-targeted enzybiotics, as well as the design of phage cocktails for current therapy. Finally, the last two chapters deal respectively with the novel methods to identify new enzybiotics and the use of modified phages to induce suicide in bacteria.

All in all, the contributors are all active researchers, involved in the topic of enzybiotics. It is hoped that the joining of different points of view, such as those reflected in this book, will help to clarify the emerging field of enzybiotics and to consolidate the idea that the therapies mediated by these compounds may contribute to the relief of pain and to the control of contagious diseases.

Santiago de Compostela, Spain   TOMAS G. VILLA
December 24, 2008   PATRICIA VEIGA-CRESPO

CONTRIBUTORS

Juan C. Alonso   Department of Microbial Biotechnology, Centro Nacional de Biotecnología, CSIC, 28049 Madrid, Spain; jcalonso@cnb.csic.es or jcalonso@cnb.uam.es

Jan Borysowski   Department of Clinical Immunology, Transplantation Institute, Warsaw Medical University, Poland; jborysowski@interia.pl

Vincent A. Fischetti   Professor and Head Laboratory of Bacterial Pathogenesis, Rockefeller University, 1230 York Avenue, New York, NY 10021, USA; vaf@rockefeller.edu OR vaf@mail.rockefeller.edu

María Gasset   Instituto de Química-Física "Rocasolano," Consejo Superior de Investigaciones Científicas, Serrano 119, 28006, Madrid, Spain; mgasset@iqfr.csic.es

Lawrence Goodridge   Department of Animal Sciences, Colorado State University, Fort Collins, CO 80523, USA; Lawrence.Goodridge@ColoState.edu

Andrzej Górski   Department of Clinical Immunology, Transplantation Institute, Warsaw Medical University, and Laboratory of Bacteriophages, L. Hirszfeld Institute of Immunology and Experimental Therapy, Wrocław, Poland; agorski@ikp.pl

Jonathan E. Schmitz   Laboratory of Bacterial Pathogenesis and Immunology, The Rockefeller University, 1230 York Avenue, Box 172, New York, NY 10065, USA; jschmitz@rockefeller.edu

Raymond Schuch   Laboratory of Bacterial Pathogenesis and Immunology, The Rockefeller University, 1230 York Avenue, Box 172, New York, NY 10065, USA

Marcelo E. Tolmasky   Center for Applied Biotechnology Studies, Department of Biological Science, College of Natural Science and Mathematics, California State University Fullerton, Fullerton, CA 92834-6850, USA; mtolmasky@Exchange.fullerton.edu

Patricia Veiga-Crespo   Department of Microbiology, Faculty of Pharmacy, University of Santiago de Compostela, Spain; patricia.veiga@usc.es

Tomas G. Villa   Department of Microbiology, Faculty of Pharmacy, University of Santiago de Compostela, and School of Biotechnology, University of Santiago de Compostela, Spain; tomas.gonzalez@usc.es

CHAPTER 1

ENZYBIOTICS AND THEIR POTENTIAL APPLICATIONS IN MEDICINE

JAN BORYSOWSKI[1] and ANDRZEJ GÓRSKI[1,2]
[1]Department of Clinical Immunology, Transplantation Institute, Warsaw Medical University, Poland
[2]Laboratory of Bacteriophages, L. Hirszfeld Institute of Immunology and Experimental Therapy, Wrocław, Poland

1. INTRODUCTION

Over the last decade, a dramatic increase in the prevalence of antibiotic resistance has been noted in several medically significant bacterial species, especially *Pseudomonas aeruginosa*, *Acinetobacter baumanii*, *Klebsiella pneumoniae*, as well as *Staphylococcus aureus*, coagulase-negative staphylococci, enterococci, and *Streptococcus pneumoniae* (Hawkey 2008). This unfavorable situation is further aggravated by a shortage of new classes of antibiotics with novel modes of action that are essential to contain the spread of antibiotic-resistant pathogens (Livermore 2004). In fact, some infectious disease experts have expressed concerns that we are returning to the pre-antibiotic era (Larson 2007). Therefore, there is an urgent need to develop novel antibacterial agents to eliminate multidrug-resistant bacteria (Breithaupt 1999). A very interesting class of novel (at least in terms of their formal clinical use) antibacterials are enzybiotics.

The term "enzybiotic" was used for the first time in a paper by Nelson et al. (2001) to designate bacteriophage enzymes endowed with bacterial cell wall-degrading capacity that could be used as antibacterial agents. While some authors suggest that this name should refer to all enzymes exhibiting antibacterial and even antifungal activity (Veiga-Crespo et al. 2007), in this chapter we will discuss only bacterial cell wall-degrading

Enzybiotics: Antibiotic Enzymes as Drugs and Therapeutics. Edited by Tomas G. Villa and Patricia Veiga-Crespo
Copyright © 2010 John Wiley & Sons, Inc.

enzymes (regardless of their source). Other names that are used with respect to enzybiotics are lytic enzymes and peptidoglycan hydrolases. The latter refers to the major mode of action of enzybiotics, that is, the enzymatic cleavage of peptidoglycan covalent bonds, which results in the hypotonic lysis of a bacterial cell. Peptidoglycan hydrolases constitute an abundant class of enzymes and may be obtained from different sources, for instance, bacteriophages (lysins) and bacteria themselves (bacteriocins and autolysins). Yet another example of well-known enzybiotics are lysozymes, including hen egg white lysozyme and human lysozyme (a list of representative enzybiotics is shown in Table 1.1).

In view of the ever-increasing antibiotic resistance of bacteria, the most important characteristics of enzybiotics are a novel mode of antibacterial action, different from those typical of antibiotics, and the capacity to kill antibiotic-resistant bacteria (Borysowski et al. 2006). Another significant feature of some lytic enzymes is the low probability of developing bacterial resistance (in some cases, the development of enzybiotic resistance results in a reduction in bacterial fitness and virulence; Kusuma et al. 2007).

The goal of this chapter is to discuss the major groups of enzybiotics, including lysins, bacteriocins, autolysins, and lysozymes, in the context of their potential medical applications.

2. LYSINS

2.1. General Features

Lysins or endolysins are double-stranded DNA bacteriophage-encoded enzymes that cleave covalent bonds in peptidoglycan (Borysowski et al. 2006; Fischetti 2008). They are naturally produced in phage-infected bacterial cells during the course of lytic cycle. At the last stage of the cycle, endolysin molecules degrade peptidoglycan, thereby causing lysis of the bacterial cell and ensuring the release of progeny virions (Young et al. 2000). The term "endolysin" was introduced to the scientific literature by F. Jacob and C. R. Fuerst to stress that enzyme molecules act on peptidoglycan from within the bacterial cell in which they are synthesized (Jacob and Fuerst 1958). In view of this, it appears that recombinant enzymes acting on the cell wall from outside the cell (e.g., those used for therapeutic purposes) should be referred to as lysins rather than endolysins. Still another name proposed to designate a lysin is "virolysin," which is intended to point out the viral origin of these enzymes (Parisien et al. 2008). However, this name has not gained popularity and is used very rarely.

TABLE 1.1. A List of Representative Enzybiotics

Enzybiotic Name	Enzybiotic Class	Source	Enzymatic Specificity	Antibacterial Range	Reference
PlyC	Lysin	Phage C1	Amidase	*S. pyogenes*, groups C and E streptococci	Nelson et al. 2006
Pal	Lysin	Phage Dp-1	Amidase	*S. pneumoniae*	Loeffler et al. 2001
Cpl-1	Lysin	Phage Cp-1	Muramidase	*S. pneumoniae*	Loeffler et al. 2003
PlyGBS	Lysin	Phage NCTC 11261	Endopeptidase muramidase	*S. agalactiae*, groups A, C, G, L streptococci	Cheng et al. 2005
Phage B30 lysin	Lysin	Phage B30	Endopeptidase muramidase	*S. agalactiae*, groups A, B, C, E, G streptococci; *E. faecalis*	Baker et al. 2006
LambdaSa1 prophage lysine	Lysin	Prophage LambdaSa1	Endopeptidase	?	Pritchard et al. 2007
LambdaSa2 prophage lysine	Lysin	Prophage LambdaSa2	Endopeptidase glucosaminidase	*S. pyogenes, S. dysgalactiae*, group E streptococci, *S. equi*, group G streptococci, *S. agalactiae*	Pritchard et al. 2007
PlyG	Lysin	Phage γ	Amidase	*B. anthracis*	Schuch et al. 2002
PlyL	Lysin	Prophage λBa02	Amidase	*B. cereus*	Low et al. 2005
				B. anthracis	
PlyPH	?	?	?	*B. anthracis*	Yoong et al. 2006
PlyB	Lysin	Phage BcpI	Muramidase	*B. anthracis*	Porter et al. 2007
Ply118	Lysin	Phage A118	Peptidase	*Listeria*	Loessner et al. 2002
Ply500	Lysin	Phage A500	Peptidase	*Listeria*	Loessner et al. 2002

3

TABLE 1.1. Continued

Enzybiotic Name	Enzybiotic Class	Source	Enzymatic Specificity	Antibacterial Range	Reference
Ply3626	Lysin	Phage Ø3626	Amidase	*C. perfringens*	Zimmer et al. 2002
PlyV12	Lysin	Phage Φ1	Amidase	*E. faecalis, E. faecium, S. pyogenes*, group B, C, E, G streptococci	Yoong et al. 2004
Lyt A	Autolysin	*S. pneumoniae*	Amidase	*S. pneumoniae*	Rodriguez-Cerrato et al. 2007
lysostaphin	Bacteriocin	*S. simulans*	Endopeptidase	*S. aureus*, coagulase-negative staphylococci	Patron et al. 1999
zoocin A	Bacteriocin	*S. equi*	Endopeptidase	*S. equi, S. pyogenes, S. mutans, S. gordonii*	Simmonds et al. 1995
hen egg white lysozyme	Lysozyme	Hen's egg white Ø3626	Muramidase	Gram-positive bacteria	Sava 1996 et al. 2002

The table does not include staphylococcal phage lysins that are discussed in Chapter 7.

The main mode of antibacterial action of lysins is the enzymatic cleavage of the covalent bonds in peptidoglycan. Depending on their enzymatic specificities, lysins fall into five major classes: N-acetylmuramoyl-L-alanine amidases, endopeptidases, N-acetylmuramidases (lysozymes), endo-β-N-acetylglucosaminidases, and lytic transglycosylases (Fig. 1.1).

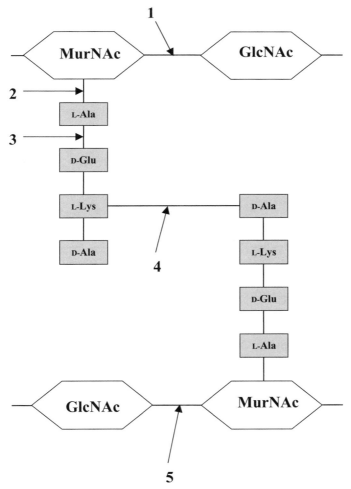

Figure 1.1. Sites of peptidoglycan cleavage by main classes of enzybiotics. This variant of peptidoglycan is typical of *S. aureus*. The backbone of peptidoglycan consists of alternating residues of N-acetylglucosamine (GlcNAc) and N-acetylmuramic acid (MurNAc). The tetrapeptide side chains branching off from N-acetylmuramic acid are cross-linked by the pentaglycine bridges. The sites of cleavage by enzybiotics with different enzymatic specificities are indicated by the numbered arrows: (1) muramidases and transglycosylases; (2) amidases; (3 and 4) endopeptidases; (5) glucosaminidases.

The majority of lysins described to date exhibit only one kind of muralytic activity, whereas relatively few possess two separate enzymatic domains (Borysowski et al. 2006).

While the main mode of antibacterial activity of lysins is based on enzymatic cleavage of peptidoglycan, it is noteworthy that some of them can also affect bacterial cells by a nonenzymatic mechanism. This mechanism relies on destabilization of the bacterial cytoplasmic membrane by amino acid sequences whose properties, especially their amphipathic secondary structure, positive charge, and hydrophobicity, are similar to those found in cationic antimicrobial peptides (CAPs). Such sequences were identified in T4 phage lysozyme and lysins encoded by *Pseudomonas aeruginosa* phages D3 and ΦKZ (Düring et al. 1999; Rotem et al. 2006). In a series of elegant experiments, these sequences were shown to be more important for T4's antibacterial activity than the enzymatic cleavage of peptidoglycan (Düring et al. 1999). As mentioned above, sequences having physicochemical characteristics typical of CAPs (X1 and Z1) are also contained within lysins encoded by two *P. aeruginosa* phages (Rotem et al. 2006). It was shown that synthetic peptides with amino acid sequences corresponding to X1 and Z1, as well as their shorter analogs, inhibited the growth of several Gram-positive bacterial species in a mechanism analogous to that of CAPs. The authors of the study suggested that endolysins of other phages could also be a source of novel antimicrobial peptides. Another unusual enzyme is the *Bacillus amyloliquefaciens* bacteriophage auxiliary lysin lys1521, whose positively charged C-terminal sequences were shown to increase the permeability of the *P. aeruginosa* outer membrane, thereby facilitating the access of the N-terminal enzymatic domain to peptidoglycan (Muyombwe et al. 1999; Orito et al. 2004).

A typical feature of lysins is their modular structure, which means that they are composed of at least two distinct domains: an N-terminal catalytic domain and a C-terminal bacterial cell wall-binding domain (Loessner et al. 2002; Loessner 2005). As mentioned above, some lysins possess two different catalytic domains. In some lysins both the catalytic and the cell wall-binding domain are indispensable for their lytic activity, while others can lyse bacteria also in their C-truncated forms, although it is the C-terminal domain that is responsible for binding to the bacterial cell wall. Interestingly, lysins were also reported to exhibit higher antibacterial activity after removing their C-terminal domains. These findings are very important because they indicate that the antibacterial activity of some lysins could be increased by simply removing their cell wall-binding domains (Borysowski et al. 2006).

Another typical feature of the vast majority of lysins described to date is a narrow antibacterial range when acting on the bacterial cell from outside. This range is usually limited to the host bacterial species of the bacteriophage encoding the given enzyme. However, it needs to be stressed that lysins are most often capable of killing the majority of strains within a given bacterial species (Fischetti 2008). For instance, Pal, an amidase encoded by *S. pneumoniae* phage Dp-1, was shown to lyse 15 out of 15 pneumococcal strains tested (Loeffler et al. 2001). Another lysin, Ply3626 of *Clostridium perfringens* bacteriophage Ø3626, could also kill all of the 48 *C. perfringens* strains tested (Zimmer et al. 2002). This feature clearly sets lysins apart from lytic phages, which are usually capable of infecting and killing only a small subset of strains within a given bacterial species. Very few lysins were reported to possess a broader antibacterial spectrum (Yoong et al. 2004).

A unique medical application of lysins may be the specific elimination of pathogenic bacterial species (e.g., *S. aureus*) colonizing mucous membranes without adversely affecting normal microflora. Such bacteria can, in some clinical settings, be a starting point for infections (Bogaert et al. 2004; Wertheim et al. 2005). Lysins could thus provide a basis for a novel strategy for preventing some bacterial infections. Furthermore, elimination of the mucosal reservoir of bacteria could contribute to containing the horizontal spread of bacterial pathogens in some communities (Fischetti 2003). Lysins appear to be better decolonizing agents than antibiotics owing to their species-specific and rapid antibacterial activity, capacity for killing antibiotic-resistant bacteria, and lower risk of developing resistance (Fischetti 2003; Cheng et al. 2005). Moreover, a considerable body of experimental data shows that lysins, in spite of their apparent immunogenicity, may also be successfully used in the treatment of systemic bacterial infections and are in this regard effective even after repeated administration (Loeffler et al. 2003; Borysowski et al. 2006).

Discussed below are lysins specific to medically significant bacterial species, including *Streptococcus pyogenes* (group A streptococci), *S. pneumoniae*, *Streptococcus agalactiae* (group B streptococci), *S. aureus*, and *Bacillus anthracis*. However, it needs to be stressed that specific lysins can be most likely obtained for any Gram-positive bacterial pathogen from dsDNA bacteriophage (Schuch et al. 2002). Gram-negative bacteria are essentially resistant to recombinant lytic enzymes due to the presence of the outer membrane (see subsection 6.2). Of particular importance is that lysins are also capable of killing antibiotic-resistant bacteria, as shown for penicillin-resistant *S. pneumoniae*

(Loeffler et al. 2001), vancomycin-resistant *Enterococcus faecalis* and *Enterococcus faecium* (Yoong et al. 2004), as well as methicillin-resistant *S. aureus* (MRSA; O'Flaherty et al. 2005) and *S. aureus* strains with reduced susceptibility to vancomycin (Rashel et al. 2007). It was also shown that lysins can act synergistically with other lytic enzymes and antibiotics (Loeffler and Fischetti 2003; Djurkovic et al. 2005; Becker et al. 2008).

2.1.1. Lysins Specific to S. pyogenes

The first and hitherto only lysin specific to *S. pyogenes* that was evaluated as a potential antibacterial agent was PlyC amidase derived from group C streptococci C1 phage (Nelson et al. 2001; Nelson et al. 2006). This enzyme is very interesting in at least two respects. First, it is the most potent lysin reported so far, its activity being over two orders of magnitude higher than those of other bacteriophage lytic enzymes. Second, PlyC is the only known multimeric lysin, while all the others are synthesized as single polypeptides. Although PlyC was first reported in 1957, it was not until 2001 that its antibacterial activity was studied in more detail both *in vitro* and *in vivo*. In fact, it is the first lysin whose activity was studied with a view to potential prophylactic or therapeutic use. *In vitro* experiments revealed that, unlike C1 phage, the enzyme lyses *S. pyogenes* most efficiently, while its activity against groups C and E is substantially lower. All 10 *S. pyogenes* strains tested were efficiently lysed by PlyC. On the other hand, the lysin practically did not act on streptococci groups B, D, F, G, L, and N or other bacterial species with the exception of *Streptococcus gordonii*, which was lysed very slowly. Such an antibacterial range appears to be very advantageous because it is essentially limited to pathogenic streptococci (groups A and C). In a murine model of oral colonization, a single dose of the lysin administered to the oral cavity of mice prior to 10^7 colony forming units (cfu) of group A streptococci resulted in significant protection from the mucosal colonization (only 28.5% of the mice that received PlyC were colonized compared with 70.5% of the animals in the control group). Importantly, in most mice that were colonized despite administration of enzyme, cfu counts remained low throughout the experiment or the bacteria were completely eliminated within 48h, whereas those in the control group increased during the same period of time. In another experiment, no streptococci were detected in oral swabs of nine heavily colonized mice 2h after administration of one lysin dose. However, in some animals recolonization was noted within 48h, which was caused most likely by bacteria previously internalized in epithelial cells of the mucous membrane. Importantly, isolated bacteria were sensitive to

PlyC, which practically rules out resistance to the enzyme's lytic activity as the reason for recolonization.

2.1.2. Lysins Specific to S. pneumoniae

Two lysins are currently being developed as potential anti-pneumococcal agents. The first is Pal amidase, encoded by the pneumococcal bacteriophage Dp-1 (Loeffler et al. 2001). *In vitro* experiments showed that Pal is capable of lysing all of the 15 clinical strains of *S. pneumoniae* tested, each of which represented a distinct serotype. Penicillin-resistant strains were lysed as efficiently as penicillin-sensitive ones. Moreover, it was found that the capsule could not block Pal's access to peptidoglycan. Apart from *S. pneumoniae*, only *Streptococcus oralis* and *Streptococcus mitis* were slightly susceptible to the enzyme's lytic activity, while six other streptococcal species belonging to the oral microflora were resistant. In a murine model of nasopharyngeal colonization, one topical dose of lysin administered to mice 42 h after pneumococci was sufficient to completely clear the bacteria from the surface of the mucous membrane. While the administration of a lower dose of Pal did not result in complete elimination of bacteria in all mice, the titers of surviving pneumococci were too low to successfully recolonize the mucous membrane.

The other anti-pneumococcal enzyme is Cpl-1 muramidase of Cp-1 phage (Loeffler et al. 2003). As was the case with Pal, the antibacterial activity of Cpl-1 is essentially specific to *S. pneumoniae*. In a murine model of nasopharyngeal colonization, a single topical dose of enzyme completely eliminated pneumococci from the mucous membrane. The high effectiveness of Cpl-1 as a topical decolonizing agent was confirmed in very interesting experiments performed on a novel murine model mimicking the natural development of secondary acute otitis media (AOM) following viral infection in children (McCullers et al. 2007). In this unique model, a pneumococcal strain engineered to express luciferase was used, which allowed monitoring infection with the use of bioluminescent imaging, was used. To evaluate the efficacy of Cpl-1, mice were colonized intranasally with bacteria and subsequently infected with influenza virus to trigger a secondary pneumococcal AOM. Administration of two topical doses of Cpl-1 resulted in complete elimination of the bacteria in 90% of the mice, while enzyme buffer administered to the mice in the control group had no effect on intranasal pneumococci. Furthermore, no mouse treated with Cpl-1 developed a secondary AOM following viral inoculation, compared with 80% of mice from the control group. It was also shown that AOM can be prevented not only by a complete elimination of colonization, but also by its partial reduction. The results of this study indicate that anti-pneumococcal lysins could provide a novel means of prophylaxis of secondary AOM in children.

The antibacterial effects of Pal and Cpl-1 were also evaluated in experimental models of different pneumococcal infections, including bacteremia, endocarditis, and meningitidis. In a murine model of pneumococcal bacteremia, a 200-μg dose of either enzymes administered to mice 1 h after inoculation with a lethal dose of multiresistant *S. pneumoniae* rescued 100% of the mice. Cpl-1 and Pal exerted a synergic effect in terms of improving survival rates of the infected mice and synergy was found with different doses and administration times of the enzymes. The antibacterial activity of the enzymes in blood was very rapid, as indicated by a sharp decrease in bacterial titers: ~4 log units 2 h after administration of 200 μg of enzyme. On the 4th and 5th day post-administration, bacteria were either undetectable or their titers in blood were very low, while the mean bacterial titer in the blood of control mice was ~10^7–10^8 cfu/mL. It was also shown that only functional lysin was capable of curing infection, while heat-inactivated enzyme did not have any positive effect. This indicates that the antibacterial effects of the studied enzymes were based on direct killing of bacteria rather than an induction of antibacterial immune response (Jado et al. 2003). The high efficacy of Cpl-1 was confirmed in another study performed on a murine model of bacteremia due to a penicillin-sensitive strain of *S. pneumoniae* (Loeffler et al. 2003).

The antibacterial effects of Cpl-1 were also evaluated in a rat model of endocarditis induced by penicillin-resistant *S. pneumoniae* (Entenza et al. 2005). In this study, two dosing regimens of Cpl-1 were compared. In the first, 16 h after pneumococcal challenge, rats received an intravenous (i.v.) bolus of 10 mg/kg of the enzyme followed by continuous infusion of 5 mg/kg/h for 6 h. This regimen resulted in only a temporary decrease in bacterial titers in blood and failed to reduce vegetation titers. In the other regimen, rats were administered an i.v. bolus of 250 mg/kg followed by continuous infusion of 250 mg/kg/h for 6 h. In this case, the bacteria were cleared from the blood within 30 min and an almost complete eradication was maintained for 6 h. Moreover, a significant decrease in vegetation bacterial titers was noted 30 min after administration of lysin. It was also found that the antibacterial effects of Cpl-1 were much more rapid than those of vancomycin with respect to decreasing bacterial titers in both blood and vegetations (differences between groups that received Cpl-1 and vancomycin were statistically significant at 6 h after administration of either drug). On the other hand, the levels of different pro-inflammatory cytokines in blood were lower in the rats that received vancomycin than in those that were administered Cpl-1.

Cpl-1 was also used in a model of experimental pneumococcal meningitis in infant rats (Grandgirard et al. 2008). Rats were inoculated with pneumococci intracisternally (i.c.) and the enzyme was administered either i.c. or intraperitoneally (i.p.). Following administration of one i.c. dose of Cpl-1 (~20 mg/kg), bacterial titers in the cerebrospinal fluid (CSF) dropped by three orders of magnitude within 30 min. Although pneumococci were essentially undetectable in the CSF for the next 2 h, their titers started to grow shortly thereafter. The half-life of Cpl-1 in the CSF was about 16 min, and the enzyme was present in the CSF for 2 h after administration of the single dose. It is very likely that the efficacy of Cpl-1 could be higher if its bioavailability in the CSF were increased (e.g., by administering repetitive doses). However, the authors failed to verify this experimentally because repetitive injections were too harmful to the rats. In another experiment, infected rats were administered one i.p. dose of Cpl-1 (200 mg/kg), which resulted in a reduction of pneumococcal titers in the CSF by 98% within 2 h. For 3 h after administration, the concentration of Cpl-1 in the CSF was within the range of 7–12 µg/mL.

2.1.3. Lysins Specific to S. agalactiae

2.1.3. Lysins Specific to **S. agalactiae**   Thus far, four lysins derived from *S. agalactiae*-specific phages have been reported. These are bacteriophage NCTC 11261 PlyGBS lysin, B30 phage lysin, and enzymes encoded by LambdaSa1 and LambdaSa2 prophages. While the enzymatic specificity of PlyGBS was not shown directly, it contains two putative catalytic domains: endopeptidase and muramidase. An interesting feature of this lysin is its relatively broad antibacterial range, encompassing, aside from *S. agalactiae*, groups A, C, G, and L streptococci (Cheng et al. 2005).

The second enzyme, phage B30 lysin, was shown to possess two separate enzymatic domains: an N-terminal cysteine, histidine-dependent aminohydrolases/peptidases (CHAP) domain (endopeptidase) and an Acm domain (muramidase) situated in the central part of the protein. Like PlyGBS, bacteriophage B30 lysin can lyse, aside from *S. agalactiae*, other bacteria, including groups A, B, C, E, and G streptococci, as well as *E. faecalis* (Pritchard et al. 2004; Baker et al. 2006).

The last two lysins, that is, those encoded by LambdaSa1 and LambdaSa2 prophages, display γ-D-glutaminyl-L-lysine endopeptidase activity, and the latter also β-D-*N*-acetylglucosaminidase activity (Pritchard et al. 2007; Donovan and Foster-Frey 2008). LambdaSa2 prophage lysin is unusual in that its two-tandem Cpl-7 cell wall-binding domains are situated not at the C-terminus of the polypeptide chain, but rather between the two enzymatic domains. In turbidity reduction assays, the enzyme was found to act potently on several streptococcal species, including *S. pyogenes*, *Streptococcus dysgalactiae*, group E streptococci, *Streptococcus equi*, and group G streptococci, while its lytic activity against *S. agalactiae* was moderate, in spite of the fact that it is encoded by an *S. agalactiae* prophage. Interestingly, a truncated form of the lysin containing the endopeptidase domain and two Cpl-7 domains had higher lytic activity than the full-length enzyme, but only against some bacteria (*S. agalactiae*, *S. dysgalactiae*, *Streptococcus uberis*, and *S. aureus*). A truncated construct containing the endopeptidase domain and one Cpl-7 domain was less active, whereas the endopeptidase domain lacking any cell wall-binding domain was virtually inactive as were all constructs containing only the glucosaminidase domain regardless of the presence or absence of the Cpl-7 domains. An interesting feature of this enzyme is that it maintains substantial lytic activity across a broad range of pH values (5.5–9.5) (Donovan and Foster-Frey 2008).

The major potential application proposed for these enzymes is intra-partum prophylaxis of early onset neonatal infections caused by *S. agalactiae* colonizing the genital tract (Pritchard et al. 2004; Cheng et al. 2005). It appears that lytic enzymes might be in several respects superior to penicillin, which is currently the first-line agent employed in intrapartum antibiotic prophylaxis. Their first advantage is a relatively higher specificity to *S. agalactiae*, especially their lack of activity against species belonging to the vaginal microflora, such as *Lactobacillus acidophilus* and *Lactobacillus crispatus*. Other characteristics of lysins favoring them over antibiotics include their rapid antibacterial activity as well as low probability of developing resistance and causing side effects. Importantly, the optimal pH values for at least some of them fall within the range typical of the human vaginal tract. While the pH optimum of B30 phage lysin (5.5–6.0) is less than the value of the normal vaginal pH (4.5), it does fall within the pH range likely to occur in women heavily colonized with *S. agalactiae* (Pritchard et al. 2004).

The only *S. agalactiae*-specific lysin whose efficacy was evaluated *in vivo* is PlyGBS (Cheng et al. 2005). In a murine model of vaginal colonization, administration of one topical dose of lysin resulted in approximately 3-log decrease in the bacterial level compared with mice in the control group. One topical dose of PlyGBS was also sufficient to significantly reduce bacterial colonization of the oropharynx mucosa. These results are very important in view of the fact that neonatal *S. agalactiae* meningitidis is likely initiated through the oropharynx. Thus it appears that *S. agalactiae*-specific lytic enzymes might be used not only to eliminate vaginal colonization in pregnant women before delivery, but also to decontaminate newborns, thereby decreasing the incidence of neonatal infections. These enzymes could be administered topically in a recombinant form or secreted in the genital tract by engineered bacteria.

2.1.4. Lysins Specific to S. aureus Thus far, several lysins encoded by *S. aureus* phages have been described, including MV-L, LysK, PlyTW, Ply187, and *S. aureus* Ø11 phage lysin (Loessner et al. 1998; Loessner et al. 1999; Navarre et al. 1999; O'Flaherty et al. 2005; Rashel et al. 2007). At least some of them can also lyse, aside from *S. aureus*, coagulase-negative staphylococci (O'Flaherty et al. 2005). Of particular importance is that they are also capable of killing MRSA and *S. aureus* strains with reduced susceptibility to vancomycin (Rashel et al. 2007). Some of them have been successfully used in experimental models of staphylococcal infections. These are discussed in detail in Chapter 7.

2.1.5. Lysins Specific to B. anthracis The first *B. anthracis*-specific lysin tested as a potential antibacterial agent was PlyG (Schuch et al. 2002). This enzyme, a putative amidase, is encoded by *B. anthracis* γ phage, which is used by the U.S. Centers for Disease Control and Prevention (CDC) in Atlanta for the identification of *B. anthracis*. It was found that the enzyme could lyse only *B. anthracis* (of all the 14 isolates tested, some were capsulated) and one *Bacillus cereus* strain (RSVF1) closely related to *B. anthracis*, while several other Gram-positive and Gram-negative bacterial species were resistant. Aside from vegetative bacterial cells, germinating *B. anthracis* spores were also susceptible to PlyG, whereas in the dormant state they were resistant. In a murine model of *B. anthracis* infection, one i.p. injection of PlyG 15 min after inoculation with a lethal dose of RSVF1 cells rescued (depending on the dose of enzyme) 68.4% or 76.9% of mice. PlyG can also be used for the specific identification of *B. anthracis* spores. In this assay, spores exposed to a germinant and lysin release adenosine triphosphate (ATP) that can be measured by means of a handheld luminometer as light emitted in the presence of luciferin/luciferase. This method was shown to be very rapid and allowed for the identification of *B. anthracis* spores within 10–60 min depending on the number of spores.

The second lysin, derived from a *B. anthracis* λ Ba02 prophage, is PlyL amidase (Low et al. 2005). This enzyme is interesting in that it displays more potent activity against *B. cereus* than *B. anthracis*. Remarkably, the lytic activity of the full-length PlyL is lower than that of its C-truncated form, most likely due to some inhibitory effects of the C-terminal domain on the N-terminal enzymatic domain, which could be relieved upon the enzyme's binding to the bacterial cell wall. Interestingly, the removal of the C-terminal domain of another lysin, Ply21 from the *B. cereus* phage TP21, had opposite effects on its capability to lyse different bacterial species. While the C-truncated enzymatic domain of Ply21 displayed higher lytic activity against *Bacillus subtilis*, its ability to lyse *B. cereus* was lower than that of the full-length enzyme (Loessner et al. 1997).

The third lysin capable of lysing *B. anthracis* is PlyPH, an enzyme of putative bacteriophage origin (Yoong et al. 2006). The antibacterial range of this enzyme, like that of PlyG, is practically restricted to *B. anthracis*. The most interesting feature of PlyPH is that it retains lytic activity over a broad range of pH values. While its maximum activity was noted between pH values of 4.5 and 8.0, partial activity was maintained between pH 4.0 and 10.5. PlyPH was also shown to display substantial antibacterial activity in a murine model of peritonitis.

The last enzyme that was reported as a potential means of preventing or treating anthrax is PlyB, a putative muramidase encoded by the BcpI bacteriophage (Porter et al. 2007). The lytic activity of PlyP against a *B. anthracis*-like strain was comparable with that of PlyG. Unlike PlyL, PlyP exhibits its maximum activity in its full-length form, while the C-truncated form is substantially less efficient.

2.1.6. Lysins Specific to Other Bacterial Species   Other lysins that might find use as antibacterial agents are Ply118, Ply500, Ply3626, and PlyV12. Ply118 and Ply500 are L-alanyl-D-glutamate peptidases encoded by *Listeria monocytogenes* phages A118 and A500, respectively (Loessner et al. 2002). The antibacterial range of both enzymes is essentially restricted to the genus *Listeria*, and they are not capable of lysing other Gram-positive or Gram-negative bacteria with the exception of *Bacillus megaterium*. It was also shown that they can lyse bacteria only in their full-length forms, while the removal of either of the two major domains resulted in a loss of lytic capacity.

Ply3626 is a putative amidase encoded by *Clostridium perfringens* bacteriophage Ø3626 (Zimmer et al. 2002). It was shown that the antibacterial range of this enzyme is restricted to the species *C. perfringens*. Interestingly, the lytic spectrum of the enzyme is much broader than that of Ø3626 phage. While the phage can infect and kill only 22% of *C. perfringens* strains, the enzyme was capable of lysing all of the 48 strains tested.

PlyV12 is a putative amidase encoded by *E. faecalis* bacteriophage Φ1 (Yoong et al. 2004). This lysin is very interesting in that it is one of the very few phage lytic enzymes possessing a broad antibacterial range. *In vitro*, PlyV12 was capable of lysing *E. faecalis* and *E. faecium* (all 15 clinical and laboratory strains were tested, including 5 vancomycin-resistant strains) as well as *S. pyogenes* and groups B, C, E, and G streptococci. Moreover, the enzyme displayed weak lytic activity against *S. aureus* and some commensal bacteria.

3. AUTOLYSINS

Another class of lytic enzymes that could be used as enzybiotics are autolysins. These are enzymes encoded by bacteria that are involved in different essential processes of bacterial cells, including cell growth and division, cell wall turnover, bacterial protein secretion, and peptidoglycan maturation (Vollmer et al. 2008). To the best of our knowledge, the first (and hitherto only) autolysin tested as a potential antibacterial agent was LytA amidase, the main autolysin of *S. pneumoniae*. In the first study aimed at evaluating the therapeutic efficacy of LytA, its antibacterial activity was compared with those of Cpl-1 lysin and cefotaxime (Rodriguez-Cerrato et al. 2007). The minimum inhibitory concentration (MIC) values of LytA, Cpl-1, and cefotaxime for a β-lactam-resistant pneumococcal isolate were 16, 32, and 4 µg/mL, respectively. In time-kill experiments, the activities of both enzymes were comparable, and much higher than that of cefotaxime. In a murine model of pneumococcal peritonitis-sepsis, LytA was essentially the most effective of the three studied agents with respect to decreasing bacterial titers in peritoneal fluid and blood.

4. BACTERIOCINS

Bacteriocins are peptides or proteins produced by bacteria to inhibit the growth of other bacteria (Nes et al. 2007). This sets them apart from autolysins, which act on the same bacterial cells in which they were produced. The bacteriocin whose antibacterial activity has been studied most thoroughly both *in vitro* and *in vivo* is lysostaphin.

4.1. Lysostaphin

Lysostaphin is discussed in more detail in Chapter 7. In this section we will present only the most important data on this enzyme and sum up the results of its use as an antibacterial agent.

Lysostaphin is an endopeptidase encoded by *Staphylococcus simulans* biovar *staphylolyticus* that specifically cleaves glycyl-glycyl bonds in the interpeptide cross-bridges of the staphylococcal peptidoglycan (Thumm and Götz 1997). Lysostaphin is very efficient in lysing *S. aureus* and can kill practically all strains of this species, including MRSA (von Eiff et al. 2003) and strains with reduced susceptibility to vancomycin (Patron et al. 1999). However, its activity against coagulase-negative staphylococci is essentially weaker due to a different amino acid composition of their cross-bridges (Kumar 2008). Aside from planktonic staphylococcal cells, lysostaphin can also specifically eliminate staphylococcal biofilms (Wu et al. 2003).

The first potential medical application of lysostaphin is the elimination of staphylococci colonizing nasal mucous membrane, which, in some clinical settings, may be a starting point for serious infections. In a cotton rat model of *S. aureus* nasal colonization, lysostaphin was shown to be more effective than mupirocin, which is currently the main antibiotic used as a decolonizing agent (Kokai-Kun et al. 2003). Another prophylactic use of lysostaphin might be prevention of catheter colonization by enzyme molecules coating their surface (Shah et al. 2004).

The second major application of lysostaphin can be the treatment of staphylococcal infections, both topical and systemic. So far, the therapeutic effectiveness of lysostaphin has been evaluated in experimental models of bacteremia, endocarditis, neonatal infections, and ocular infections, especially endophthalmitis and keratitis (Patron et al. 1999; Dajcs et al. 2000; Dajcs et al. 2001; Kokai-Kun et al. 2007; Oluola et al. 2007). Essentially, these studies revealed that lysostaphin can efficiently kill bacteria *in vivo* without causing any serious side effects. It is also noteworthy that in some experiments, lysostaphin was found to be more effective than antibiotics (Climo et al. 1998). Importantly, it was shown that specific antibodies do not completely neutralize, but rather moderately reduce, lysostaphin's antibacterial activity *in vivo*, which suggests that the enzyme could exert substantial antibacterial activity even after repeated injection (Climo et al. 1998; Dajcs et al. 2002). It was also found that lysostaphin can exert a synergistic antibacterial activity with other lytic enzymes, cationic antimicrobial peptides, and some antibiotics (Polak et al. 1993; Graham and Coote 2007; Becker et al. 2008). Development of resistance to lysostaphin, at least in some cases, can result in an increase in bacterial sensitivity to antibiotics and a reduction in their fitness and virulence (Kusuma et al. 2007).

4.2. Other Bacteriocins

Another bacteriocin whose antibacterial activity was studied in more detail and which could be used as an antibacterial agent is zoocin A, produced by *S. equi* ssp. zooepidemicus 4881 (Akesson et al. 2007). It is composed of an N-terminal catalytic domain of putative endopeptidase activity and a C-terminal cell wall-binding domain (Lai et al. 2002). It was shown that zoocin A-susceptible streptococcal species include, aside from *S. equi*, *S. pyogenes*, *Streptococcus mutans*, and *S. gordonii*. It is worth noting that all five *S. pyogenes* strains tested were extremely sensitive to zoocin A (MIC ≤ 31.5 ng/mL). Some other streptococcal species, especially *S. oralis* and *S. rattus*, are not susceptible to the enzyme's lytic activity (Akesson et al. 2007). Interestingly, zoocin A was found to be a penicillin-binding protein (PBP). In this regard, the enzyme was shown to bind penicillin covalently, to possess a weak β-lactamase activity, and to contain motifs typical of other PBPs. Furthermore, incubation of zoocin A with penicillin decreased its enzymatic activity (Heath et al. 2004). Zoocin A was shown to be capable of killing *S. mutans* in a triple-species plaque model (this species is involved in the pathogenesis of dental caries; Simmonds et al. 1995).

5. LYSOZYMES

Lysozymes, or *N*-acetylmuramidases, are hydrolases that specifically cleave the β-1,4 glycosidic linkages between *N*-acetylmuramic acid and *N*-acetylglucosamine in peptidoglycan (Jolles and Jolles 1984). Lysozymes are produced by cells of many different animal species, plants, insects, bacteria, and viruses. Based on their amino acid sequences and structural features, lysozymes are divided into several main subfamilies (Masschalck and Michiels 2003). In the human organism, lysozyme is produced by cells of the immune system (polymorphonuclear granulocytes, monocytes, macrophages) and is found in different biological fluids and tissues, including tears, urine, milk, saliva, liver, cartilage, and skin. However, the best known and most often used lysozyme is hen egg white lysozyme (Jolles and Jolles 1984; Masschalck and Michiels 2003).

Lysozymes are unique enzybiotics in that they exert not only antibacterial activity, but also antiviral, anti-inflammatory, anticancer, and immunomodulatory activities (Sava 1996). They are also the only peptidoglycan hydrolases that have been used on a larger scale in humans for the past several decades.

Although the best known mode of antibacterial action of lysozyme is based on the enzymatic cleavage of peptidoglycan, in fact it can also kill bacteria by some nonenzymatic mechanisms. First, lysozyme, in view of its cationic nature, can activate bacterial autolytic enzymes (autolysins). The second nonenzymatic mechanism by which lysozyme can kill bacteria is cytoplasmic membrane destabilization resulting from the removal of divalent ions from the membrane surface. In fact, some studies show that nonenzymatic mechanisms may be more important for killing bacteria than the enzymatic cleavage of peptidoglycan (Masschalck and Michiels 2003).

Generally, lysozyme is capable of killing only Gram-positive bacteria, while Gram-negative bacteria are resistant owing to the presence of the outer membrane. However, several exceptions to this rule have been reported, including both lysozyme-resistant Gram-positive bacteria (e.g., some strains of *S. aureus* and *E. faecalis*) and lysozyme-sensitive Gram-negative bacteria (e.g., *Capnocytophaga gingivalis*). It is also worth mentioning that several modifications of the lysozyme molecule have been developed to enable the enzyme to kill Gram-negative bacteria. These are essentially based on coupling lysozyme to molecules facilitating the penetration of the outer membrane (e.g., fatty acids and hydrophobic peptides) (Ibrahim et al. 2002; Masschalck and Michiels 2003).

For the past several decades, lysozyme has been used, often combined with antibiotics, in the prophylaxis and treatment of different bacterial infections, including pharyngitis, tonsillitis, dysentery, and wound infections (Sava 1996). More recently, patents for several lysozyme applications have been applied for or issued (Donovan 2007). These include the use of lysozyme formulated as a gel for topical treatment of wounds, the treatment of acne using different formulations of the enzyme, the prophylaxis of infections due to skin piercing, and the use of aerosolized lysozyme for the treatment of tracheitis, pneumonia, amyglalitis, and faucitis. Another interesting application of lysozyme is the use of its mutant to neutralize the activity of a lysozyme inhibitor produced by *Treponema pallidum*. Lysozyme has also been used as a component of oral health products (e.g., mouthwashes; Tenovuo 2002; Gil-Montoya et al. 2008), taking advantage of its capacity to kill different oral bacteria. A recent study showed that lysozyme can be utilized as a carrier allowing specific delivery of antibiotic molecules to bacterial cells (Hoq et al. 2008).

6. IMPORTANT ASPECTS OF ENZYBIOTIC THERAPY

6.1. Resistance

An important feature of some lytic enzymes, especially lysins, is the low risk of developing resistance. This results likely from the fact that lysins interact with those components of the cell wall that are necessary for bacterial viability (Borysowski et al. 2006; Fischetti 2008). For instance, the receptor for *S. pneumoniae*-specific lysins is choline, an essential component of the pneumococcal cell wall. In fact, in none of the hitherto conducted studies have any lysin-resistant bacteria been identified (Loeffler et al. 2001). In two separate studies, both *S. pneumoniae* and *B. cereus* failed to develop resistance to lysin even following repeated exposure to low doses of enzyme (the same results were obtained in experiments carried out on solid media and in liquid cultures; Loeffler et al. 2001; Schuch et al. 2002). Furthermore, *B. cereus* remained sensitive to lysin even after exposure to mutagens that rapidly induced mutations resulting in resistance to novobiocin and streptomycin (Schuch et al. 2002). On the other hand, it was found that bacteria can be less susceptible to lysins during the stationary phase, probably owing to some changes in the cell wall composition (Borysowski et al. 2006).

While no cases of bacterial resistance to lysins have been reported as yet, four mechanisms inducing resistance to lysostaphin have been identified (Shaw et al. 2005; Gründling et al. 2006; Kusuma et al. 2007). These are described in more detail in Chapter 7. Interestingly, in some cases the development of resistance to lysostaphin in MRSA results in an increase in susceptibility to methicillin and other antibiotics as well as a reduction in bacterial fitness and virulence (Kusuma et al. 2007).

Moreover, two general mechanisms mediating bacterial resistance to lysozyme have been reported. The first is based on the modifications of some peptidoglycan components (either O-acetylation of *N*-acetylmuramic acid residues by O-acetyltransferase or deacetylation of *N*-acetylglucosamine residues by *N*-acetylglucosamine deacetylase; Vollmer and Tomasz 2000; Bera et al. 2005). Such modified residues restrict the access of lysozyme to its substrate. The other mechanism involves the production of a lysozyme inhibitor (Binks et al. 2005).

6.2. Gram-negative Bacteria

While in Gram-positive bacteria peptidoglycan is easily accessible to recombinant lytic enzymes from outside the cell, in Gram-negative bacteria it is protected by the outer membrane that is impermeable to macromolecules (Vaara 1992). Therefore, Gram-negative bacteria are essentially resistant to lytic enzymes. However, the results of some studies indicate that these bacteria can also be killed by recombinant lytic enzymes

(Masschalck and Michiels 2003; Borysowski et al. 2006). For instance, two lysins were reported to be capable of killing Gram-negative bacteria (Düring et al. 1999; Orito et al. 2004). Moreover, different modifications of enzyme molecules were developed to enable lytic enzymes to penetrate the outer membrane (Ibrahim et al. 2002).

6.3. Immunogenicity

One of the major factors that can decrease the efficacy of protein therapeutics is the induction of a humoral immune response (De Groot and Scott 2007). However, a number of studies have consistently shown that specific antibodies do not completely block, but at most moderately reduce, the antibacterial activity of lytic enzymes (Borysowski et al. 2006; Fischetti 2008). For instance, in a murine model of pneumococcal bacteremia it was shown that a second dose of Cpl-1 administered to mice i.p. 10 days after the first dose can also cure mice from infection (Jado et al. 2003). Similar results were obtained by Loeffler et al. (2003), who found that Cpl-1 had comparable efficacy in terms of decreasing bacterial titers in the blood in naive mice and mice that were administered three i.v. doses of the enzyme 4 weeks earlier. In the same study it was shown that preincubation of Cpl-1 with hyperimmune rabbit serum for 10 or 60 min resulted in only a slight decrease in the enzyme's lytic activity *in vitro* (Loeffler et al. 2003). The most likely explanation for these unexpected findings is the very high affinity of lysins to their receptors on the bacterial cell wall (Fischetti 2008). These results are very important because they indicate that the apparent immunogenicity of lysins may not considerably decrease their therapeutic efficacy following repeated systemic administration. There are also data showing that specific antibodies do not completely neutralize, but rather to some extent decrease, the therapeutic efficacy of lysostaphin *in vivo* (Climo et al. 1998; Dajcs et al. 2002).

It is also worth mentioning that the immunogenicity of lytic enzymes can be considerably reduced by coupling enzyme molecules to polyethylene glycol (PEG), as shown for lysostaphin. However, it has not yet been shown whether modified lysostaphin with better pharmacokinetic features indeed has a higher therapeutic efficacy *in vivo* (Walsh et al. 2003).

6.4. Safety

The major mode of the antibacterial action of lytic enzymes relies on the enzymatic cleavage of peptidoglycan, which is an exclusive component of bacterial cells. Therefore, lytic enzymes are not likely to adversely affect mammalian cells, at least not directly (Fischetti 2003; Borysowski et al. 2006). To the best of our knowledge, in no study involving the administration of a lytic enzyme to experimental animals have any serious side effects been found. For instance, no signs of toxicity were detected following repeated administration of lysins to mice regardless of the route of administration (Fischetti 2003; Loeffler et al. 2003). However, it is noteworthy that some glycylglycine endo-peptidases, especially lysostaphin, were shown to degrade elastin, which has a high content of glycine residues (Park et al. 1995). This suggests that lysostaphin (and perhaps also other peptidases) could cleave other proteins present in mammalian tissues. It remains to be verified whether or not this activity may translate into any side effects of lytic enzymes.

Another important aspect of the safety of enzybiotic therapy is the possibility of the release of different pro-inflammatory components from bacterial cells being lysed. These components include especially endotoxin, peptidoglycan, as well as teichoic and lipoteichoic acids. Theoretically, massive release of these components could lead to septic shock and multiple organ failure (Nau and Eiffert 2002). However, many experimental studies have consistently shown that even massive bacteriolysis during treatment of systemic bacterial infections, including bacteremia, does not result in any serious side effects. For instance, practically no side effects were found following treatment of bacteremic animals with lysins and lysostaphin, which rapidly reduced bacterial titers in blood (Jado et al. 2003; Loeffler et al. 2003; Kokai-Kun et al. 2007).

Moreover, side effects associated with enzybiotic therapy might occur following the massive release of preformed bacterial toxins from the cytoplasm of bacteria during bacteriolysis. In fact, autolysins of some bacterial species may be involved in the pathogenesis of infections in the mechanism based on the release of different toxins. For instance, autolysins of *Clostridium difficile* may be involved in the release of toxin A and toxin B (Dhalluin et al. 2005). This potential effect should also be taken into account in the discussion about the safety of enzybiotic therapy.

7. CONCLUDING REMARKS

Many experimental studies performed both *in vitro* and *in vivo* have shown that enzybiotics constitute highly effective antibacterial agents.

In view of the dramatic and continuing increase in the prevalence of multidrug-resistant bacteria, the most important features of enzybiotics are their novel mode of action and the capability to kill antibiotic-resistant bacteria. Moreover, at least for some lytic enzymes, the risk of developing resistance is relatively lower than for traditional antibiotics. While unmodified enzybiotics essentially lyse only Gram-positive bacteria, some modifications were developed that enable them to also kill Gram-negative bacteria.

The potential medical applications of enzybiotics include different forms of prophylaxis and treatment of bacterial infections. For instance, some lytic enzymes were shown (in animal models) to be very effective in killing bacteria colonizing mucous membranes upon topical administration. These enzymes could be employed as a unique means of prophylaxis based on clearing bacteria that present a potential starting point for infections. Many experimental studies have also shown that lytic enzymes are efficacious in the treatment of systemic infections, including bacteremia, even in immunized animals.

In view of the unique therapeutic capabilities they provide, lytic enzymes definitely deserve a wider attention of the medical community.

ACKNOWLEDGMENTS

This work was supported by the Ministry of Science and Higher Education Grant No. 2 P05B 012 30 and Warsaw Medical University Intramural Grants 1MG/W1/08 and 1MG/N/2008.

REFERENCES

Akesson M., M. Dufour, G. L. Sloan, and R. S. Simmonds (2007) *FEMS Microbiology Letters* **270**, 155–161.

Baker J. R., C. Liu, S. Dong, and D. G. Pritchard (2006) *Applied and Environmental Microbiology* **72**, 6825–6828.

Becker S. C., J. Foster-Frey, and D. M. Donovan (2008) *FEMS Microbiology Letters* **287**, 185–191.

Bera A., S. Herbert, A. Jakob, W. Vollmer, and F. Götz (2005) *Molecular Microbiology* **55**, 778–787.

Binks M. J., B. A. Fernie-King, D. J. Seilly, P. J. Lachmann, and K. S. Sriprakash (2005) *Journal of Biological Chemistry* **280**, 20120–20125.

Bogaert D., R. de Groot, and P. W. M. Hermans (2004) *Lancet Infectious Diseases* **4**, 144–154.

Borysowski J., B. Weber-Dabrowska, and A. Gorski (2006) *Experimental Biology and Medicine* **231**, 366–377.

Breithaupt H. (1999) *Nature Biotechnology* **17**, 1165–1169.

Cheng Q., D. Nelson, S. Zhu, and V. A. Fischetti (2005) *Antimicrobial Agents and Chemotherapy* **49**, 111–117.

Climo M. W., L. R. Patron, B. P. Goldstein, and G. L. Archer (1998) *Antimicrobial Agents and Chemotherapy* **42**, 1355–1360.

Dajcs J. J., E. B. H. Hume, J. M. Moreau, A. R. Caballero, B. M. Cannon, and R. J. O'Callaghan (2000) *Investigative Ophthalmology Visual Science* **41**, 1432–1436.

Dajcs J. J., B. A. Thibodeaux, E. B. H. Hume, X. Zheng, G. D. Sloop, and R. J. O'Callaghan (2001) *Current Eye Research* **22**, 451–457.

Dajcs J. J., B. A. Thibodeaux, D. O. Girgis, M. D. Shaffer, S. M. Delvisco, and R. J. O'Callaghan (2002) *Investigative Ophthalmology Visual Science* **43**, 3712–3716.

De Groot A. S. and D. W. Scott (2007) *Trends in Immunology* **28**, 482–490.

Dhalluin A., I. Bourgeois, M. Pestel-Caron, E. Camiade, G. Raux, P. Courtin, M.-P. Chapot-Chartier, and J.-L. Pons (2005) *Microbiology* **151**, 2343–2351.

Djurkovic S., J. M. Loeffler, and V. A. Fischetti (2005) *Antimicrobial Agents and Chemotherapy* **49**, 1225–1228.

Donovan D. M. (2007) *Recent Patents on Biotechnology* **1**, 113–122.

Donovan D. M. and J. Foster-Frey (2008) *FEMS Microbiology Letters* **287**, 22–33.

Düring K., P. Porsch, A. Mahn, O. Brinkmann, and W. Gieffers (1999) *FEBS Letters* **449**, 93–100.

Entenza J. M., J. M. Loeffler, D. Grandgirard, and V. A. Fischetti (2005) *Antimicrobial Agents and Chemotherapy* **49**, 4789–4792.

Fischetti V. A. (2003) *Annals of New York Academy of Sciences* **987**, 207–214.

Fischetti V. A. (2008) *Current Opinion in Microbiology* **11**, 393–400.

Gil-Montoya J. A., I. Guardia-Lopez, and M. A. Gonzales-Moles (2008) *Gerodontology* **25**, 3–9.

Graham S., P. J. Coote (2007) *Journal of Antimicrobial Chemotherapy* **59**, 759–762.

Grandgirard D., J. M. Loeffler, V. A. Fischetti, and S. L. Leib (2008) *The Journal of Infectious Diseases* **197**, 1519–1522.

Gründling A, D. M. Missiakas, and O. Schneewind (2006) *Journal of Bacteriology* **188**, 6286–6297.

Hawkey P. M. (2008) *Journal of Antimicrobial Chemotherapy* **62 (Suppl. 1)**, 1–9.

Heath L. S., H. E. Heath, P. A. LeBlanc, S. Rochelle Smithberg, M. Dufour, R. S. Simmonds, and G. L. Sloan (2004) *FEMS Microbiology Letters* **236**, 205–211.

Hoq M. I., K. Mitsuno, Y. Tsujino, T. Aoki, and H. R. Ibrahim (2008) *International Journal of Biological Macromolecules* **42**, 468–477.

Ibrahim H. R., T. Aoki, and A. Pellegrini (2002) *Current Pharmaceutical Design* **8**, 671–693.

Jacob F. and C. R. Fuerst (1958) *Journal of General Microbiology* **18**, 518–526.

Jado I., R. López, E. García, A. Fenoll, J. Casal, P. García, and Spanish Pneumococcal Infection Study Network (2003) *Journal of Antimicrobial Chemotherapy* **52**, 967–973.

Jolles P. and J. Jolles (1984) *Molecular and Cellular Biochemistry* **63**, 165–189.

Kokai-Kun J. F., S. M. Walsh, T. Chanturiya, and J. J. Mond (2003). *Antimicrobial Agents and Chemotherapy* **47**, 1589–1597.

Kokai-Kun J. F., T. Chanturiya, and J. J. Mond (2007) *Journal of Antimicrobial Chemotherapy* **60**, 1051–1059.

Kumar J. K. (2008). *Applied Microbiology Biotechnology* **80**, 555–561.

Kusuma C., A. Jadanova, T. Chanturiya, and J. F. Kokai-Kun (2007) *Antimicrobial Agents and Chemotherapy* **51**, 475–482.

Lai A. C.-Y., S. Tran, and R. S. Simmonds (2002) *FEMS Microbiology Letters* **215**, 133–138.

Larson E. (2007) *Annual Review of Public Health* **28**, 435–447.

Livermore D. H. (2004) *Clinical Microbiology and Infection* **10 (Suppl. 4)**, 1–9.

Loeffler J. M., S. Djurkovic, and V. A. Fischetti (2003) *Infection Immunity* **71**, 6199–6204.

Loeffler J. M. and V. A. Fischetti (2003) *Antimicrobial Agents and Chemotherapy* **47**, 375–377.

Loeffler J. M., D. Nelson, and V. A. Fischetti (2001) *Science* **294**, 2170–2172.

Loessner M. J. (2005) *Current Opinion in Microbiology* **8**, 480–487.

Loessner M. J., S. Gaeng, and S. Scherer (1999) *Journal of Bacteriology* **181**, 4452–4460.

Loessner M. J., S. Gaeng, G. Wendlinger, S. K. Maier, and S. Scherer (1998) *FEMS Microbiology Letters* **162**, 265–274.

Loessner M. J., K. Kramer, F. Ebel, and S. Scherer (2002) *Molecular Microbiology* **44**, 335–349.

Loessner M. J., S. K. Maier, H. Daubek-Puza, G. Wendlinger, and S. Scherer (1997) *Journal of Bacteriology* **179**, 2845–2851.

Low L. Y., C. Yang, M. Perego, A. Osterman, and R. C. Liddington (2005) *Journal of Biological Chemistry* **280**, 35433–35439.

Masschalck B. and C. W. Michiels (2003) *Critical Reviews in Microbiology* **29**, 191–214.

McCullers J. A., A. Karlström, A. R. Iverson, J. M. Loeffler, and V. A. Fischetti (2007) *PLoS Pathogens* **3**, e28.

Muyombwe A., Y. Tanji, and H. Unno (1999) *Journal of Bioscience and Bioengineering* **88**, 221–225.

Nau R. and H. Eiffert (2002) *Clinical Microbiology Reviews* **15**, 95–110.

Navarre W. W., H. Ton-That, K. F. Faull, and O. Schneewind (1999) *Journal of Biological Chemistry* **274**, 15847–15856.

Nelson D., L. Loomis, and V. A. Fischetti (2001) *Proceedings of the National Academy of Sciences USA* **98**, 4107–4112.

Nelson D., R. Schuch, P. Chahales, S. Zhu, and V. A. Fischetti (2006) *Proceedings of the National Academy of Sciences USA* **103**, 10765–10770.

Nes I. F., D. B. Diep, and H. Holo (2007) *Journal of Bacteriology* **189**, 1189–1198.

O'Flaherty S., A. Coffey, W. Meaney, G. F. Fitzgerald, and R. P. Ross (2005) *Journal of Bacteriology* **187**, 7161–7164.

Oluola O., L. Kong, M. Fein, and L. E. Weigman (2007) *Antimicrobial Agents and Chemotherapy* **51**, 2198–2200.

Orito Y., M. Morita, K. Hori, H. Unno, and Y. Tanji (2004) *Applied Microbiology and Biotechnology* **65**, 105–109.

Parisien A., B. Allain, J. Zhang, R. Mandeville, and C. Q. Lan (2008) *Journal of Applied Microbiology* **104**, 1–13.

Park P. W., R. M. Senior, G. L. Griffin, T. J. Broekelmann, M. S. Mudd, and R. P. Mecham (1995) *International Journal of Biochemistry Cell Biology* **27**, 139–146.

Patron R. L., M. W. Climo, B. P. Goldstein, and G. L. Archer (1999) *Antimicrobial Agents and Chemotherapy* **43**, 1754–1755.

Polak J., P. Della Latta, and P. Blackburn (1993) *Diagnostic Microbiology and Infectious Disease* **17**, 265–270.

Porter C. J., R. Schuch, A.J. Pelzek, A. M. Buckle, S. McGowan, M. C. J. Wilce, J. Rossjohn, R. Russell, D. Nelson, V. A. Fischetti, and J. C. Whisstock (2007) *Journal of Molecular Biology* **366**, 540–550.

Pritchard D. G., S. Dong, J. R. Baker, and J. A. Engler (2004) *Microbiology* **150**, 2079–2087.

Pritchard D. G., S. Dong, M. C. Kirk, R. T. Cartee, and J. R. Baker (2007) *Applied and Environmental Microbiology* **73**, 7150–7154.

Rashel M., J. Uchiyama, T. Ujihara, Y. Uehara, S. Kuramoto, S. Sugihara, K. Yagyu, A. Muraoka, M. Sugai, K. Hiramatsu, K. Honke, and S. Matsuzaki (2007) *The Journal of Infectious Diseases* **196**, 1237–1247.

Rodriguez-Cerrato V., P. Garcia, G. del Prado, E. Garcia, M. Gracia, L. Huelves, C. Ponte, R. Lopez, and F. Soriano (2007) *Journal of Antimicrobial Chemotherapy* **60**, 1159–1162.

Rotem S., I. Radzishevsky, R. T. Inouye, M. Samore, and A. Mor (2006) *Peptides* **27**, 18–26.

Sava G. (1996) *EXS* **75**, 433–449.

Schuch R., D. Nelson, and V. A. Fischetti (2002) *Nature* **418**, 884–889.

Shah A., J. Mond, and S. Walsh (2004 *Antimicrobial Agents and Chemotherapy* **48**, 2704–2707).

Shaw L. N., E. Golonka, G. Szmyd, S. J. Foster, J. Travis, and J. Potempa (2005) *Journal of Bacteriology* **187**, 1751–1762.

Simmonds R. S., J. Naidoo, C. L. Jones, and J. R. Tagg (1995) *Microbial Ecology in Health and Disease* **8**, 281–292.

Tenovuo J. (2002) *Oral Diseases* **8**, 23–29.

Thumm G. and F. Götz (1997) *Molecular Microbiology* **23**, 1251–1265.

Vaara M. (1992) *Microbiology Reviews* **56**, 395–411.

Veiga-Crespo P., J. M. Ageitos, M. Poza, and T. G. Villa (2007) *Journal of Pharmaceutical Sciences* **96**, 1917–1924.

Vollmer W., B. Joris, P. Charlier, and S. Foster (2008) *FEMS Microbiology Reviews* **32**, 259–286.

Vollmer W. and A. Tomasz (2000) *Journal of Biological Chemistry* **275**, 20496–20501.

von Eiff C., J. F. Kokai-Kun, K. Becker, and G. Peters (2003) *Antimicrobial Agents and Chemotherapy* **47**, 3613–3615.

Walsh S., A. Shah, and J. Mond (2003) *Antimicrobial Agents and Chemotherapy* **47**, 554–558.

Wertheim H. F., D. C. Melles, M. C. Vos, W. van Leeuwen, A. van Belkum, H. A. Verbrugh, and J. L. Nouwen (2005) *Lancet Infectious Diseases* **5**, 751–762.

Wu J. A., C. Kusuma, J. J. Mond, and J. F. Kokai-Kun (2003) *Antimicrobial Agents and Chemotherapy* **47**, 3407–3414.

Yoong P., R. Schuch, D. Nelson, and V. A. Fischetti (2004) *Journal of Bacteriology* **186**, 4808–4812.

Yoong P., R. Schuch, D. Nelson, and V. A. Fischetti (2006) *Journal of Bacteriology* **188**, 2711–2714.

Young R. Y., I.-N. Wang, and W. D. Roof (2000) *Trends in Microbiology* **8**, 120–128.

Zimmer M., N. Vukov, S. Scherer, and M. J. Loessner (2002) *Applied and Environmental Microbiology* **68**, 5311–5317.

CHAPTER 2

ADVANTAGES AND DISADVANTAGES IN THE USE OF ANTIBIOTICS OR PHAGES AS THERAPEUTIC AGENTS

PATRICIA VEIGA-CRESPO[1] and TOMAS G. VILLA[1,2]

[1]Departament of Microbiology, Faculty of Pharmacy, University of Santiago de Compostela, Spain
[2]School of Biotechnology, University of Santiago de Compostela, Spain

1. INTRODUCTION

1.1. Antibiotic Resistance Phenomena

The discovery of penicillin by Sir Alexander Fleming and its subsequent industrial exploitation marked the beginning of the antibiotic era, and the pharmaceutical industry immediately began to produce penicillin for clinical use (Sulakvelidze 2005). For years, the development of new natural, synthetic, or semisynthetic antibiotics was continuous. The investigative methods for discovering new agents, such as cephalosporin C-producing organisms (Powers 2004), were based on observation. Antibiotics have traditionally been recovered from nature, and nearly all of today's antibiotics are versions of natural ones. The developers continued to work on chemical modifications of previously existing antibiotics. Some of these new generations of antibiotics are more effective against some diseases than previous generations. In the last thirty years, the new agents approved by the Food and Drug Administration (FDA) belonged to existing classes, such as cephalosporins, mainly β-lactams. These modified antibiotics basically use the same mechanisms as the preceding ones, making it easy for bacteria to develop resistance to drugs. The chemical synthesis of entirely created

Enzybiotics: Antibiotic Enzymes as Drugs and Therapeutics. Edited by Tomas G. Villa and Patricia Veiga-Crespo

TABLE 2.1. Milestones in Antimicrobial Chemotherapy

Year	
1929	Fleming discovered penicillin.
1935	Gerhard Domagk developed sulphonamides.
1939	Discovery and purification of gramicidin by René Dubos
1941	The United States commenced commercial production of penicillin
1944	Amynoglycosides
1945	Golden age of antibiotics
1949	Chloramphenicol
1950	Tetracyclines
1952	Macrolides
1953	Multidrug-resistant dysentery bacilli in Japan. Drug-resistant tuberculosis (TB)
1956	Glycopeptides
1957	Rifamycins
1959	Nitromidiazoles
1960	Methicillin-resistant *S. aureus*
1962	Quinolones
1968	Trimethoprim
1980s	Major drug companies scaled down antibiotic discovery programs
2000	Linezolid launched—first new class of agents in 30 years Oxazalidinones
2003	Lipopeptides

antibiotics is reduced to the fluoroquinolones, a group of broad-spectrum antibiotics based on the chemistry of nalidixic acid (Table 2.1).

Over the past few decades, the number of antibiotics has risen in parallel to the emergence of resistant microorganisms. Lately, however, owing to the ever-increasing number of new resistant strains, the microorganisms are winning the race. As an example, between 2000 and 2003 the number of antibiotics approved by the FDA was very low (Table 2.2). The need for antibiotics or antibacterials will increase in the future due to factors such as aging population, immunocompromised patients, bacterial resistance, and organ transplantation. However, research on the development of new antibiotics has been decreasing.

Changes within the pharmaceutical industry may be contributing to the decrease in new antimicrobials. The major pharmaceutical companies have stopped research in the development of new antibiotic drugs (Talbot et al. 2006). One of the reasons for this cessation is lower profits relative to other drugs such as those for heart disease, mental disorders, or hypertension. In 2000, amoxicillin-clavulanate was the only antibiotic in the list of the top 20 prescription drugs (Kreling et al. 2001). In the last few years, larger pharmaceutical firms have merged to form even larger entities, and these require larger profits to sustain themselves. The withdrawal of the larger entities opens the

TABLE 2.2. New Antimicrobial Agents Approved by the U.S. Food and Drug Administration in 2003

Generic Name	Manufacturer	Indications for Use
Moxifloxacin hydrochloride	Bayer Corporation	Modification of community-acquired pneumonia indication to include penicillin-resistant strains of *Streptococcus pneumoniae*
Gatifloxacin	Allegan	Treatment of bacterial conjunctivitis
Moxifloxacin	Alcon	Treatment of bacterial conjunctivitis
Gemifloxacin mesylate	LG Life Sciences	Infections caused by susceptible strains of designated microorganisms in acute bacterial exacerbations in chronic bronchitis and community-acquired pneumonia
Levofloxacin	Ortho-McNeil	Treatment of chronic bacterial prostatitis due to *Escherichia coli, Enterococcus faecalis* or *Staphylococcus epidermidis*
Ciprofloxacin and dexamethasone	Alcon	Treatment of acute otitis media in pedriatic patiens due to *Staphylococcus aures, S. pneumoniae, Haemophilus influenzae, Moraxella catarrhalis*, or *Pseudomona aeruginosa*
Linezolid	Pharmacia & Upjohn	Treatment of diabetic foot infections
Gemifloxacin mesylate	GeneSoft Pharmaceuticals	Treatment of infections caused by susceptible strains of designated microorganisms in acute bacterial exacerbations of chronic bronchitis and community-acquired pneumonia, including those caused by multidrug-resistant strains of *S. pneumoniae*
Ciprofloxacin	Bayer Corporation	For complicated urinary tract infections and acute uncomplicated pyelonephritis
Daptomycin	Cubist Pharmaceuticals	For treatment of complicated skin and skin structure infections caused by susceptible strains *of S. aureus* (including methicillin resistant), *Streptococcus pyogenes, Streptococcus agalactiae, Streptococcus dysgalactiae* spp. *eauisimilisor E. faecalis* (vancomycin susceptible only)

It can be observed that the major difference between the approved ones is the formulaton of the active principles.
Source: Adapted from the U.S. Food and Drug Administration Approvals (2004).

field to smaller firms, which need comparatively smaller profits to survive.

Today, the strategy for obtaining new antibiotics is very laborious, although it can be carried out in different ways. The traditional method involves screening large libraries of compounds and huge numbers of producer organisms. The modification of natural compounds can also be attempted, but the newest approach is to find new targets in bacteria and fungal cells, followed by the design of new compounds that will act against them. To better understand resistance phenomena, experiments of reverse engineering or genotyping the resistant strains are currently under way (Hiramatsu et al. 1997; Struble and Gill 2006).

Bacterial antibiotic resistance can be acquired mainly by two mechanisms: a result of genetic events causing variations in the bacterial genome or by horizontal gene transfer between bacteria. This results in an accumulation of genetic elements that contribute to create clones with multiresistant properties (Livermore 2003, 2004). The acquirement of resistance follows a sigmoid distribution, with a lag phase before resitance, a quick increase in the proportion of resistant bacteria, and finally a third phase in which the proportion of resistant bacteria reaches an equilibrium (Austin and Anderson 1999).

Several major programs have been deployed to monitor antimicrobial resistance phenomena: the National Nosocomial Infection Surveillance System, Intensive Care Antimicrobial Resistance Epidemiology (ICARE), SENTRY Antimicrobial Surveillance Program, Community Network for Epidemiological Surveillance and Control of Communicable Diseases from the European Union, European Antimicrobial Reistance Surveillance System (EARSS), and so on (Diekema et al. 1999; Fridkin et al. 1999; Sahm et al. 1999; Gales et al. 2001). The aim of these programs is to provide a body of information that will further highlight the global nature of the problem of resistance. In the last decades, an overuse of antibiotics has been carried out. Around half of the antibiotic use for humans is based on incorrect indications, mostly viral infections. The World Health Organization (WHO; 2001) defines the appropiate use of antibiotics as "the cost-effective use of antibiotics, which maximizes clinical therapeutic effect while minimizing both drug-related toxicity and the development of antibiotic resistance."

Resistance phenomena represent not only an important health-care issue but also a huge economic problem, with an estimated cost of about $4000 million per year (Workshop Summary 1998). To curb the problem, the European Union and the U.S. government, the Infectious Diseases Society of America (IDSA), and WHO have established directives aimed at coordinating the fight against antibiotic resistance and the rational use of antibiotics (European Union Council 1999; U.S. Centers for Disease Control et al. 2000; Infecious Diseases Society of America 2004; Nordberg et al. 2005). However, these initiatives do not seem to be sufficient.

Methicillin-resistant *Staphylococcus aureus* caused nearly 60% of nosocomial infections in 2001; 30% of *Streptococcus pneumoniae* strains are penicillin-resistant; and vancomycin-resistant *Enterococcus faecium* and *Enterococcus faecalis* are increasingly seen. Other important groups of resistant strains are the opportunistic pathogens in immuno-compromised patients such as multidrug-resistant *Stenotrophomonas maltophilia* or *Acinetobacter* species (Gales et al. 2001).

Antibiotic resistance phenomena are also a problem in cattle care and veterinary medicine. Thus, mastitis—caused mainly by *Streptococcus agalactiae, S. aureus, Streptococcus uberis*, and *Streptococcus dysgalactiae*—generates losses estimated at between $1700 million and $2000 million per year (Donovan et al. 2006). Also, the use of subtherapeutic concentrations of antibiotics in animal feed has been extensive. This practice has promoted the development of resistant strains from farms to humans by food chain or direct contact (Endtz et al. 1991).

1.2. Phages

Phages are the most abundant and diverse biological agents. Phages have been detected in all environments; however, they are more abundant in the water column of freshwater and marine habitats.

Phages are cassified by the International Committee on Taxonomy of Viruses (ICTV) according to morphology and nucleic acid. Bacteriophage genomes consist of either single- or double-stranded DNA or DNA with a high variable size and may be circular or lineal. Bacteriophages have been classified into 12 families (Table 2.3). The tailed phages are comprised into Myoviridae, Siphoviridae, and Podoviridae, and they represent over 95% of the total number of identified phages.

Phages may undergo two life cycles: the lysogenic cycle, in which the phage DNA is inserted into the bacterial chromosome, and the lytic type, in which the phages uses the cellular machinary to replicate their own DNA and assemble its components. After assembly, the progeny lyse the membrane and the bacterial cell wall, thus killing the bacterial host.

The lytic cycle consists of five main steps: attachment, injection of phage DNA into the host, halting of the host molecular processes, and

TABLE 2.3. Classification of Bacteriophages

Family	Nucleic Acid	Main Characteristics and Members
Myoviridae	dsDNA	Contractile tails. T4, T2, or Mu bacteriophage
Siphoviridae	dsDNA	No-contractile tails. λ bacteriophge
Podoviridae	dsDNA	Short tails. P22 or T7 phages
Tectiviridae	dsDNA	Double capsid, one internal lipid. Bam35 or PRD1 phages, Thermus phage 37–61
Plasmaviridae	dsDNA	Pleomorphic. Infect *Mycoplasma*
Corticoviridae	dsDNA	Present of internal lipids. Infect *Pseudomona*
Microviridae	ssDNA	Icosaedric capsid with spikes. G4 and ΦX174 bacteriophage
Inoviridae	ssDNA	Filamentous phages. M13 phage
Cysaviridae	ssRNA	Infect *Pseudomona*
Leviviridae	ssRNA	Icosaedric capsid; bind to F pilus. MS2 and Qβ coliphages
Lipothrixviridae	dsDNA	Also contain DNA binding protein. Infect members of Archea
Fuselloviridae	dsDNA	Infect members of Archea

beginning of the replication of phage DNA as well as the components of the phage capsid, assembly of the phage-progen, and subsequent release. The lysogenic cycle is more complicated than the lytic one; however, the start of the cycle is similar. In this case, viral DNA is inserted into the bacterial chromosome and host cells can replicate normally for many generations. When lysogenic induction occurs, the phage DNA initiates replication of its components for release from host cell (Fig. 2.1).

1.3. A Brief Background

More than a century ago, Hankin (1896) reported that the waters of the Ganges and Juma in India had a marked antibacterial against cholera infection by *Vibrio cholerae*. This antibacterial action could be destroyed by simply boiling. He suggested that an unidentified substance was responsible for limiting the spread of cholera epidemics. He studied the effect of "this antibacterial action" on *V. cholerae*, but he did not go too deep into resolving the problem. When Twort (1915) and D'Hérelle (1917) finally discovered phages in 1915 and 1917, respectively, their application in the treatment of bacterial infections was immediate. D'Hérelle and his coworkers successfully applied bacteriophage therapy to treat dysentery in Paris as early as 1918 (Summers 1999). In 1921, Bruynoghe and Maisin (1921) reported for the first time the use of bacteriophages

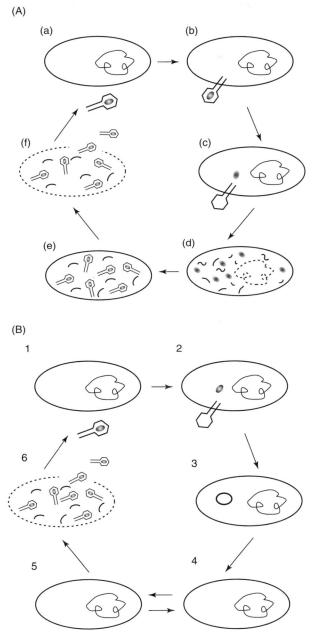

Figure 2.1. Life cycles of phages. (A) The lytic cycle can be summarized in the next steps: recognition of cell type (a); adsoption of the bacterial cell (b); injection of phage DNA (c); (d) arrest of the cell machinery and synthesis of phage components; assembly and maduration of phage particles (e); and, finally, liberation of mature phage particles. (B) The lysogenic cycle begins as the lytic one (1, 2, and 3 steps). However, the phage DNA is integrated in bacterial chromosome (4). This association is stable during cell generations until the phage DNA is induced (5). Then, the lytic cycle occurs (6).

Figure 2.2. Fathers of phage therapy. a: A young Frederick Twort; b: a young Felix D'Hérelle; c: Joseph Maisin; and d: Richard Bruynoghe. Together with Ernest Hankin, the observations of these men and their ability to go further led to the development of phage therapy.

to treat staphylococcal skin disease (Fig. 2.2). The potential applications for bacteriophages in medicine were quickly explored (Table 2.4). The reviews of early therapeutic applications of phages concluded that the results were variable, but in many cases were sucessful in controlling infections.

TABLE 2.4. First Applications of Phage Therapy

Treatment of	
Vibrio cholorea	Hankin 1896
Staphylococcus and *Bacillus anthracis*	Bruynoghe and Maisin 1921
Bacillary dysentery	Beckerich and Hauduroy 1922
Thypoid and parathyphoid fever	Beckerich and Hauduroy 1922
Bacillus dysenteriae "Shiga" and Bacillys dysenteriae "Flexner"	Beckerich and Hauduroy 1922
Typhoid fever	Davison 1922
Pyelonephritis	Courcoux 1922
Purperal pyelocystitis caused by *Bacillus coli*	Spence and McKinley 1924
Staphylococcus aureus	Spence and McKinley 1924
V. cholorea	Morrison 1932

The bacteriophages were applied not only in human and veterinary therapy. At the same time the first investigations against human pathogens were carried out, the bacteriophages were being employed in plant pathogen control (Mallmann and Hermstreet 1924; Kotila and Coons 1925; Link 1928).

The success of the initial bacteriophage-based treatment led pharmaceutical companies to start commercializing drugs known as Bacté-coli-phage, Bacté-rhino-phage, and so on. The preparations had been elaborated by D'Hérelle and coworkers (Summers 1999). However, the controversial results, as well as the advances made in antibiotic therapies, condemned phage-based preparations to oblivion in Western countries.

When phage-based therapy against cholera was revised, it was observed that it was better than the absence of any treatment, although it was not proven to be more efficient than vaccination and the observance of strict hygienic practices (Pollitzer 1959).

Over ensuing years, research into phages continued in different ways. Thus, while in Western countries the main topics of research focused on the new field of molecular genetics, the study of phages as therapeutic agents continued in Eastern-bloc countries, the main centers of knowledge in this field being the Hirszfeld Institute in Poland and the Eliava Institute in Georgia Republic (Sulakvelidze et al. 2001). The beginings of molecular biology and phages are intimately related (Table 2.5). The development of the investigation of phages led to the establishment of the actual molecular techniques such as gene cloning and DNA manipulation.

Studies concerning the use of bacteriophages for therapy were centered on the prophylaxis and treatment of bacterial infections in humans,

TABLE 2.5. Studies of Genetic of Phages and Its Influence on Molecular Biology Development

Year		
1928–1929	First observations of lysogeny	Wollman 1928; Burnett and Mckie 1929
1936	Chemical composition of phages (50% nucleic acids/50% proteins)	Schlesinger 1936
Early 1940s	Phage physiology and relationships with the host during lysis	Ellis and Delbruck 1939; Delbruck 1940a, b
1943	First images of tailed bacteriophages	Luria et al. 1943
1943	Mutual exclusion principle and fluctuation test	Luria and Delbruck 1943
1949	First recombinational maps	Hershey and Rotman, 1949
1950–1952	Role of bacteriophage in genetic transfer	Lederberg et al. 1951; Zinder and Lederberg 1952
1952	Only the nucleic acid entries in the bacteria during infection	Hershey and Chase, 1952
1952	Host-induced modification of phage	Luria and Human, 1952;
1953	Final demonstration of lysogeny	Lwoff 1953
1955–1963	Discovery of lethal mutations in λ and T4 phages and its implication on phage replication cycles	Benzer 1955; Cambell 1961; Epstein et al. 1963
1966	Discovery of the methylation of DNA	Gefter et al. 1966

and target pathogens and illnesses were indeed very diverse (Table 2.6). For example, Slopek et al. (1983, 1984, 1985a, b, c) employed phages for the treatment of gastrointestinal tract, skin, head, and neck infections caused by *Staphylococcus, Pseudomonas, Escherichia coli, Klebsiella*, and *Salmonella*. Other authors worked hard to treat other sicknesses such as intestinal dysbacteriosis (Litvinova et al. 1978), lung infections (Meladze et al. 1982), or infectious allergoses (Sakandelidze 1991). Unfortunately, these and other works were published in their respective languages, with no or little diffusion in Western countries. This, together with political isolation behind the "Iron Curtain," meant that their results were unavailable to the rest of the world.

Meanwhile, in the West, the λ and T2/T4 phages were proposed and used as vehicles for displaying proteins in their coatings, because they exhibited proteins on their capsids (Smith 1985); they were employed to build gene libraries of proteins with practical applications (Clark and March 2006); used as potential delivery vectors in genetic therapy; and employed as vaccine delivery vehicles (Clark and March 2004).

While in the East the investigation about phage therapies went on, the main application of phages in the West was the phage typing of

TABLE 2.6. List of the Main Therapeutic Applications of Phages in Eastern Block

Clinical Diagnosis	Etiology	Full Recovery	Marked Improvement	No Effect
Septicemia	*Staphylococcus aureus, Escherichia coli, Klebsiella, Proteus, Pseudomonas*	93 (87.7%)	8 (7.5%)	5 (4.7%)
Purulent otitis media	*S. aureus, Klebsiella, Pseudomonas*	28 (88.4%)	3 (9.09%)	2 (6.06%)
Purulent meningitis	*S. aureus, E. coli, Klebsiella, Proteus, Pseudomonas*	10 (100%)	—	—
Varicose ulcers of lower extremities	*S. aureus, E. coli, Klebsiella, Proteus, Pseudomonas*	47 (61.03%)	21 (27.2%)	9 (11.6%)
Mucopurulent chronic bronchitis, laryngitis, rhinitis	*S. aureus, E. coli, Klebsiella, Proteus, Pseudomonas*	224 (82.6%)	46 (16.9%)	1 (0.3%)
Bronchopneumonia, empysema	*S. aureus, E. coli, Klebsiella, Proteus, Pseudomonas*	47 (82%)	—	10 (18%)
Pleuritis with fistula	*S. aureus, Escherichia coli, Klebsiella, Proteus, Pseudomonas*	42ʼ (86%)	5 (10%)	2 (4%)
Suppurative peritonitis	*S. aureus, E. coli, Klebsiella, Enterobacter, Proteus, Pseudomonas*	60 (91%)	5 (8%)	1 (0.15%)
Urinary tract infections	*S. aureus, E. coli, Klebsiella, Proteus, Pseudomonas*	59 (75.6%)	9 (11.5%)	10 (12.8%)
Furunculosis	*S. aureus*	90 (100%)	—	—
Decubitus with infection	*S. aureus, E. coli, Klebsiella, Proteus, Pseudomonas*	13 (81%)	—	3 (1
Pyogenic arthritis and myositis	*S. aureus, E. coli, Klebsiella, Proteus, Pseudomonas*	17 (89%)	—	2 (11%)
Osteomyelitis of the long bones	*S. aureus, E. coli, Klebsiella, Proteus, Pseudomonas*	38 (95%)	2 (5%)	—

TABLE 2.6. *Continued*

Clinical Diagnosis	Etiology	Full Recovery	Marked Improvement	No Effect
Suppurative osteitis after bone fractures	*S. aureus, E. coli, Klebsiella, Proteus, Pseudomonas*	37 (90%)	4 (10%)	—
Suppurative osteitis after bone fractures	*S. aureus, E. coli, Klebsiella, Proteus, Pseudomonas*	37 (90%)	4 (10%)	—
Pyogenic infections of burns	*S. aureus, E. coli, Klebsiella, Proteus, Pseudomonas*	42 (86%)	7 (14%)	—
Pyogenic postoperative infection	*S. aureus, E. coli, Klebsiella, Pseudomonas*	29 (83%)	6 (17%)	—
Chronic suppurative fistulas	*S. aureus, E. coli, Klebsiella, Proteus, Pseudomonas*	168 (93%)	12 (7%)	—
Suppurative sinusitis	*S. aureus, E. coli, Klebsiella, Proteus, Pseudomonas*	38 (83%)	3 (7%)	5 (11%)
Purulent mastitis	*S. aureus, E. coli*	41 (93.1%)	3 (6.8%)	—
Total		1123 (85.9%)	134 (10.2%)	50 (3.8%)

bacteria. Bacteriphage typing was one of the first techniques for the identification and classification of clinical strains (Parker, 1972). In 1942, a set of phages was established for the epidemiological typing of *S. aureus* (Fisk 1942). Phage-typing systems have been described for *E. coli* 0157, *Listeria*, and *Campylobacter* (Khakhria et al. 1990; Loessner 1990; Owen et al. 1990). Phages had been used for the confirmatory diagnosis of *V. cholerea* 01 or to detect non-enteropathogen-specific fecal contamination in water (Dutka et al. 1987).

The cessation of investigations into phage therapies in the West was due to two main reasons: the advent of broad-spectrum antibiotics and the unreliable and inconsistent initial results. However, it is now clear that that the initial lack of consistency was due to poor knowledge of phage biology (Summers 1999; Sulakvelidze et al. 2001). The potential of bacteriophages as therapeutic agents in the West was first evaluated in fields such as aquaculture, plague control in agriculture, and veterinary medicine (Heuer et al. 2002; Goodridge 2004; Higgins 2005; Donovan et al. 2006). The success of phage therapy during the treatment of septicemia and meningitis in chickens and calves or the possibility of using bacteriophages as biocontrol agents to reduce *Salmonella* in poultry products has been reported (Barrow et al. 1998; Huff et al. 2005). Phages have also been proposed as an alternative to antibiotic sprays to control bacterial infections in high-value crops (Holtzman 2003; Leverentz et al. 2004).

These advances and the increase in antibiotic resistance phenomena evidently changed the perspective of the use of bacteriophages in Western countries.

Advances in molecular techniques allowed researchers to focus phage-based therapies on two main aspects: the administration of whole phages and the administration of phage lytic enzymes. As can be gathered from several chapters of this book, when applied externally as purified recombinant proteins, lysins are capable of degrading peptidoglycan. In this context, lysozymes play a relevant role in the control of bacterial populations in a huge variety of environments and applications.

Bacteriophages can be used as food additives and food preservatives; in the production of cheese and wine; in medical uses, such as eye drops; in toothpaste; and so on (Burman et al. 1991; Sherman et al. 1994; Ibrahim et al. 2002; Touch et al. 2003; Delfini et al. 2004). The FDA has approved using bacteriophages on cheese to control *Listeria monocytogenes*, categorizing phages as "generally recognized as safe" (GRAS) in 2006 (http://www.cfsan.fda.gov/~rdb/opa-g198.html). In 2007, these phages were classified as GRAS for their use on all food products (http://www.cfsan.fda.gov/~rdb/opa-g218.html).

1.4. Actual Fungal Therapies

Fungal infections have been increasing over the past several years, particularly in immunocompromised patients. The development of active antifungal principles has been slower than that of antibacterial agents, mainly because of their toxicity to host cells; that is, the targets of the main antifungal agents are also present in host cells. The exception to this rule is the fungal cell wall. The main components of this cellular structure are chitin, which consists of an N-acetyl-D-glucosamine polymer with 1,4-β-D-linkages organized in microfibrils, and β-1,3-glucan, formerly known as "yeast cellulose," which is almost as rigid as chitin and is present in nearly all varieties of yeasts. Both polysaccharides are the target of chitinases and glucanases, respectively. These enzymes have two main natural functions: normal morphogenetic events related to cell wall growth, and participation as a defense mechanism in both plants and lower animals. Chitinases and glucanases are therefore the "enzybiotic alternative" for treating fungal infections.

Current antifungal drugs have different targets, such as cytochrome P450-dependent enzymes, ergosterol, the mitotic cycle, or squalene epoxidase. However, these targets are also present in host cells. Azoles can cause liver toxicity; amphotericin B can cause phlebitis or renal dysfunction; and so on.

The development of antifungal drugs has only recently focused on the cell wall, an essential component of fungal cells and one exclusive to fungi. Currently, there are two drugs whose targets are on the fungal cell wall: nikkomycin, a chitin synthase inhibitor, and caspofungin, a glucan synthesis-interfering compound (www.doctorfungus.org/thedrugs/medical.htm).

The most accurate definition of enzybiotics refers to a group of phage-associated enzymes that are produced actively during the phage lytic cycle. If the definition is extended, the antifungal glucanases and chitinases can also be included in this definition. Accordingly, the term "enzybiotic" is wider than first thought and should include all enzymes, regardless of their origin, able to act on microbial cells, causing or contributing to their death.

2. PHAGE ENZYMES

2.1. Structure of Lytic Enzymes from Phages

The lytic cycle of phages can be described as a tightly regulated number of events that occur in the blink of an eye. During bacteriophage infection, the control of cycle timing and the lysis of the host cell are crucial. In these events, two proteins are essential: holins and lysins.

Phage lysins can be classified according to the site of cleavage where they act: N-acetylmuramidases or lysozymes, endo-β-N-acetylglucosaminidases, transglycosylases, endopeptidases, and N-acetylmuramoyl-L-alanine amidases (Borysowski et al. 2006). The members of the lysin family are often chimaeric proteins, with a well-conserved catalytic domain fused to a largely divergent binding domain directed toward species or strains (Lopez et al. 1997). The cell wall-binding domain is notably divergent and can distinguish discrete epitopes present within the cell wall, mainly carbohydrates or teichoic acids (Nelson et al. 2006). The binding between a lysin and the bacterial cell wall is not a random event and takes place through the C-terminus (Jado et al. 2003). Within the specificity-exhibiting lysins, the Pal case merits closer analysis. Pal is an amidase with lytic activity against pneumococci. The protein has a catalytic domain with no homology at all with other pneumococcal amidases, but its binding domain is conserved and it exhibits the typical choline-binding motif of these lysins (Varea et al. 2004).

Some lysins may show more than one catalytic domain: PlyGBS from GBS phage NCTC 11621, B30 bacteriophage, and Φ11 phage (Navarre et al. 1999; Pritchard et al. 2004; Cheng et al. 2005). Other lysins, such as the lysin from T7 bacteriophage, show only the catalytic domain (Cheng et al. 1994). Some lysins, in contrast, may contain introns that split the gene (Fischetti 2005). There are multimeric lysins, such as PlyC from streptococcal C_1 bacteriophage. This enzyme is built of chains: a heavy chain and a light chain, both encoded by different genes (Nelson et al. 2006).

Generally, lysins lack secretory signals; holins are responsible for the formation of holes in the cellular membrane so that lysins can reach the peptidoglycan layer. However, these rules have some exceptions, as is for example the case of Lys44 from *Oenococcus oeni* phage fOp44, or Lys from *Lactobacillus plantarum* phage Φgle, which show a secretory signal at the N-terminus (Sao-Jose et al. 2000; Kakikawa 2002).

Holins are classified in two classes according to their transmembrane domains (TMD). Class I holins have three TMD while class II holins have two TMD (Young 2002). Within each class, there are holins from phages able to act on Gram-positive bacteria, and others that exert their action on Gram-negative bacteria. All holins share some characteristics such as a high content of basic amino acids at the C-terminus and the presence of a short and polar N-terminus (Wang et al. 2000). Holins can be thought of as molecular clocks, because they control the timing of lysis. Many of the holin genes show the dual-start motif and thus encode two proteins: holin and anti-holin (Park et al. 2006). The interaction between both is unknown, but no doubt their relationship represents the real clock bacteriophage after infection. The oligomerization process of holins remains unknown, although recently it has been described that holin from bacteriophage λ forms rings of at least two size classes, containing approximately 72 S105 monomers (Savva et al. 2008).

The typical organization of lysis genes in bacteriophages is in clusters, although there are some cases, such as bacteriophage P1, in which the holin gene is not clustered with the lysozyme gene (Lee et al. 2006).

There are other phages, such as ΦX174 or Qβ, that produce proteins that block the synthesis of peptidoglycan. These phages encode neither an endolysin nor a holin, and it is only the lack of coordination in the morphogenesis of the cell wall that leads to weakness in the wall (Hatfull 2001).

2.2. Phages and Therapy

Early work using phages as therapeutic tools afforded ambiguous results. This, plus the advances in antibiotic research, led many researchers to leave the field. However, the reported initial failures were due to specific issues: (i) a poor understanding of bacterial and phage ecology, (ii) incorrect selection of phages for therapy, (iii) the use of a single phage in illnesses caused by more than one bacteria, (iv) the emergence of resistant bacteria after treatment with phages, (v) problems with the titer of phage preparations used, (vi) incorrect preparation and administration of phages, (vii) incorrect identification of the causal agent, (viii) the absence of selection of specific phages *in vitro* against the target bacteria before use *in vivo* models, and (ix) a lack of knowledge about the release of endotoxins as a result of the lysis of bacteria (Kutter 2001). All these pitfalls can now be solved.

Antibacterial phage therapy still has limitations that hamper its application: (i) the phages' host range, (ii) the requirement of prior identification of the causal agent of the illness, (iii) sterilization of the phage preparation without damaging the phages, (iv) the immune response to phages inside the human body, and (vi) the pharmacokinetics of phage treatment (Hermoso et al. 2007).

Not all phages can be used for therapy. Of the two major classes of phages, lytic and lysogenic, only the former are good candidates for use as therapeutic agents. Lysogenic phages integrate their genome inside the host genome, and this can lead to the transfer of virulence genes from one host to another via transduction.

The phenomenon of resistance to bacteriophages can be solved more easily than antibiotic resistance. It has been estimated that for each bacterial cell there are 10 bacteriophage particles. Thus, the development of new alternatives must be easier and cheaper, and since phages and bacteria have been evolving over million of years together, this should facilitate the overall research. Recent experiments have shown that the mutation rate of *E. coli* to phage T1 resistance is 1.4×10^{-8} to 4.1×10^{-8}, a lower rate than when antibiotic resistance is evaluated. Recently, phages able to infect Archeabacteria have been discovered (Stettler et al. 1995; Luo et al. 2001).

Since the bacteriophage-target cell is a receptor-specific union, the side effects are fewer than in antibiotic therapy. Since bacteriophages will destroy only their natural host, the cause of the normal microflora will not be affected. The ideal antibiotic should effectively kill the pathogens responsible for infection and simultaneously cause as little disturbance as possible to the normal flora of the individual. But an antibiotic such as this does not exist. Yet after oral administration of a four-phage cocktail (Bruttin and Brussow 2005), human volunteers show no alteration whatsoever of normal microbiota. It has been demonstrated that it is possible to combine phages and bifidobacteria for the treatment of dysentery in immunosuppressed leukemia patients. In this way phages deplete the pathogenic bacteria while bifidobacteria provide a source of new normal flora (Sulakvelidze et al. 2001).

The side effects due to cell death are the same in antibiotic and phage therapies, but it has been possible to engineer phages for "death without lysis" (Westwater et al. 2003). These engineered phages could offer a very elegant solution for eliminating toxin-producing bacteria such as *Clostridium difficile*.

One of the major criticisms in bacteriophage therapy is the need to identify the causal agent before treatment is begun. However, some phages are highly specific, while others are extremely broad in their host range (Skurnik and Strauch 2006). Recently, the FDA has approved a cocktail of bacteriophages for use against *L. monocytogenes* in ready-to-eat meat and poultry products (Fischetti et al. 2006).

The application of whole phage particles is not limited to the direct use of phages as active principles; they can also be used as vehicle vaccines in the form of immunogenic peptides to coat proteins or as DNA vaccines (Clark and March 2004). Rapid diagnostic kits based on phages have been developed: for example, the fast diagnostic kit for *Mycobacterium tuberculosis* antibiotic resistance is based on an engineered phage that expressed luciferase and detection with a dental X-ray film (Riska et al. 1999). Drug delivery systems based on viral capsids have also been developed (Kovacs et al. 2007).

A major pitfall in phage therapy is the pharmacokinetics of the preparations used. This is a complex problem because phages are live organisms and are self-replicating. The first consideration to be taken into account is how phages can be stored. It has been shown that lyophilized phages show greater stability and activity than liquid preparations (Brussow 2005). An advantage in the formulation of whole phages is that these are quite resistant to environmental conditions, thus allowing a variety of different formulations. It is difficult to obtain the exact compositions of the different product preparations and it is also difficult to know the final doses inside the organism after colonization of the target bacteria. It is therefore mandatory to develop mathematical models that will be able to account for their behavior inside the organism since models of *in vitro* growth do not reflect the *in vivo* patterns (Payne and Jansen 2003; Weld et al. 2004). Whole-phage particle administration is the only medicine that multiplies, so it is important to know what it is happening inside the human body.

Another aspect to be taken into account is how phages reach target bacteria. Experiments with mice have shown that phages are able to cross the blood-brain barrier (Brussow 2005), and similar results have been obtained in chickens (Barrow et al. 1998). Phages are more effective than antibiotics in areas of the body with poor blood circulation (Kutter 2001). Phages can not only reach the site of infection inside the organism, but have also been reported to cause the lysis of intracellular pathogens. The effectiveness of phages against *M. tuberculosis* and *Mycobacterium avium* within macrophages has been reported (Broxmeyer et al. 2002), and it should be kept in mind that *M. tuberculosis* and *M. avium* have been reported to be the most important opportunistic pathogens in immunocompromised patients such as those with AIDS. Multidrug-resistant strains have also been described.

Bacteriophage therapies are not effective against dormant spores because these have barriers that protect the peptidoglycan layer, such as the cortex. However, such barriers soon disappear, 10 min after germination, after which the bacteriophages are able to infect and destroy the pathogen (Schuch et al. 2002).

The Southwest Regional Wound Care Centre in Texas has been using phages in therapy to treat antibiotic-resistant infections (Clark and March 2006). Biodegradable patches impregnated with bacteriophages have also been used in Georgia (former Soviet State) to treat patients with chronic infections (Fischetti et al. 2006). The Phage International company has developed a product called PhagoBioDerm, a biodegradable polymer impregnated with bacteriophages (PyoPhage, BioPharm-L, Georgia), antibiotics (ciprofloxacin, benzocaine), and proteolytic enzymes (α-chymotrypsin), which can be used for both prophylaxis and therapy (www.phageinternational.com). This product showed very promising results when it was assessed in the treatment of infected venous stasis ulcers and other poorly healing wounds, where antibiotics are unable to penetrate because of poor wound vascularization. The study also demonstrated a fair degree of efficacy in the eradication of multidrug-resistant *S. aureus* in patients with skin damage (Markoishvili et al. 2002; Jikia et al. 2005). Additionally, there is a version available called PhageDent, which was formulated for periodontal applications (Shishniashvili 1999). The largest consumer of phage preparations has been the Soviet military. During the Georgian civil war in the early 1990s, soldiers carried sprays with phages against *S. aureus*, *E. coli*,

Pseudomona aeruginosa, *Streptococcus pyogenes*, and *Proteus vulgaris* (Stone 2002).

Bacteriophage components have been explored to develop tools to control processes such as milk fermentation and cheese ripening; indeed, even transgenic cattle able to secrete a recombinant lysin-like hydrolase into their milk to protect against *S. aureus* mastitis have been investigated (Fischetti et al. 2006). The use of cocktails of phages against pathogens in cattle has also been evaluated and their use as biocontrols has been demonstrated (O'Flynn et al. 2004).

In phage therapy, one of the major fears is the safety of the preparations, together with issues such as whether phages are able to act in mammalian cells or whether they are able to integrate their genes in human chromosomes. Nevertheless one consideration must be taken into account and that is that humans consume phages continuously in fermented food such as yogurt, sauerkraut, and salami, such that phages must probably be almost innocuous (Brussow 2005). Indeed, phages have even been described as contaminants in live polio vaccine preparations (Huff et al. 2003).

Humans are exposed to phages from the very moment of birth. It has been reported that contamination-free water contains up to 2×10^8 phages/mL, and they are present in the skin, urine, mouth, and many other parts of the body (Bergh et al. 1989; Yeung and Kozelsky 1997; Bachrach et al. 2003).

The most extensive safety trials in humans were undertaken on Staphage Lysate by Delmont Laboratories (Swarthmore, PA). This was administered to humans intranasally, topically, orally, subcutaneously, and intravenously. After 12 years, minor side effects were observed (Sulakvelidze and Kutter 2005). At the Institute of Immunology and Experimental Medicine, in Poland, several administration routes were tested: in oral and aerosol form, and even as infusions for rectal or surgical wounds. Intravenous administration was rejected, whereas it was observed that access through the digestive system was a fast and effective route of administration (Slopek et al. 1983; Weber-Dabrowska et al. 1985).

When immunity was considered, side effects were not observed. However, phages are antigenic and they can be cleared by the reticuloendothelial system. To resolve this problem, long-circulating phages might be employed (Merril et al. 1996; Vitiello et al. 2005).

Summarizing the foregoing, it might be said that the major problems involved in the reintroduction of phage therapy would be the biology itself of phages, and hence the difficulties in obtaining approved use. However, the alarming increase in the numbers of antibiotic-resistant bacteria and the trend against the use of antimicrobials in the food production chain seem to recommend a reappraisal of the use of phage-based therapies. In 2000, the American National Institutes of Health released a challenge-grant announcement that specifically called for proposals using phage therapies for emerging and resistant infections, placing special focus on vancomycin-resistant enterococci and multi-drug-resistant staphylococci.

A final question merits consideration. Would it be justifiable to patent a phage? More importantly, would the pharmaceutical industry be willing to invest the money in research and development into such matters and accept limited patent protection?

The use of phages has several advantages: (i) phages are self-repli-cating; (ii) they are more specific than antibiotics and hence do not cause damage to the normal microbiota; (iii) they have few side effects; (iv) they are alternatives for patients with allergies to antibiotics; (v) they have low production costs; (vi) they might be used for prophylaxis in hospital settings; (vii) their administration can be carried out through different routes; (viii) they may exhibit synergistic effects with antibiot-ics or even other medicines; (ix) whole-phage particles are able to replicate at the site of infection and so concentration occurs only where they are needed; and (x) the search for and development of new alter-natives against resistant bacteria is faster than the development of new antibiotics.

Evidently, antibiotics have certain advantages over bacteriophages, the main one being that they show a broad spectrum of action, such they can be used when the exact nature of the disease-causing bacteria is unknown. However, this wasteful use has also been responsible for the increase in antibiotic resistance.

2.3. Therapy with Phage Enzymes

The development of molecular techniques has allowed a new approach to bacteriophage-based therapy: the use of purified lytic enzymes or holins alone without the need of the whole phage particle. Indeed, in this sense, the efficacy of lysins against a variety of causal agents has been demonstrated.

The cloned enzyme PlyG from phage γ can kill *Bacillus anthracis* strains, such that when it is administered *in vivo* the enzyme readily protects mice from *B. anthracis* infections (Schuch et al. 2002).

The success of lytic enzymes has been demonstrated against infec-tions caused by vancomycin-resistant *E. faecium* and vancomycin-resis-tant *E. faecalis* (Biswas et al. 2002; Yoong et al. 2004).

For example, it has been possible to swap different lysin domains with different bacterial and catalytic domains, resulting in a new enzyme that cleaves different bonds in the peptidoglycan, but with the same specificity (Fischetti 2005). Bacteriophages show normal specificity toward the bacterial host. However, bacteriophages such as P1 are able to inject their DNA into a broad range of Gram-negative bacteria (Schoolnik et al. 2004). The lysin plyGBS, isolated from the GBS bacteriophage NCTC 11261, shows different lytic activity within the different groups of streptococci and exhibits enzymatic activity against other species such as *Streptococcus salivarius, Streptococcus gordonii*, and *Streptococcus mutans* (Cheng et al. 2005). This enzyme can be used in treatment against infections caused by *S. agalactiae* located in the vagina or oropharynx. The virion p68 produces the p17 lysin, effective against *S. aureus* (Takac and Blasi 2005). The lysins φ13 and φ6 from bacteriophages φ13 and φ6 have been shown to be efficient against *Pseudomonas syringae* (Dangelavicius et al. 2005). LysK, cloned from broad-host range staphylococcal phage K, may have therapeutic efficacy against methicillin-resistant *S. aureus* (O'Flaherty et al. 2005).

It has been possible to observe *in vitro* synergistic effects between combinations of different lytic activities against *S. pneumoniae* (Loeffler and Fischetti 2003) and *in vivo* (Jado et al. 2003). A very interesting study demonstrated the synergistic effect between lysine and antibiotics useful against *S. pneumoniae* (Djurkovic et al. 2005).

The activity of lysins from phages against caries-causing agents has also been demonstrated. Phage lysins showed a pH of action that was more effective than other types of lysin against *Streptococcus oralis* present in dental plaque. The ability of lysins to act on antibiotic-resistant bacterial biofilms, such as that formed by *S. aureus* (Sass and Bierbaum 2007), has also been demonstrated.

Nevertheless, when exogenous lytic enzymes are considered for therapy, a new problem must be taken into account. Most lysins lack a secretory signal to enable them to cross the cellular membrane. In the case of Gram-positive bacteria, exogenous lysins would have no difficulty in reaching the peptide-glycan layer. However, in Gram-negative bacteria the cell wall is more complicated, mainly due to the presence of the outer membrane, which hampers the access of lysins to the peptidoglycan layer. Thus, further research is needed to solve this problem.

During the treatment of hyperimmunized rabbits with lysins, it was discovered that no lysin-neutralizing antibodies could be recovered (Fischetti 2006). The enhancement of the lytic activity of lysozyme against *L. monocytogenes* using immunoparticles has also been reported (Yang et al. 2007).

Recently, the importance of holins and their possible applications has been brought to the attention of researchers. Recent studies have addressed the cytotoxic activity of the holin from λ bacteriophage against mammary cancer cells *in vivo*, suggesting its potential therapeutic use in cancer therapy (Agu et al. 2005).

3. FUNGAL ENZYMES

3.1. Glucanases

Fungal cell walls contain two major highly resistant polysaccharides, namely, 1,3-β-glucan and chitin, and these must be worked upon by endogenous lysins when normal morphogenetic process such as cell wall elongation or yeast budding occurs. When either of these endogenous lysins acts uncoordinatedly with biosynthetic enzymes the 1,3-β-glucan-glucanohydrolases (E.C. 3.2.1.x) are responsible for the hydrolysis of the glucoside linkage either at the nonreducing end (exo-acting enzymes, E.C. 3.2.1.58) or in the middle of the polysaccharide chain (endo-acting: E.C. 3.2.1.39). There are two main groups, depending on the type of the active amino acid involved at their active site: aspartic acid or glutamic acid (Ring et al. 1988; Sinnot 1990). There are 88 glycosyl hydrolase families (www.expasy.ch/cig-bin/lists?glycosid.txt). Family 55 is formed only by filamentous fungal 1,3-β-glucanases and its components are characterized by displaying two complete β-helix domains, which are exclusive to this family (Kawai et al. 2006). Glucanases may be used as food additives, although adverse reactions have been reported (Coenen et al. 1995).

β-1,6-glucan is also a fundamental component of fungal cell walls. Recently, it has been described that the *Candida albicans* KR5 gene-null mutants exhibit a lower adhesion capacity to human epithelial cells and are avirulent for mice (Herrero et al. 2004). This gene is involved in β-1,6-glucan synthesis, and hence degradation of the β-1,6-glucan of the fungal cell wall is one of the major possible targets in antifungal therapies.

3.2. Chitinases

Chitinases (E.C. 3.2.1.14) hydrolyze the β-1,4-glycosidic linkage between *N*-acetylglucosamine residues. They are classified within families 18 and 19 of the glycosylhydrolase families (www.expasy.ch/cig-bin/lists?glycosid.txt). They can thus be considered as part of the innate first-immune response against a chitin-containing pathogen not only in plants but also in vertebrates (Elias et al. 2005). The human chitinase AMCase is involved in host defense and and may be used in the food-processing industry (Boot et al. 2001).

Chitinases have a broad range of applications, such as in the pharmaceutical industry, protoplast isolation, control of pathogenic fungi, and the treatment of chitinous wastes (Dahiya et al. 2005). Several American companies are now developing transgenic plants with chitinase or glucanase genes to protect wheat and barley plants against fungal infections such as *Fusarium*-caused head blight, which results in important economic losses (Dahleen et al. 2001).

The use of chitinases was suggested in treatment against fungal diseases many years ago (Pope and Davis 1979), but little research has been carried out in this respect. However, the compounds generated in chitin degradation offer some interest for medical and industrial applications. For example, chitin and chitosan are used in bioremediation, in paper production, as food additives, in vaccine development, as treatment for burns, and as anticlotting agents (Felt et al. 1998; Drozd et al. 2001; Singla and Chawla 2001; Howard et al. 2003).

4. OTHER ENZYMES

Although there are many enzymes that can be considered to be enzybiotics, such as phage lytic enzymes, glucanases, and chitinases, the list must be extended to include any medically important microbial enzyme preparation, regardless of its origin (Biziulevicius et al. 2008). We believe that another group of enzymes should be taken into account, such as those that can act as helpers for enzybiotics.

Within the list of helper enzymes, one important group is that containing enzymes able to degrade the exopolysaccharides needed for biofilm formation. Biofilm is an important barrier that decreases the efficacy of treatment against pathogenic bacteria to a huge extent. In this sense, phages could represent a powerful tool against biofilms. Thus, it has been demonstrated that phage T4 can infect and replicate within *E. coli* biofilms and disrupt biofilm morphology by killing bacterial cells (Corbin et al. 2001). Recently, bacteriophage T7 has been modified to express Dispersin B, an enzyme produced by *Actinobacillus actinomycetemcomitans* (Lu and Collins 2007). This engineered phage is able to degrade biofilm, meaning that it offers a good solution against species such as *Staphylococcus* and *E. coli*. Also, exogenously applied Dispersin B is able to reduce the biofilms of species such as *E. coli, Staphylococcus epidermidis, Pseudomonas fluorescens, Yersinia pestis, Salmonella enterica* serovar Typhimurium, and *P. aeruginosa* (Itoh et al. 2005).

Dispersin B is an example of a non-bacteriophage-originated enzyme that should be considered an enzybiotic, or perhaps better as a helper. Another example is alginate lyase. This enzyme is expressed by species such as *Azotobacter chroococcum, Bacillus circulans*, and *P. aeruginosa* (Hensen and Nakamura 1985; Schiller et al. 1993; Peciña et al. 1999). Alginate is one of the main components of the secretions of *P. aeruginosa* in cystic fibrosis patients. Alginate protects microorganisms from phagocytes and antibiotics. Alginate lyase is able to degrade alginate and can thus help to eradicate bacteria.

Another important example is lysosthaphin, an endopeptidase produced naturally by *Staphylococcus simulans*. This enzyme had been approved in the former Soviet Union for treatment of gastrointestinal and gynecological diseases and its efficacy has been tested (Biziulevicius and Zukaite 1999). It has also been demonstrated that lysostaphin disrupts *S. aureus* and *S. epidermidis* biofilms on artificial surfaces (Wu et al. 2007).

Zoocin A is a peptidoglycan hydrolase produced by *Streptococcus equi* spp. *zooepidemicus* 4881, which targets a number of pathogenic streptococci such as *S. equi, S. pyogenes*, and *S. mutans*. Its capacity to bind to and hydrolyse the peptideglycan layer of several strains of *Streptococcus* has been tested *in vitro*, with the observation of its high potential as an enzybiotic (Akesson et al. 2007).

Aspergillus giganteus is an imperfect ascomycete fungus that secretes an antifungal protein (AFP). This protein is a good candidate for antifungal therapy because it has a low molecular weight and is able to inhibit the growth of filamentous fungi, especially *Fusarium* and *Aspergillus* species, but does not affect the growth of bacteria or yeast (Meyer and Stahl 2002).

Finally, the group of enzymes that block the peptideglycan layer must be taken into account. This group of enzymes that block the peptideglycan layer synthesis lidered by that of Qβ phage must be revised and considered as a possible therapeutic alternative (Bernhardt et al. 2001).

Together with these strategies for finding new therapeutic alternatives, it will be necessary to develop new bacterial vaccines to control the spread of microorganisms.

ACKNOWLEDGMENTS

The authors wish to express their appreciation to the Ramon Areces Foundation of Madrid for partly furnishing the laboratory where some experiments reported here were conducted. Patricia Veiga-Crespo also thanks the Xunta de Galicia and the University of Santiago de Compostela for financial support from the Anxeles Alvariño Grants.

REFERENCES

Agu C. A., R. Klein, S. Schwab, M. Konig-Schuster, P. Kodajova, M. Ausserlechner, B. Binishofer, U. Blasi, B. Salmons, W. H. Günzburg, and C. Hohenadl (2005) *Journal of Gene Medicine* **8**, 229–241.

Akesson A., M. Dufour, G. L. Sloan, and R. S. Simmonds (2007) *FEMS Microbiology Letters* **270**, 155–161.

Austin D. J. and R. M. Anderson (1999) *Philosophical Transactions of the Royal Society of London B Biological Sciences* **354**, 441–446.

Bachrach G., M. Leizerovici-Zigmond, A. Zlotkin, R. Naor, and D. Steinberg (2003) *Letters in Applied Microbiolgy* **36**, 50–53.

Barrow P., M. Lowell, and A. Berchieri (1998) *Clinical and Diagnostic Laboratory Immunology* **5**, 294–298.

Beckerich A. and P. Hauduroy (1922) *Comptes Rendus Des Seances de la Societe de Biologie et de ses Filiales* **86**, 168.

Benzer S. (1955) *Proceedings of the National Academy of Sciences of United States of America* **41**, 344.

Bergh O., K. Y. Borsheim, G. Bratbak and M. Heldal (1989) *Nature* **340**, 467–468.

Bernhardt T. G., I.-N. Wang, D. K. Struck, and R. Young (2001) *Science* **292**, 2326–2329.

Biswas B., S. Adhya, P. Washart, B. Paul, A. N. Trostel, B. Powell, R. Carlton, and C. R. Merril (2002) *Infection and Immunity* **70**, 204–210.

Biziulevicius G. A., G. Biziuleviciene, and J. Kazlauskaite (2008) *Journal of Pharmacy and Pharmacology* **60**, 531–532.

Biziulevicius G. A. and V. Zukaite (1999) *International Journal of Pharmaceutics* **189**, 43–55.

Boot R. G., E. F. Blonmaart, E. Swart, B. G. Van der Vlugt, N. Bijl, and C. Moe (2001) *Journal of Biological Chemistry* **276**, 6770–6778.

Borysowski J., B. Weber-Dabrowska, and A. Gorski (2006) *Experimental Biology and Medicine* **231**, 366–377.

Broxmeyer L., D. Sosnowska, E. Miltner, O. Chacon, D. Wagner, J. McGarvey, R. G. Barletta, and L. E. Bermudez (2002) *The Journal of Infectious Diseases* **186**, 1155–1160.

Brussow H. (2005) *Microbiology* **151**, 2133–2140.

Bruttin A. and H. Brussow (2005) *Antimicrobial Agents Chemotherapy* **49**, 2874–2878.

Bruynoghe R. and J. Maisin (1921) *Comptes Rendus Des Seances de la Societe de Biologie et de ses Filiales* **85**, 1120–1121.

Burman L. G., G. Lundblad, P. Cammer, R. Fange, M. Lundborg, and P. Soder (1991) *Lakartidningen* **88**, 3665–3668.

Burnett F. M. and M. Mckie (1929) *The Australian Journal of Experimental Biology and Medical Science* **6**, 277–284.

Cambell A. (1961) *Virology* **27**, 340–348.

Clark J. R. and J. B. March (2004) *FEMS Immunology and Medical Microbiology* **40**, 21–26.

Clark J. R. and J. B. March (2006) *Trends in Biotechnology* **24**, 212–218.

Cheng Q., D. Nelson, S. Zhu, and V. A. Fischetti (2005) *Antimicrobial Agents Chemotherapy* **49**, 111–117.

Cheng X., X. Zhang, J. W. Pflugrath, and F. W. Studier (1994) *Proceedings of the National Academy of Sciences of United States of America* **91**, 4034–4038.

Coenen T. M. M., A. C. M. Schoenmakers, and H. Verhagen (1995) *Food Chemistry* **95**, 859–866.

Corbin B. D., R. J. McLean, and G. M. Aron (2001) *Canadian Journal of Microbiology* **47**, 680–684.

Courcoux A. (1922) *Bull de la Societe Medicale des Hospitaux, Paris* **46**, 1151.

Dahiya N., R. Tewari, and G. S. Hoondal (2005) *Applied Microbiology and Biotechnology Onlinefirst,* 1–10.

Dahleen L. S., P. A. Okubara, and A. E. Blechl (2001) *Crop Science* **41**, 628–637.

Dangelavicius R., V. Cvirkarte, A. Gaidelyte, E. Bakiene, R. Gabrenaite-Verkhovskaya, and D. H. Bamford (2005) Journal of Viriology **79**, 5017–5026.

Davison W. C. (1922) *American Journal of Diseases of Children* **23**, 531.

Delbruck M. (1940a) *Journal of General Physiology* **23**, 643–660.

Delbruck M. (1940b) *Journal of General Physiology* **23**, 631–642.

Delfini C., M. Cersosimo, V. Del Prete, M. Strano, G. Gaetano, A. Pagliara, and S. Ambro (2004) *Journal of Agricultural and Food Chemistry* **52**, 1861–1866.

Diekema D. J., M. A. Pfaller, R. N. Jones, G. V. Doern, P. L. Winokur, A. C. Gales, H. S. Sader, K. Kugler, and M. Beach (1999) *Clinical Infectious Diseases* **29**, 595–607.

D'Hérelle F. (1917) *Les Comptes rendus de l'Académie des Sciences* **165**, 373–375.

Djurkovic S., J. M. Loeffler, and V. A. Fischetti (2005) *Antimicrobial Agents Chemotherapy* **47**, 1225–1228.

Donovan D. M., S. Dong, W. Garrett, G. M. Rousseau, S. Moineau, and D. G. Pritchard (2006) *Applied and Environmental Microbiology* **72**, 2988–2996.

Drozd N., A. Sher, V. Makarov, L. Galbraikh, G. Vikhoreva, and I. Gorbachiova (2001) *Thrombosis Research* **102**, 445–455.

Dutka B. J., A. E. Shaarawi, M. T. Matins, and P. S. Sanchez (1987) *Water Research* **21**, 1107–1025.

Elias J. A., R. J. Homer, Q. Hamid, and C. G. Lee (2005) *Journal of Allergy and Clinical Immunology* **116**, 497–500.

Ellis E. L. and M. Delbruck (1939) *Journal of General Physiology* **22**, 365.

Endtz H. P., G. J. Ruijs, B. van Klingeren, W. H. Jansen, T. vand der Reyden, and R. P. Mouton (1991) *Journal of Antimicrobial Chemotherapy* **27**, 199–208.

Epstein R. H., A. Bolle, and C. M. Steinberg (1963) *Cold Spring Symposium of Quant Biology* **16**, 413.

European Union Council (1999). *Resolution 8.7.* C195 17.7.1999

Felt O., P. Buri, and R. Gurny (1998) *Drug Development and Industrial Pharmacy* **24**, 979–993.

Fischetti V. A. (2005) *Trends in Microbiology* **13**, 491–496.

Fischetti V. A. (2006) *BMC Oral Health* **6**, S16.

Fischetti V. A., D. Nelson, and R. Schuch (2006) *Nature Biotechnology* **24**, 1508–1511.

Fisk R. T. (1942) *The Journal of Infectious Diseases* **71**, 153–160.

Fridkin S. K., C. D. Steward, J. R. Edwards, E. R. Pryor, J. E. McGowan Jr., L. K. Archibald, R. P. Gaynes, and F. C. Tenover (1999) *Clinical Infectious Diseases* **29**, 245–252.

Gales A. C., R. N. Jones, K. R. Forward, J. Liñares, H. S. Sader, and J. Verhoel (2001) *Clinical Infectious Diseases* **32**, 104–113.

Gefter M., R. Hausmann, M. Gold, and J. Hurwitz (1966) *Journal of Biological Chemistry* **241**, 1995–2004.

Goodridge L. D. (2004) *Trends in Biotechnology* **22**, 384–385.

Hankin E. H. (1896) *Annals de l'Institute Pasteur* **10**, 511.

Hatfull G. F. (2001) *Science* **22**, 2263–2264.

Hensen, J. B. and L. K. Nakamura (1985) *Applied and Environmental Microbiology* **49**, 1019–1021.

Hermoso J. A., J. L. García, and P. García (2007) *Current Opinion in Microbiology* **10**, 461–472.

Herrero A. B., P. Magnelli, M. Mansour, S. M. Levitz, H. Bussey, and C. Abeijon (2004) *Eukaryotic Cell* **3**, 1423–1429.

Hershey A. D. and M. Chase (1952) *Journal of General Physiology* **36**, 31–56.

Hershey A. D. and R. Rotman (1949) *Genetics* **34**, 44–50.

Heuer H., R. M. Kroppenstedt, J. Lottman, G. Berg, and K. Smalla (2002) *Applied and Environmental Microbiology* **68**, 1325–1335.

Higgins J. P., S. E. Higgins, K. L. Guenther, W. Huff, A. M. Donoghue, D. J. Donoghue, and B. M. Hargis (2005) *Poultry Science* **84**, 1141–1145.

Hiramatsu K., N. Aritaka, H. Hanaki, S. Kawasaki, Y. Hosoda, S. Hori, Y. Fuckichi, and I. Kobayashi (1997) *The Lancet* **350**, 1670–1673.

Holtzman D. (2003) *ASM News* **69**, 489–490.

Howard M. B., N. A. Ekborg, R. M. Weiner, and S. W. Hutcheson (2003) *Journal of Industrial Microbiology and Biotechnology* **30**, 627–635.

Huff W. E., G. R. Huff, N. C. Rath, J. M. Balog, and A. M. Donoghue (2005) *Poultry Science* **84**, 655–659.

Ibrahim H. R., T. Aoki, and A. Pellegrini (2002) *Current Pharmaceutical Design* **8**, 671–693.

Infectious Diseases Society of America (2004) *Bad Bugs, No Drugs* July (7).

Itoh Y., X. Wang, B. J. Hinnebusch, J. F. Preston III, and T. Romeo (2005) *Journal of Bacteriology* **187**, 382–387.

Jado I., R. Lopez, E. Garcia, A. Fenoll, J. Casal, and P. Garcia (2003) *Journal of Antimicrobial Chemotherapy* **52**, 967–973.

Jikia D., N. Chkhaidze, E. Imedashvilli, I. Mgaloblishvli, F. Tsitlanadze, R. Katsarava, J. G. Morris, and A. Sulakvelidze (2005) *Clinical and Experimental Dermatology* **30**, 23–26.

Kakikawa M. (2002) *Gene* **299**, 227–234.

Kawai R., K. Iigarashi, and M. Samejima (2006) *Biotechnology Letters* **28**, 365–371.

Khakhria R., D. Duck, and H. Lior (1990) *Epidemiology and Infection* **105**, 511–520.

Kotila J. E. and G. H. Coons (1925) *Michigan Agricultural Experiment Station Technical Bulletin* **67**, 357–370.

Kovacs E. W., J. M. Hooker, D. W. romanini, P. G. Holder, K. E. Berry, and M. B. Francis (2007) *Bioconjugate Chemistry* **18**, 1140–1147.

Kreling D. H., D. A. Mott, J. B. Wiederholt, J. Lundy, and L. Levitt (2001) Prescription Drug Trends. A Chartbook Update. Menlo Park, CA: The Henry J. Kaiser Family Foundation.

Kutter E. (2001) *Phage Therapy: Bacteriophage as Natural Self-limiting Antibiotics*, 1st ed. Bangalore, India: AstraZeneca Research Foundation India.

Lederberg E. M. (1951) *Genetics* 36, 560.

Lee C., J. Lin, T. Chow, Y. Tseng, and S. Weng (2006) *Protein Expression and Purification* **50**, 229–237.

Leverentz B., W. S. Conway, W. Janisiewicz, and M. J. Camp (2004) *Journal of Food Protection* **67**, 1682–1686.

Link G. K. (1928) *The Newer Knowledge of Bacteriology and Immunology.* Chicago: University of Chigago Press, 560–606.

Litvinova A. M., V. M. Chtetsova, and I. G. Kavtreva (1978) *Voprosy okhrany materinstva i detstva* **9**, 42–44.

Livermore D. (2003) *Clinical Infectious Diseases* **36**, S11–S36.

Livermore D. (2004) *Nature Reviews in Microbiology* **2**, 73–78.

Loeffler, J. M. and V. A. Fischetti (2003) *Antimicrobial Agents Chemotherapy* **47**, 375–377.

Loessner M. J. (1990) *Deutsche Milchwirtschaft* **33**, 1115–1118.

Lopez R., E. Garcia, P. Garcia, and J. L. Garcia (1997) *Microbial Drug Resistance* **3**, 199–211.

Lu T. K. and J. J. Collins (2007) *Proceedings of the National Academy of Sciences of United States of America* **104**, 11197–11202.

Luo Y., P. Pfister, T. Leisinger, and A. Wasserfallen (2001) *Journal of Bacteriology* **183**, 5788–5792.

Luria S. E. and M. Delbruck (1943) *Genetics* **28**, 491–511.

Luria S. E., M. Delbruck, and T. F. Anderson (1943) *Journal of Bacteriology* **46**, 57–67.

Luria S. E. and M. L. Human (1952) *Journal of Bacteriology* **64**, 557–562.

Lwoff A. (1953) *Bacteriology Reviews* **17**, 269–337.

Mallmann W. L. and C. J. Hermstreet (1924) *Agricultural Research* **28**, 599–602.

Markoishvili K., G. Tsitlanadze, R. Katsavara, J. G. Morris, and A. Sulakvelidze (2002) *International Journal of Dermatology* **41**, 453–458.

Meladze G. D., M. G. Mebuke, N. S. Chkhetia, N. I. Kiknadze, G. G. Koguashvili, I. I. Timoshuk, N. G. Larionova, and G. K. Vasadze (1982) *Grudnaia khirurgiia* **1**, 53–56.

Merril C. R., B. Biswas, R. Carlton, N. C. Jensen, G. J. Creed, S. Zullo, and S. Adhya (1996) *Proceedings of the National Academy of Sciences of United States of America* **93**, 3188–3192.

Meyer V. and U. Stahl (2002) *Currents in Genetics* **42**, 36–42.

Morrison J. (1932) *Bacteriophage in the Treatment and Prevention of Cholera.* London: HK Lewis & Co.

Navarre W. W., H. Ton-That, K. F. Faull, and O. Schneewind (1999) *Journal of Biology Chemistry* **274**, 15847–15856.

Nelson D., R. Schuch, P. Chahales, and V. A. Fischetti (2006) *Proceedings of the National Academy of Sciences of United States of America* **103**, 10765–10770.

Nordberg P., D. L. Monnet, and O. Cars (2005) Background Document for the WHO Project: Priority Medicines for Europe and the World, a Public Health Approach to Innovation. http://mednet3.who.int/prioritymeds/report/index.htm (accessed August 9, 2005).

O''Flaherty S., A. Coffey, W. Meaney, F. F. Fitzgerald, and R. P. Ross (2005) *Journal of Bacteriology* **187**, 7161–7164.

O'Flynn G., R. P. Ross, G. F. Fitzgerald, and A. Coffey (2004) *Applied and Enviromental Microbiology* **70**, 3417–3424.

Owen R. J., J. Hernandez, and F. Bolton (1990) *Epidemiology and Infection* **105**, 265–275.

Park T., D. K. Struck, J. F. Deaton, and R. Young (2006) *Proceedings of the National Academy of Sciences of the United States of America* **103**, 19713–19718.

Parker M. T. (1972) Phage-typing of *Staphylococcus aureus*, pp. 1–28. *In* J. R. Norris and D. W. Ribbons (ed.), *Methods in microbiology*, vol. **7B**. Academic Press, London, United Kingdom.

Payne R. J. and V. A. Jansen (2003) *Clinical Pharmacokinetics* **42**, 315–323.

Peciña A., A. Pascual, and A. Paneque (1999) *Journal of Bacteriology* **181**, 1409–1414.

Pollitzer R. (1959) *Cholera*. Geneva: World Health Organization.

Pope A. and D. Davis (1979) *Postgraduate Medical Journal* **55**, 674–679.

Powers J. H. (2004) *Clinical Microbiological Infection* **10**, 23–31.

Pritchard D. G., S. Dong, J. R. Baker, and J. A. Engler. (2004) *Microbiology* **150**, 2079–2087.

Ring M, D. E. Bader, and M. Hubber (1988) *Biochemical and Biophysical Research Communications* **152**, 1050–1055.

Riska P. F., Y. Su, S. Bardarov, L. Freundlich, G. Sarkis, G. Hatfull, C. Carriere, V. Kumar, J. Chan, and W. R. Jacobs (1999) *Journal of Clinical Microbiology* **37**, 1144–1149.

Sahm D. F., M. K. Marsilio, and G. Piazza (1999) *Clinical Infectious Diseases* **29**, 259–263.

Sakandelidze V. M. (1991) *Vrachebnoe delo* **3**, 60–63.

Sao-Jose C., R. Parreira, G. Vieira, and M. A. Santos (2000) *Journal of Bacteriology* **182**, 5823–2831.

Sass, P. and G. Bierbaum (2007) *Applied and Enviromental Microbiology* **73**, 347–352.

Savva C. G., J. S. Dewey, J. Deaton, R. L. White, D. K. Struck, A. Holzenburg, and R. Young (2008) *Molecular Microbiology* **69**, 784–793.

Schiller N. L., S. R. Monday, C. M. Boyd, N. T. Keen, and D. E. Ohman (1993) *Journal of Bacteriology* **175**, 4780–4789.

Schlesinger M. (1936) *Nature* **138**, 508–509.

Schoolnik G. K., W. C. Summers, and J. D. Watson (2004) *Nature Biotechnology* **22**, 505.

Schuch R., D. Nelson, and V. A. Fischetti (2002) *Nature* **418**, 884–889.

Sherman C. A., K. L. Jury, and M. J. Gasson (1994) *Applied and Envioroment Microbiology* **60**, 3063–3073.

Shishniashvili T. E. (1999) *Medical Journal of Australia* **2**, 71–78.

Singla A. and M. Chawla (2001) *Journal of Pharmacy and Pharmacology* **53**, 1047–1067.

Sinnot M. L. (1990) *Chemistry Review* **90**, 1171–1202.

Skurnik M. and E. Strauch (2006) *International Journal of Medical Microbiology* **296**, 5–14.

Slopek S., I. Durlakowa, B. Weber-Dabrowska, A. Kucharewicz-Krukowska, M. Dabrowski, and R. Bisikiewicz (1983) *Archivum Immunologiae et Therapie Experimentalis* **31**, 267–291.

Slopek S., I. Durlakowa, B. Weber-Dabrowska, M. Dabrowski, and A. Kucharewicz-Krukowska (1984) *Archivum Immunologiae et Therapie Experimentalis* **32**, 317–335.

Slopek S., A. Kucharewicz-Krukowska, B. Weber-Dabrowska, and M. Dabrowski (1985a) *Archivum Immunologiae et Therapie Experimentalis* **33**, 219–240.

Slopek S, A. Kucharewicz-Krukowska, B. Weber-Dabrowska, and M. Dabrowski (1985b) *Archivum Immunologiae et Therapie Experimentalis* **33**, 241–259.

Slopek S., A. Kucharewicz-Krukowska, B. Weber-Dabrowska, and M. Dabrowski (1985c) *Archivum Immunologiae et Therapie Experimentalis* **33**, 261–273.

Smith G. P. (1985) *Science* **228**, 1315–1317.

Spence R. C. and E. B. McKinley (1924) *Southern Medical Journal* **17**, 563.

Stettler R., C. Thurner, D. Stax, L. Meile, and T. Leisinger (1995) *FEMS Microbiol Letters* **132**, 85–89.

Stone R. (2002) *Science* **298**, 728–731.

Struble J. M. and R. T. Gill (2006) *Antimicrobial Agents and Chemotherapy* **50**, 2506–2515.

Sulakvelidze A. (2005) *Drugs Discovery Today* **12**, 807–809.

Sulakvelidze A., Z. Alavidze Jr., and J. G. Morris (2001) *Antimicrobial Agents Chemotherapy* **45**, 649–659 .

Sulakvelidze A. and E. Kutter (2005) Bacteriophage therapy in humans. In *Bacteriophages: Biology and Applications*, ed. E. Kutter and A. Sulakvelidze. Boca Raton, FL: CRC Press.

Summers W. C. (1999) *Felix D'Hérelle and the Origins of Molecular Biology*, 1st ed. New Haven, CT: Yale University Press.

Takac M. and U. Blasi (2005) *Antimicrobial Agents Chemotherapy* **49**, 2934–2940.

Talbot G. H., J. Bradley, J. E. Edwards, D. Gilbert, M. Scheld, and J. G. Bartlett (2006) *Clinical Infectious Diseases* **42**, 657–668.

Touch V., S. Hayakawa, K. Fukada, Y. Aratani, and Y. Sun (2003) *Journal of Agricultural and Food Chemistry* **51**, 5154–5161.

Twort F. W. (1915) *Lancet* **ii**, 1241.

U.S. Centers for Disease Control, Food and Drug Administration, and U.S. National Institutes of Health (2000) www.cdc.gov/drugresistance/action-plan/aractionplan.pdf (accessed July 2009).

U.S. Food and Drug Administration Approvals (2004). *Antimicrobial Agents and Chemotherapy* **48**, 1438–1439.

Varea J., B. Monterroso, J. Saiz, C. Lopez-Zumel, J. L. Garcia, J. Laynez, P. Garcia, and M. Menendez (2004) *Journal of Biological Chemistry* **279**, 43697–43707.

Vitiello C. L., C. R. Merril, and S. Adhya (2005) *Virus Research* **114**, 101–103.

Wang I. N., D. L. Smith, and R. Young (2000) *Annual Review of Microbiology* **54**, 799–825.

Weber-Dabrowska B., M. Dabrowski, and S. Slopek (1985c) *Archivum Immunologiae et Therapie Experimentalis* **35**, 261–273.

Weld R. J., C. Butts, and J. A. Heinemann (2004) *Journal of Theorical Biology* **227**, 1–11.

Westwater C., L. M. Kasman, D. A. Schofield, P. A. Werner, J. W. Dolan, M. G. Schmidt, and J. S. Norris (2003) *Antimicrobial Agents Chemotherapy* **47**, 1301–1307 .

Wollman E. (1928) *Bulletin Institute Pasteur* **26**, 1–14.

Workshop Summary (1998). *Antimicrobial Resistance: Issues and Options.* Washington, DC: Institute of Medicine, National Academy Press.

World Health Organization (2001) *WHO Global Strategy for Containment of Antimicrobial Resistance.* Geneva: WHO. http://www.who.int/entity/csr/resources/publicatios/drugresist/en/EGlobal_Strat.pdf (accessed July 2009).

Wu J. A., C. Kusuma, J. J. Mond, and J. F. Kokai-Kun (2007) Antimicrobial Agents and Chemotherpay **47**, 3407–3414.

Yang H., L. Qu, A. Wimbrow, X. Jiang, and Y.-P. Sun (2007) *Journal of Food Protection* **70**, 1844–1849.

Yeung M. K. and C. S. Kozelsky (1997) *Plasmid* **37**, 141–153.

Yoong P., R. Schuch, D. Nelson, and V. A. Fischetti (2004) *Journal of Bacteriology* **186**, 4808–4812.

Young R. (2002) *Journal of Molecular Microbiology and Bioctehnology* **4**, 21–36.

Zinder N. D. and J. Lederberg (1952) *Journal of Bacteriology* **64**, 679.

CHAPTER 3

ENZYBIOTICS AS SELECTIVE KILLERS OF TARGET BACTERIA

JUAN C. ALONSO[1] and MARCELO E. TOLMASKY[2]

[1]Department of Microbial Biotechnology, Centro Nacional de Biotecnología, CSIC, Madrid, Spain

[2]Center for Applied Biotechnology Studies, Department of Biological Science, College of Natural Science and Mathematics, California State University Fullerton, CA

1. INTRODUCTION

Disinfection is the reduction of the load of pathogens from the environment with the aim of decreasing the risk of disease. It is commonly accepted that Joseph Lister introduced the first "modern" disinfectant, phenol (known at that time as carbolic acid), which was used to decontaminate surgical wounds. Interestingly, the irritation produced in the hands of surgeons by this disinfectant led to the first use of rubber gloves by surgeons. There are several classes of disinfectants available, including chlorine compounds such as bleach, iodine and iodophores, aldehydes, phenols, quaternary ammonium compounds, oxidizing agents, and biguanidines. Although in theory they should have low toxicity to humans and be able to reduce microbial contamination by several orders of magnitude, in practice they have a certain inherent level of danger and in many cases they exhibit less than ideal reduction of the load of pathogens.

Immediately following the tragic events of September 11, 2001, in the United States, a bioterrorist act was committed at multiple locations (Atlas 2002; Schmid and Kaufmann 2002). Between September and November of that year, the perpetrator(s) intentionally mailed *Bacillus anthracis* spores to several sites, creating terror and panic in a broad population and causing five deaths and a total of 22 infections

Enzybiotics: Antibiotic Enzymes as Drugs and Therapeutics. Edited by Tomas G. Villa and Patricia Veiga-Crespo

(at least 18 confirmed). The attacks served as a wake-up call, and many industrialized countries are now considered extremely vulnerable to future terrorist attacks with biological weapons, or bioweapons, which could result in a large number of casualties and panic in the population and could also dramatically affect the economy.

To discern the nature of the threat of bioweapons and the appropriate responses to them requires great attention to the biological characteristics of the bioweapon and the panic that it can generate. To reduce their potentially devastating effects an effective policy to clean up contaminated area is essential.

It is believed that the person(s) responsible for the 2001 anthrax outbreak mailed at least seven letters containing anthrax spores, resulting in over 20 contaminated sites. The team assigned to clean up the sites, confronting the unprecedented challenge of dealing with weapon-grade *B. anthracis* spores, was forced to rely on methods that were untested in the types of environments that were affected, which included office buildings and a US Postal Service Processing and Distribution Center. The Hart Senate Office Building was fumigated and remained closed for a total of 96 days (http://www.avma.org/onlnews/javma/mar02/s031502c.asp). The cleanup of this building required the collection of about 6000 samples and generated more than 30 tons of waste. As of December 3, 2002, there were still several contaminated locations, and the cost of the cleanup had reached $750 million. By the completion date, the cost for decontamination of all buildings had climbed even higher. Clearly, the issues of toxicity, time, and resources indicate that decontaminating environments after a bioterrorist attack is a critical component of the response and one that urgently needs further research.

While disinfectants are considered effective if they reduce the load of pathogens by several orders of magnitude, new standards must be applied for the cleanup of accidental or intentional contamination with biological weapons in which all microorganisms should be physically destroyed. New disinfectants to be used in these cases should reduce the threat to negligible levels, have low toxicity to humans, be easy to obtain and to remove, cause as little as possible damage to structures, and act as quickly as possible. Their affordability should be also considered.

Among the new ideas for generating novel disinfectants is the utilization of bacteriophages (Hermoso et al. 2007; Yacoby et al. 2007; Parisien et al. 2008) or their murein hydrolases (enzybiotics; Schuch et al. 2002; Cheng and Fischetti 2007; see Chapters 2 and 5). Murein hydrolases are grouped in two classes: autolysins, bacterium-coded enzymes that play a variety of roles such as cell wall biosynthesis, cell separation, cell

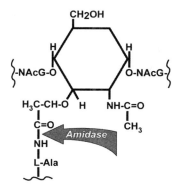

Figure 3.1. *N*-acetylmuramoyl-L-alanine amidase activity. The diagram shows the relevant portion of the peptidoglycan structure and the enzymatic activity. In *B. cereus* the GlcNAc (*N*-acetyl-D-glucosamine) is mostly deacetylated (Loessner et al. 1997).

adhesion, and virulence, which are secreted from the bacterial cytoplasm (Smith et al. 2000; Parisien et al. 2008; Vollmer et al. 2008), and endolysins, bacteriophage-encoded enzymes that reach and subsequently cleave the peptidoglycan, through membrane lesions formed by holins. This process results in lysis of the bacterial cell and the release of mature viral particles (Loessner et al. 1997; Navarre et al. 1999; Smith et al. 2000; Wang et al. 2000; Parisien et al. 2008; also see Chapters 4–8). Endolysins have a modular structure and include domains responsible for amidase, peptidase, and *N*-acetylglucosaminidase activities. In addition they posses a cell wall-specific targeting domain that is commonly found at the C-terminal end (Loessner et al. 1999; Navarre et al. 1999; Fig. 3.1). An interesting characteristic of some of these enzymes is that they induce cell lysis when added to the appropriate bacterial cells from the outside (Loessner et al. 1997; Loeffler et al. 2001; Nelson et al. 2001; Schuch et al. 2002).

B. anthracis belongs to the *Bacillus cereus* group of bacteria, which comprises closely related Firmicute organisms with highly divergent virulence properties (Hoffmaster et al. 2004). It was shown that the *B. anthracis* phage γ-encoded endolysin was capable of degrading peptidoglycan when applied externally to *B. anthracis* or to the closely related *B. cereus* strain RSVF1, resulting in a rapid lysis of the cell (Schuch et al. 2002). Furthermore, several other endolysins have been shown to induce cell lysis when added to the appropriate cells from the outside. Different endolysins were found to have different degrees of specificity. The group B streptococcal B30 bacteriophage endolysin is active against all β-haemolytic streptococci tested, including groups A, B, C, E, and G streptococci, with different efficiencies (Pritchard et al. 2004). *Enterococcus faecalis* phage φ1 endolysin PlyV12 is active against groups B and C streptococci, and *Staphylococcus aureus* (Yoong et al. 2004). *Lactobacillus helveticus* phage φ0303 protein Mur-LH is active against at least 10 bacterial species (Deutsch et al. 2004). The φ11 endolysin hydrolyzed heat-killed staphylococci as well as staphylococcal biofilms, untreated staphylococcal mastitis pathogens, *S. aureus*, and coagulase-negative staphylococci (Donovan et al. 2006, Sass and Bierbaum 2007). The murein hydrolase of the bacteriophage φ3626 dual lysis system was active against all tested *C. perfringens* strains (Zimmer et al. 2002). The gp144 endolysin protein from the *Pseudomonas aeruginosa* phage φKZ showed *in vitro* lytic activity against *P. aeruginosa* cells and degraded purified peptidoglycan of Gram-negative bacteria (Paradis-Bleau et al. 2007). In other cases endolysins showed little killing activity when added exogenously to the viral host, for example, the *Listeria monocytogenes* bacteriophage N-acetylmuramoyl-l-alanine amidase endolysin, Ply511, was unable to control *L. monocytogenes* growth (Turner et al. 2007).

The discovery that endolysins from bacteriophages that have Firmicutes as hosts can also mediate hydrolysis of the peptidoglycan from the outside of the cells led to their use as agents to kill bacteria (Borysowski et al. 2006; Chapter 7). Although the bactericidal action of endolysins when added from the outside seems to be mainly confined to Firmicutes, their potential as antimicrobials for a large number of bacteria led to the coinage of the term enzybiotics (Nelson et al. 2001; Jado et al. 2003; Veiga-Crespo et al. 2007; see Chapters 2 and 5). This term is currently also used for all enzymes that are able to mediate killing of microbial cells regardless of their origin (Veiga-Crespo et al. 2007; see Chapter 4).

2. SPECIFIC DISEASES ASSOCIATED WITH BIOLOGICAL WEAPONS

In theory, any infectious agent could be engineered for use as a biological weapon. Experts in this field believe that bacteria of the Firmicutes (*B. anthracis* and *Clostridium botulinum*, C. perfringens, S. aureus, etc.) and of the γ-proteobacteria phylum (*Yersinia pestis* and *Francisella tularensis*) are among the pathogens most likely to be used in a bioterrorist attack.

The disease caused by *B. anthracis*, anthrax, has been in existence for hundreds of years and still occurs naturally in both animals and humans in many parts of the world, including Asia, southern Europe, sub-Saharan Africa, and parts of Australia. There are three forms of disease caused by anthrax: cutaneous anthrax, inhalation anthrax, and gastrointestinal anthrax. Cutaneous anthrax is the mildest form and usually gets cured in about 6 weeks, whereas the most severe form is inhalational anthrax, which can go unrecognized until it is too late for effective treatment (Mock and Fouet 2001).

Cutaneous anthrax is characterized by a red-brown spot that becomes larger and surrounded by redness. The center develops into a fluid-filled blister that eventually ruptures, forming an ulceration that dries up with a black crust. A series of images of the lesions caused by cutaneous anthrax can be seen at the Centers for Disease Control and Prevention (CDC) website (http://www.bt.cdc.gov/agent/anthrax/anthrax-images/cutaneous.asp). Lymph nodes swell and the disease can be accompanied by muscle pain, headache, fever, nausea, and vomiting. Mortality of untreated cutaneous anthrax is about 20%, but the rate is reduced to 1% when the infected person is given appropriate treatment. Intestinal anthrax is currently rare and is acquired by consumption of contaminated undercooked meat. The earlier symptoms include loss of appetite, fever, and bloody diarrhea followed by spreading throughout the bloodstream. Fatality rate is between 25% and 60% of the cases and treatment success strongly depends on how soon treatment starts. Inhalation anthrax, also known as pulmonary, pneumonic, or respiratory anthrax, is the most serious form of the infection with a mortality rate of almost 100%. Spore-bearing particles reach the alveolar spaces where spores are ingested by macrophages. Although some of the spores are lysed, those that survive are transported via lymphatics to mediastinal lymph nodes, where germination occurs in a period of up to 60 days (Friedlander et al. 1993; Inglesby et al. 1999). Inhalation anthrax starts with flu-like symptoms; after a few days the second stage of the disease starts abruptly, and the patient worsens feeling severe respiratory distress with sudden fever, dyspnea, diaphoresis, and shock. The disease continues to evolve and the patient may fall into a coma followed by death. As many as 50% of the patients develop hemorrhagic meningitis, delirium, and obtundation. A summary of the symptoms and evolution of bioterrorism-related inhalation anthrax victims has been published (Inglesby et al. 1999; Jernigan et al. 2001). To identify the presence of *B. anthracis* in the United States, the Department of Health and Human Services, CDC, engages its partners in the Laboratory Response Network (LRN), which was established by the same institution to rapidly identify threat agents in 1999 and is a collaborative partnership and multilevel system linking state and local public-health laboratories. Local clinical laboratory testing is confirmed at state and large metropolitan public health laboratories. The CDC conducts the definitive or highly specialized testing for major threat agents. Samples must be collected in a form suitable for testing, which is a two-step process. The first test may show positive results within 2 h, but only for samples that contain large numbers of cells. However, the result must be confirmed with a second test, which requires more sophisticated laboratory facilities and takes longer (usually 1–3 days; data from http://www.bt.cdc.gov/agent/anthrax/faq/labtesting.asp).

   C. botulinum produces botulism, a rare but serious paralytic illness caused by a neurotoxin known as the botulinum toxin, which has seven different serotypes (A-G; Dembek et al. 2007). The botulinum toxin can be inhaled or ingested via contaminated food or water (Arnon et al. 2001). There are three main kinds of botulism: foodborne, wound, and infant botulism. The first is caused by eating food that contains the botulism toxin; the second occurs when a wound is infected with *C. botulinum*; and the third occurs when an infant ingests the spores of the botulinum bacteria, which can be found in dirt and dust, or which can contaminate honey. This form of the illness mostly affects babies who are between 3 weeks and 6 months old. The spores germinate in the intestines and release the toxin. All three forms can be fatal. The botulinum toxin is synthesized as a 150 kD polypeptide that is subsequently activated by proteolysis to form a heavy and a light chain that remain linked by a disulfide bond (Lacy et al. 1998; Stenmark et al. 2008). The toxin causes muscle paralysis, which translates in a variety of symptoms such as double vision, blurred vision, drooping eyelids, slurred speech, difficulty swallowing, dry mouth, and muscle weakness. These symptoms progress to cause paralysis of the arms, legs, trunk, and respiratory muscles. An aerosolized or foodborne botulinum toxin weapon would cause acute symptoms 12–72 h after exposure. An effective response to an intentional release of the toxin include rapid diagnosis and treatment with antitoxin (Arnon et al. 2001; Kman and Nelson 2008).

   Y. pestis is the causative agent of plague, an infectious disease of animals and humans, and it is probably the first recorded utilization of a bioweapon when in 1347 the Tartar army catapulted infected corpses over the Caffa city walls. Plague is transmitted between rodents by rodent fleas or to people through infected rodent fleabites or through direct contact with infected animal tissue. There are three main forms of plague in humans: bubonic, pneumonic, and septicemic. Bubonic plague, the most common form of the disease, usually starts with a bite of an infected flea that inoculates bacterial cells. After 2–8 days the patient develops symptoms that include abrupt onset of fever, chills, headache, and weakness. Painful proximal lymphadenopathy develops, most commonly in the inguinal, axillary, or cervical areas. Lymph nodes become enlarged, matted, and associated with extensive, overlying edema (buboes). Gastrointestinal symptoms such as nausea, vomiting, and abdominal pain may occur. The body defenses become over-whelmed by the large number of bacteria, which leave the lymph nodes and enter the bloodstream from which they may colonize other organs. Then, the secondary septicemia usually results in sepsis and death. In about 5%–15% of the cases, when the disease reaches the terminal stage, bacterial cells spread to the lungs, causing secondary pneumonic plague. The pneumonic infection is rare but is the most likely presenta-tion of a bioterrorist attack as a consequence of deliberate aerosoliza-tion of bacteria that would mimic the contact with respiratory droplets from an infected person. The symptoms develop 1–6 days following exposure and consist of a bronchopneumonia characterized by fever, cough, dyspnea, and serosanguineous sputum. From the lungs, the bacterial cells spread to the blood, causing symptoms of septicemic plague. Suspicion of a deliberate attack should be raised by the rapid progression of a lethal pneumonia in a group of previously healthy people from the time of respiratory exposure to death, which ranges from 2 to 6 days. The septicemia arises secondary to bubonic plague or following a fleabite without the development of buboes. Septicemia is associated with disseminated intravascular coagulation, necrosis of small blood vessels, purpuric skin lesions, and acral gangrene. Death is the result of shock and multiple organ failure.

F. tularensis causes a highly infectious disease called Tularemia. There are two predominant subspecies: *F. tularensis tularensis* (type A), which is the most virulent and is found in North America, and *F. tularensis palaearctica* (type B), which occurs in Asia, Europe, and North America. A large number of animal species throughout the world are susceptible to *F. tularensis*, which can be recovered from water, soil, and vegetation. Humans can be infected through the skin, mucous membranes, gastrointestinal tract, and lungs. Clinical manifes-tations depend on the route of entry and the virulence of the agent. There are seven forms of tularemia in humans: ulceroglandular, pneu-monial, oculoglandular, glandular, oropharyngeal, typhoidal, and sepsis. The ulceroglandular accounts for 60%–80% of all naturally occurring cases. Infection typically results from handling an infected carcass or follows the bite of an infected arthropod or another animal.

With the onset of generalized symptoms, a papule appears at the inoculation site; it quickly becomes pustular, ulcerates, and may develop an eschar. Regional lymphadenopathy develops; it may suppurate and rupture. In the oculoglandular case, ulceration occurs on the conjunctiva, accompanied by chemosis, vasculitis, and regional lymphadenitis, whereas in the glandular case lymphadenopathy and generalized symptoms without an ulcer were observed. Pneumonial tularemia can result from direct inhalation of contaminated aerosols (naturally occurring or deliberately disseminated, as in a bioterrorist attack) or be secondary to hematogenous spread from another site. One or more of the following is present: pharyngitis, bronchiolitis, pleuropneumonitis, or hilar lymphadenitis. In the correct clinical setting, the presence of nodular infiltrates with a pleural effusion should suggest either tularemia or plague pneumonia. In a substantial number of patients, pulmonary signs may be minimal or absent, and generalized constitutional symptoms may predominate. Oropharyngeal tularemia is acquired by ingesting contaminated water or food, or occasionally by inhaling contaminated droplets. The patient may develop stomatitis, but more commonly develops exudative pharyngitis or tonsillitis with ulceration. Cervical or retropharyngeal lymphadenopathy may also occur. Typhoidal tularemia is a fatal form of the disease with the microorganisms disseminating rapidly to multiple organs by hematogenous spread. It has a gastrointestinal manifestation, including abdominal pain and diarrhea. In tularemia sepsis the early symptoms are fever, abdominal pain, diarrhea, and vomiting. It may progress to septic shock with complications of the systemic inflammatory response including disseminated intravascular coagulation, adult respiratory distress syndrome, and multiple organ failure.

3. THE PARADIGM: AN ENZYBIOTIC ATTACHED TO NONPATHOGENIC NONPROLIFERATING BACTERIA

In case of a bioterrorist attack with a bacterial agent such as any of those described in the previous section or an accidental release, one critical issue will be the decontamination of the areas affected. In this section we describe a methodology using innocuous live organisms and enzybiotics.

The capacity of a phage-encoded murein hydrolase to specifically kill bacteria was first documented by Krause (1957), and demonstrated with a purified enzyme by Vincent Fischetti's group (Nelson et al. 2001). Table 3.1 shows a list of bacteria that could be used as bioweap-

TABLE 3.1. Target Bacteria (Selected Agents) and Potential Bacteriophages

Selected Agents[a]	Bacteriophage Family(ies)[b]
Bacillus anthracis	Siphoviridae
Brucella abortus	Podoviridae morphotype C1
Brucella melitensis	Podoviridae morphotype C1
Brucella suis	Podoviridae morphotype C1
Burkholderia mallei[c]	Myoviridae morphotype A1, A2
Burkholderia pseudomallei[d]	Myoviridae morphotype A1, A2
Clostridium perfringens	Siphoviridae morphotype B1, Inoviridae, Myoviridae morphotype A1, A2
Clostridium botulinum	Siphoviridae morphotype B1, Inoviridae, Myoviridae morphotype A1, A2
Coxiella burnetii	Myoviridae morphotype A1, A2
Francisella tularensis	Myoviridae morphotypes A1, A2
Rickettsia prowazekii	Podoviridae morphotype C1
Rickettsia prowazekii	Podoviridae morphotype C1
Rickettsia rickettsii	Podoviridae morphotype C1
Staphylococcus aureus	Siphoviridae morphotype B1, B2
Yersinia pestis	Myoviridae morphotypes A1, A2

[a]http://www.cdc.gov/od/sap/docs/salist.pdf
[b]http://www.mansfield.ohio-state.edu/~sabedon/names.htm; http://www.phage.org; http://www.ebi.ac.uk/genomes/phage.html
[c]Formerly *Pseudomonas mallei.*
[d]Formerly *Pseudomonas pseudomallei.*

ons and bacteriophages that could serve as sources of endolysins (enzybiotics).

The enzybiotics could be used directly or attached to microorganisms, the latter approach offers advantages over the direct used of the purified enzyme (Table 3.2). Different enzybiotics attached to microorganisms could be suitable candidates to design "a la carte" disinfectants suitable for treating accidentally or intentionally contaminated environments.

These enzybiotics include many desirable characteristics:

- easy to store and to scale up
- easy to deliver
- effective
- gentle to the environment
- nonhazardous and non dangerous
- specie-specific
- affordable
- easy to remove

TABLE 3.2. Comparative of Purified Enzybiotic Enzyme and Enzybiotic Attached to a Nonproliferating Bacteria

Purified Enzybiotics	Enzybiotic Attached to a Bacteria
a. Bacterial cells bearing a plasmid-borne enzybiotic gene must be over-expressed.	a. Bacterial cells bearing a plasmid-borne enzybiotic fused to a membrane protein are grown.
b. Enzybiotic must be purified or partially purified, concentrated, and lyophilized.	b. Different cell culture exposing different enzybiotic are lyophilized and ready to be used.
c. Large amounts of enzybiotic that target different bioweapons need to be produced.	c. Not applicable
d. Large amounts of enzybiotic need to be stored at low temperatures or need to develop a system to keep their activity at room temperature.	d. Strains can be kept in a few small vials at −80°C (minimal storage room in a freezer) or in a lyophilized (at room temperature).
e. If there are any activity problems before its use it can be time-consuming to produce a new protein batch.	e. Production is very fast (a bacterial culture). There is no risk of inactivation because there will be new cultures every time there is a need to use them.
f. Not applicable	f. Once the bacterial cells are no longer needed (after cleanup), they can be eliminated by inducing a block in cell proliferation.

The vast majority of virulence factors or toxin genes described in bacteria with bioweapon potential, namely *B. anthracis*, *Clostridium tetani*, *C. botulinum*, and *Y. pestis*, are plasmid-borne (Mock and Fouet 2001; Bruggemann 2005; Huang et al. 2006). Therefore, to avoid frequent horizontal gene transfer during the cleanup step, the enzybiotic should be housed by a nonproliferating distantly related bacteria to block the possibility of plasmid mobilization and subsequent establishment. Once the enzybiotic bacteria are released to the environment to act as disinfectant, their growth must be tightly controlled by introducing multiple auxotrophies and check-points elements (e.g., toxin-antitoxin) that should block cell proliferation.

Previous work using tripartite fusions consist of (i) the signal sequence and first nine N-terminal amino acids of the mature major *Escherichia coli* lipoprotein, (ii) the amino acids 46–159 of the outer membrane protein OmpA, and (iii) the amino acid sequence of proteins of interest showed that the protein can be transported through the membrane and become anchored to the external surface of the cell and keep enzymatic activity (Georgiou et al. 1993; Francisco and Georgiou 1994). Tripartite fusion systems expressing beta-lactamase, the entire Cex exoglucanase from *Cellulomonas fimi*, the Cex cellulose-binding domain, and a single-chain Fv antibody fragment were generated using this scheme and all of them were active and expressed on the surface of *E. coli* (Francisco et al. 1992; Francisco et al. 1993a, b).

4. THE CONSTRUCTION OF AN *E. COLI* STRAIN THAT INTERFERES WITH *B. ANTHRACIS* PROLIFERATION

A construct based on the system described in the previous section was designed to fuse the hybrid *lpp-ompA* gene to an enzybiotic-coding gene (e.g., *plyG*) and expressed on the surface of *E. coli* (Fig. 3.2). Taking advantage of the seminal work showing that PlyG, the murein hydrolase coded for by the phage γ, induces lysis of *B. anthracis* cells when added to the culture (Schuch et al. 2002; see Chapter 5), we used

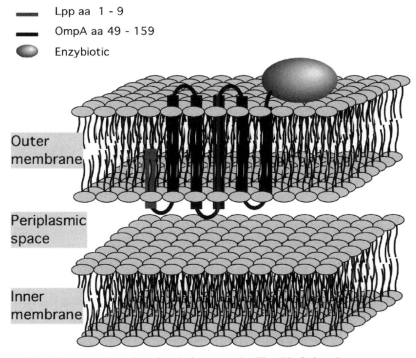

Figure 3.2. Diagram of a tripartite fusion protein. The PlyG fused to a presentation protein (Lpp fused to OmpA) renders the fused *lpp-ompA-plyG* gene. The Lpp-OmpA-PlyG protein is exposed on the cell's surface. The contact of the target bacteria with the fused PlyG protein will produce cell lysis. Redrawn from Francisco et al. (1992).

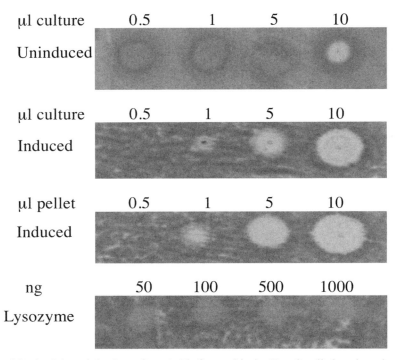

Figure 3.3. Activity of the Lpp-OmpA-PlyG enzybiotic. *E. coli* cells bearing plasmid-borne *lpp-ompA-plyG* gene were grown in Luria-Bertani broth (LB) rich medium until $OD_{560} \sim 0.4$ at $37\,°C$. Then the cultured was divided and one of the aliquots was induced by addition of IPTG. The cultures were incubated for 60 min. Aliquots of induced or uninduced cell cultures or of the resuspended pellet were spotted on a loan of *B. cereus* RSVF1 cells. The indicated amounts of lysozyme were spotted in a control experiment.

this enzyme as a paradigm to determine if it feasible to anchor a murein hydrolase to the surface of *E. coli* and generate a strain that can kill the target bacteria (Fig. 3.2). The concept of a phage-encoded PlyG endolysin anchored to the external surface of the cell (Lpp-OmpA-PlyG) to kill bacteria by breaking peptidoglycan bonds of the specific bacterial pathogen cell wall has been proven to be valid. Exposure of a target bacterium, *B. cereus* RSVF1, to the enzybiotic-producing strain resulted in enhanced loss of viability (Tolmasky and Alonso 2007, patent pending; Fig. 3.3).

To ensure that the released enzybiotic-borne genetically modified organism cannot proliferate in the environment different traits were introduced. The enzybiotic strains were auxotrophic for various amino acids and the strain showed addiction to a xenobiotic compound (isopropyl-β-D-thio-galactoside [IPTG]). The expression of an antitoxin (e.g., ε antitoxin) was IPTG-controlled and in its absence its expression was turned off, allowing the free toxin (e.g., ζ toxin) to block cell proliferation (Camacho et al. 2002; Meinhart et al. 2003). It has been recently shown that when the ζ toxin is freed from the ε antitoxin, it induces a reversible block of cells proliferation (Lioy et al. 2006).

Once a target bacterium is selected, a bacteriophage endolysin known to effectively lyse it when added from the outside should be identified. If a target bacterium and specific bacteriophages are well cataloged, the appropriate lytic enzyme gene sequence will be easily identified and obtained for use in a hybrid gene sequence. The lytic enzyme of bacteriophages that infect potential target bacteria are then produced and analyzed to ensure that it will be active when supplied from the outside of the target bacterium. If no bacteriophage is known for use with the described bacterium, a bacteriophage can be isolated from natural sources.

The combined action of enzybiotics with mild chemicals agents or physical treatments may provide a tool to control and to decontaminate areas polluted by biological weapons or pathogenic or toxic bacteria that have been accidentally released.

5. PERSPECTIVE

Apocalyptic scenarios from large-scale biological attacks have been predicted using comprehensive computer models (Wein et al. 2003). One of the pressing goals to be able to confront the 21st century's public-health challenges is the development of an armamentarium of new chemotherapeutic agents to efficiently disinfect accidentally or intentionally contaminated areas. Three unconnected strategies facilitate the achievement of this goal. One of these is the thorough understanding of the specific components of the potential bacterial bioweapon's membranes and cell wall, which can help in the development of sensitive biosensors as well as efficient bactericidals. Second, the development of mild physical or chemical agents with enhanced capacity to treat polluted areas. Finally, the development of comprehensive computer models of accidentally or intentionally released biological weapons and sensitive methods to detect the bioweapon as soon as possible. So far the cleanup of contaminated areas has been achieved using hazardous chemicals and/or physical disinfectants. An alternative paradigm consists of creating biological agents that can achieve the cleanup but are mild to the environment and innocuous to humans. This strategy relies on the recent progress in the development of enzybiotics, in bacteria distantly related to the target with the aim of avoiding frequent horizontal gene transfer of plasmid-borne toxins, as efficient bactericidal agents and the methodologies to manipulate bacterial cells, and does not exclude their utilization in concert with mild chemical and/or physical disinfectants.

ACKNOWLEDGMENTS

We are very grateful to S. Ayora for the critical reading of the manuscript. This work was supported by grants BFU2006–01062 from the Ministerio de Ciencia e Innovación, Dirección General de Investigación to JCA, and from California State University Fullerton to MET.

REFERENCES

Arnon S. S., R. Schechter, T. V. Inglesby, D. A. Henderson, J. G. Bartlett, M. S. Ascher, E. Eitzen, A. D. Fine, J. Hauer, M. Layton, S. Lillibridge, M. T. Osterholm, T. O'toole, T., G. Parker, T. M. Perl, P. K. Russell, D. L. Swerdlow, and K. Tonat (2001) *JAMA* **285**, 1059–1070.

Atlas R. M. (2002) *International Microbiology* **5**, 161–167.

Borysowski J., B. Weber-Dabrowska, and A. Gorski (2006) *Experimental Biology and Medicine (Maywood)* **231**, 366–77.

Bruggemann H. (2005) *Current Opinion in Microbiology* **8**, 601–605.

Camacho A. G., R. Misselwitz, J. Behlke, S. Ayora, K. Welfle, A. Meinhart, B. Lara, W. Saenger, H. Welfle, and J. C. Alonso (2002) *Biological Chemistry* **383**, 1701–1713.

Cheng Q. and V. A. Fischetti (2007) *Applied Microbiology and Biotechnology* **74**, 1284–1291.

Dembek Z. F., L. A. Smith, and J. M. Rusnak (2007) *Disaster Medicine and Public Health Preparedness* **1**, 122–134.

Deutsch S. M., S. Guezenec, M. Piot, S. Foster, and S. Lortal (2004) *Applied and Environmental Microbiology* **70**, 96–103.

Donovan D. M., M. Lardeo, and J. Foster-Frey (2006) *FEMS Microbiology Letters* **265**, 133–139.

Francisco J. A., R. Campbell, B. L. Iverson, and G. Georgiou (1993a) *Proceedings of the National Academy of Sciences USA* **90**, 10444–10448.

Francisco J. A., C. F. Earhart, and G. Georgiou (1992) *Proceedings of the National Academy of Sciences USA* **89**, 2713–2717.

Francisco J. A. and G. Georgiou (1994) *Annals of the New York Academy of Sciences* **745**, 372–382.

Francisco J. A., C. Stathopoulos, R. A. Warren, D. G. Kilburn, and G. Georgiou (1993b) *Biotechnology (NY)* **11**, 491–495.

Friedlander A. M., S. L. Welkos, M. L. Pitt, J. W. Ezzell, P. L. Worsham, K. J. Rose, B. E. Ivins, J. R. Lowe, G. B. Howe, and P. Mikesell (1993) *The Journal of Infectious Diseases* **167**, 1239–1243.

Georgiou G., H. L. Poetschke, C. Stathopoulos, and J. A. Francisco (1993) *Trends in Biotechnology* **11**, 6–10.

Hermoso J. A., J. L. Garcia, and P. Garcia (2007) *Current Opinions in Microbiology* **10**, 461–472.

Hoffmaster A. R., J. Ravel, D. A. Rasko, G. D. Chapman, M. D. Chute, C. K. Marston, B. K. De, C. T. Sacchi, C. Fitzgerald, L. W. Mayer, M. C. Maiden, F. G. Priest, M. Barker, L. Jiang, R. Z. Cer, J. Rilstone, S. N. Peterson, R. S. Weyant, D. R. Galloway, T. D. Read, T. Popovic, and C. M. Fraser (2004) *Proceedings of the National Academy of Sciences USA* **101**, 8449–8454.

Huang X. Z., M. P. Nikolich, and L. E. Lindler (2006) *Clinical Medicine and Research* **4**, 189–99.

Inglesby T. V., D. A. Henderson, J. G. Bartlett, M. S. Ascher, E. Eitzen, A. M. Friedlander, J. Hauer, J. Mcdade, M. T. Osterholm, T. O'toole, G. Parker, T. M. Perl, P. K. Russell, and K. Tonat (1999) *JAMA* **281**, 1735–1745.

Jado I., R. Lopez, E. Garcia, A. Fenoll, J. Casal, and P. Garcia (2003) *Journal of Antimicrobial Chemotherapy* **52**, 967–973.

Jernigan J. A., D. S. Stephens, D. A. Ashford, C. Omenaca, M. S. Topiel, M. Galbraith, M. Tapper, T. L. Fisk, S. Zaki, T. Popovic, R. F. Meyer, C. P. Quinn, S. A. Harper, S. K. Fridkin, J. J. Sejvar, C. W. Shepard, M. Mcconnell, J. Guarner, W. J. Shieh, J. M. Malecki, J. L. Gerberding, J. M. Hughes, and B. A. Perkins (2001) *Emerging Infectious Diseases* **7**, 933–944.

Kman N. E. and R. N. Nelson (2008) *Emergency Medicine Clinics of North America* **26**, 517–547, x-xi.

Krause R. M. (1957) *The Journal of Experimental Medicine* **106**, 365–384.

Lacy D. B., W. Tepp, A. C. Cohen, B. R. Dasgupta, and R. C. Stevens (1998) *Nature Structural & Molecular Biology* **5**, 898–902.

Lioy V. S., M. T. Martin, A. G. Camacho, R. Lurz, H. Antelmann, M. Hecker, E. Hitchin, Y. Ridge, J. M. Wells, and J. C. Alonso (2006) *Microbiology* **152**, 2365–2379.

Loeffler J. M., D. Nelson, and V. A. Fischetti (2001) *Science* **294**, 2170–2172.

Loessner M. J., S. Gaeng, and S. Scherer (1999) *Journal of Bacteriology* **181**, 4452–4460.

Loessner M. J., S. K. Maier, H. Daubek-Puza, G. Wendlinger, and S. Scherer (1997) *Journal of Bacteriology* **179**, 2845–2851.

Meinhart A., J. C. Alonso, N. Strater, and W. Saenger (2003) *Proceedings of the National Academy of Sciences USA* **100**, 1661–1666.

Mock M. and A. Fouet (2001) *Annuals Reviews in Microbiology* **55**, 647–671.

Navarre W., H. Ton-That, K. Faull, and O. Schneewind (1999) *Journal of Biological Chemistry* **274**, 15847–15856.

Nelson D., L. Loomis, and V. A. Fischetti (2001) *Proceedings of the National Academy of Sciences USA* **98**, 4107–4112.

Paradis-Bleau C., I. Cloutier, L. Lemieux, F. Sanschagrin, J. Laroche, M. Auger, A. Garnier, and R. C. Levesque (2007) *FEMS Microbiology Letters* **266**, 201–209.

Parisien A., B. Allain, J. Zhang, R. Mandeville, and C. Q. Lan (2008) *Journal of Applied Microbiology* **104**, 1–13.

Pritchard D. G., S. Dong, J. R. Baker, and J. A. Engler (2004) *Microbiology* **150**, 2079–2087.

Sass P. and G. Bierbaum (2007) *Applied and Environmental Microbiology* **73**, 347–352.

Schmid G. and A. Kaufmann (2002) *Clinical Microbiology and Infection* **8**, 479–488.

Schuch R., D. Nelson, and V. A. Fischetti (2002) *Nature* **418**, 884–889.

Smith T. J., S. A. Blackman, and S. J. Foster (2000) *Microbiology* **146**, 249–262.

Stenmark P., J. Dupuy, A. Imamura, M. Kiso, and R. C. Stevens (2008) *PloS Pathogenesis* **4**, e1000129.

Tolmasky M. E. and J. C. Alonso (2007) Method of making active biological containment factors for use in selectively killing target bacteria. 11/651,292, patent pending.

Turner M. S., F. Waldherr, M. J. Loessner, and P. M. Giffard (2007) *Systematic and Applied Microbiology* **30**, 58–67.

Veiga-Crespo P., J. M. Ageitos, M. Poza, and T. G. Villa (2007) *Journal of Pharmaceutical Science* **96**, 1917–1924.

Vollmer W., B. Joris, P. Charlier, and S. Foster (2008) *FEMS Microbiology Reviews* **32**, 259–286.

Wang I. N., D. L. Smith, and R. Young (2000) *Annuals Reviews in Microbiology* **54**, 799–825.

Wein L. M., D. L. Craft, and E. H. Kaplan (2003) *Proceedings of the National Academy of Sciences USA* **100**, 4346–4351.

Yacoby I., H. Bar, and I. Benhar (2007) *Antimicrobial Agents Chemotherapy* **51**, 2156–2163.

Yoong P., R. Schuch, D. Nelson, and V. A. Fischetti (2004) *Journal of Bacteriology* **186**, 4808–4812.

Zimmer M., N. Vukov, S. Scherer, and M. J. Loessner (2002) *Applied and Environmental Microbiology* **68**, 5311–5317.

CHAPTER 4

PHYLOGENY OF ENZYBIOTICS

PATRICIA VEIGA-CRESPO[1] and TOMAS G. VILLA[1,2]
[1]Department of Microbiology, Faculty of Pharmacy, University of Santiago de Compostela, Spain
[2]School of Biotechnology, University of Santiago de Compostela, Spain

1. INTRODUCTION

Enzymes are now increasingly being used in different industrial fields, not only as final products but also as biocatalysts (Schoemaker et al. 2003), and it is therefore of crucial importance to have an in-depth knowledge of their characteristics in order to improve the success of their applications. In this sense, rational protein design is necessary if better efficacy is to be achieved in the shortest time possible.

There are several strategies for modifying the activity of a given enzyme. In the last decade, the most employed technique has been the so-called direct evolution of enzymes (Williams et al. 2004; Hibbert and Dalby 2005; Chatterjee and Yuan 2006).

It is necessary to understand the natural evolution of enzymes in order to obtain better, rationalized, and fast-modified enzymes, and the best source of knowledge about these must surely come from studies of the relationships among the groups of enzymes of interest. In the case of this book, interest is focused on enzybiotics. Knowledge of the evolution of natural enzymes should provide insight into how we should act when resistance phenomena take place and should help us program rational uses of enzybiotics. Additionally, such a scenario could facilitate the discovery of potential pitfalls that might arise during enzybiotic-based therapies.

2. PHAGE ENZYMES EVOLUTION

Phages and bacteria have been evolving together for the last 2–3 billion years. Bacteria direct their efforts to making themselves resistant to attack by phages, whereas phages try to render the efforts of bacteria unfruitful. There is a considerable body of evidence supporting the hypothesis that host and phage murein hydrolases share a common ancestry, and interestingly, in some organisms it is evident that the proteins have co-evolved by exchanging their functional domains.

2.1. Holins

The role of holins is associated with the collapse of the membrane potential and permeabilization of the bacterial membrane (Young 1992). The holins are expressed in later stages of phage infection, forming a pore in the cell membrane and allowing the lysins to gain access to the cell wall peptidoglycan, which results in the release of the phage progeny. In general, lysins do not have signal sequences that would enable them to move through the membrane. Instead, this movement is controlled by holins (Wang et al. 2000) and lysins are accumulated in the cytoplasm during phage development (Smith et al. 1998). At specific times, holin is inserted into the membrane, thus disrupting it and affording the access for the lysins to the peptidoglycan layer (Fischetti 2005). Studies based on mathematical models have shown that there is an optimal lysis time. The mathematical model using the Marginal Value Theorem has been employed successfully to calculate optimal lysis times according to fitness, and environmental and genetic factors (Bull et al. 2004). When optimal timing has been reached, the fitness of the infection is maximized (Wang et al. 1996; Wang 2006). Thus, holins must be under strong evolutionary pressure because of their role as molecular clocks.

Holins are extremely diverse and are found in many unrelated sequence families with at least three membrane topologies, suggesting that they may have evolved from many different origins. Holins are classified in different classes and families on the basis of the number of transmembrane domains contained in their sequence. The members of class I exhibit three transmembrane domains, whereas those of class II show two transmembrane domains. The major class I holin is phage λ holin whereas in class II it is phage 21 holin. Most genes for classes I and II encode two proteins: the holin and its anti-holin. This is so because these genes show the dual-star motif as in the case of the λ phage (Park et al. 2006). Within each class, there are holins from phages able to act on Gram-positive bacteria and other holins that exert their action on Gram-negative bacteria. Nevertheless, all holins share some characteristics, such as a high content of basic amino acids in the C-terminal domain, and the presence of a short and polar N-terminal domain (Wang et al. 2000). Better knowledge of the different sequences of the holins and the implications of the different motifs is therefore essential.

A phylogentic tree protein sequence-based was elaborated with known holin sequences employing the Neighbor-Joining algorithm with Kimura corrections. When the tree was analyzed, it was observed that holin sequences occur not only in phages and prophages but also in bacterial chromosomes such as those of *Pseudomonas aeruginosa*, *Pseudomonas putida*, *Salmonella typhimurium*, and species of the genera *Serratia* and *Yersinia*. This was probably due to the presence of the remains of phage DNA sequences that were integrated in the bacterial chromosome during lysogenic cycles.

2.2. Lysins

Lysins are often chimaeric proteins, with a well-preserved catalytic domain fused to a divergent binding domain (Wang et al. 2000). Normally, high-affinity binding is directed toward species or strains (Lopez et al. 1997).

Lysins are named on the basis of the linkages they split. Thus, readers will find names such as N-acetylmuramidases or lysozymes, glucosaminidases, amidases, and endopeptidases (Stojkovic and Rothman-Denes 2006). This classification is made according to their catalytic domains and lysins may occasionally have more than one catalytic domain; such as the PlyGBS lysine from the GBS phage NCTC 11261, which shows one endopeptidase domain and one muramidase domain, both separated by a short linker (Cheng et al. 2005).

Although rarely, some lysins (i.e., *Streptococcus* bacteriophages) may even contain introns (Fischetti 2005). Normally, lysins are defective in secretion signals inside their sequence and, as indicated, holins are responsible for their access to the peptidoglycan layer. However, there are lysins such as the one from P1 phage or lamboid phage 21 that have an N-terminal signal-anchor release (SAR) sequence (Xu et al. 2005). These enzymes therefore use the host sec system. The SAR sequence has a high frequency of non-hydrophobic residues such as Gly, Ala, Ser, and Thr (Park et al. 2006).

The binding between a lysin and the bacterial cell wall is not a random event. When different lytic activities against *Streptococcus pneumoniae* were analyzed, it was observed that the union between the enzyme and the cell wall was always accomplished through the choline residues present in the teichoic acids of the cell wall (Jado et al. 2003). The entity responsible for the recognition and union is a six-repeat motif located at the cell wall-anchoring domain (Hermoso et al. 2003). It was also observed that a minimum of four repeats of choline-binding motif was necessary for efficient binding to the cell wall (Lopez and Garcia 2004). It was proposed that the amino alcohol would serve as an element of selective pressure to preserve the substrate-recognition domain.

Recently we have elaborated a protein sequence-based phylogenetic tree with known lysin sequences of bacteriophages able to recognize the amino alcohol motif. The tree was elaborated employing the Neighbor-Joining algorithm with Kimura corrections (Fig. 4.1). When the tree was analyzed, it was observed that lysin sequences were present in phage and bacteria genomes. The presence of lysins in bacteria occurs because these enzymes are necessary during normal cell wall extension, the separation of daughter cells, cell motility, and so on. It has been suggested that endogenous lytic cell wall enzymes would be related to the irreversible effect of β-lactam antibiotics (Tomasz 1979). The high similarity found between cells and phage lysins can be taken as a proof of the selective pressure to preserve substrate-recognition domains. It is noteworthy that the lysins able to act against *Streptococcus* species are not directly related to any other lysin. This property must be due to the aforementioned presence of the six-repeat motif located at the C-terminus binding domain. The high similarity between the genes of phages infecting *Streptococcus pyogenes* and *S. pneumoniae* suggests a frequent genetic interchange between both species and the recent divergence from a common ancestor phage (Obregon et al. 2003).

Interesting proof of the above selective pressure in lysin-binding domains is the enzyme Pal. This is an amidase with lytic activity against pneumococci. The protein exhibits the typical dual organization of lysins: a catalytic domain with amidase activity and a binding domain with a six-repeat motif. The catalytic domain has no homology with the rest of the *Streptococcus* amidases. However, the binding domain is well conserved (Varea et al. 2004). It is possible that a natural recombination process, a catalytic domain from an unknown precursor and the binding domain acquired from *S. pneumoniae*, is the origin of this enzyme.

Analysis of the phylogenetic relationships between phage and bacterial genomes suggests that the former may have evolved by recombi-

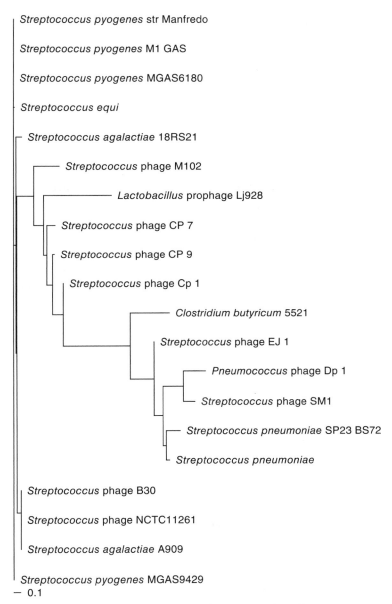

Figure 4.1. Phylogenetic tree of lysins able to act against *Streptococcus* genus.

national re-assortment of genes and by the acquisition of novel genetic elements (Hendrix et al. 2000).

Glycoside hydrolase family 24 (GH24) comprises enzymes with only one known activity. This family includes λ phage lysozyme and

Escherichia coli endolysin. Lysozymes (EC 3.2.1.17) are widespread in plants and animals, where they constitute a natural defense mechanism against bacterial pathogens. Lysozymes are divided into types according to their sequence similarity and three-dimensional structure (Jolles 1996). These types are c-type (chicken-type lysozyme), including the stomach lysozyme and insect lysozyme; goose-type lysozyme (g-type); plant lysozyme; bacterial lysozyme; and T4 phage-lysozyme (phage-type). c-Type lysozymes have been isolated from many vertebrates and, interestingly, also from insect species (Qasba and Kumar 1997).

When a phylogenetic analysis of GH24 members was carried out employing the Neighbor-Joining algorithm with Kimura corrections (Fig. 4.2), an absence of lysozymes able to recognize the amino alcohol motif was observed, thus confirming the fact that both motifs are poorly related. This analysis sheds light on the relationship between the lysozymes of enterobacteria and their phages. Genera such as *Synechococcus, Pseudomonas, and Salmonella* or different strains of *Escherichia* are extensively represented in the tree, and this in turn suggests deep relationship in the coevolution of lysozymes from bacteria and phages. Upon examining a variety of known peptideglycan structures, it is clear that bacteria attempt to modify their structures to avoid invasion by phages; however, in turn phages evolve their genome in order to proceed with their biological cycles. Recently, a new family of murein hydrolases has been reported. This new family is based on the sequence of coliphage N4 N-acetylmuramidase (pfam05838.4; Stojkovic and Rothman-Denes 2006). The main characteristic of this family is the presence of an EGGY (Glutamic acid-Glycine-Glycine-Tyrosine) motif near the N-terminus that contains a glutamic acid residue essential for its enzymatic activity. The sequence of this enzyme still lacks any statistically significant sequence similarity with any previously characterized muramidases.

Among the lysozymes, some curious mechanisms are worthy of note. Thus, some lysozymes stimulate the autolysin activity of bacteria (Iacono et al. 1985). Others, such as human lysozyme and hen egg-white lysozyme (HEWL), show a bactericidal mechanism without peptideglycan hydrolysis (Laible and Germaine 1985; Ibrahim et al. 2001; Masschalck et al. 2002). Finally, lysozyme from the fungus *Chalaropsis punctulata* shows lysozyme activity together with β-1,4-N,6-O-diacetylmuramidase activity.

As indicated above, the most common animal lysozyme belongs to the c-type group. Some differences, however, can be observed between monogastric animals and ruminants. Monogastric animals possess a single gene for c-lysozyme, mainly involved in defense against bacterial infections (Yu and Irwin 1996). However, ruminants, such as cows and

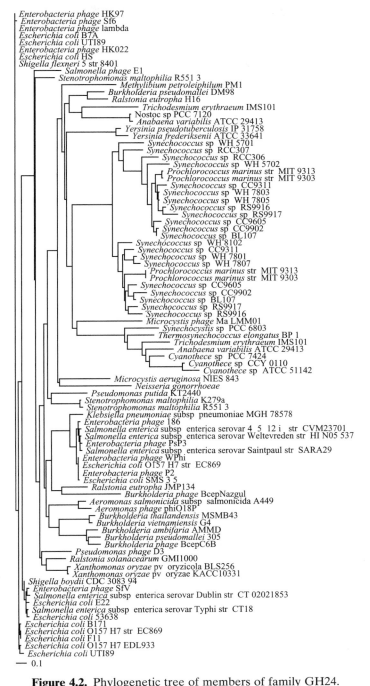

Figure 4.2. Phylogenetic tree of members of family GH24.

sheep, have multiple lysozyme genes and at least four that code for gastric lysozyme with a clear digestive role (Irwin and Wilson 1989). The recruitment of lysozymes as digestive enzymes has occurred at least in three vertebrate groups: ruminants, leaf-eating monkeys, and leaf-eating birds. This recruitment could have occurred through gene duplication (Irwin 1995).

2.3. Common Ancestral Precursor

Along evolution, the amino acid sequence of a protein is more variable and changes more rapidly than its tertiary structure. On the other hand, structural homology in the absence of sequence homology could be attributed to convergent rather than divergent evolution.

The structural similarities between T4 phage lysozyme and HEWL suggest that the two proteins have arisen by divergent evolution from a common precursor (Weaver et al. 1984; Weaver et al. 1985; Thunnissen et al. 1995). Indeed, X-ray crystallography reveals the correspondence of goose-type lysozyme (GEWL) with T4 lysozyme and HEWL and a common ancestral precursor, even though their amino acid sequences appear to be unrelated. The structure of GEWL has striking similarities to the lysozymes from hen egg-white and bacteriophage T4. However, some parts of GEWL resemble HEWL while other parts correspond only to the phage enzyme. The nature of the structural correspondence strongly suggests that all three lysozymes evolved from a common precursor (Grütter et al. 1983).

Lysozymes have undergone conformational changes both at the global and at the local levels. It has been possible to observe displacements of helices relative to each other and the corresponding helices may have increased or decreased in length. The glutamic acid residues of the active site are essential for lytic activity, but the position of the residue can be displaced in the sequence, that is, the glutamic acid of GEWL is at position 73, in HEWL the position is 35, and Glu11 for T4 lysozyme. The rest of the amino acid residues of the active site are poorly conserved. Phage P22 lysozyme shows a sequence homology of 26% with the lysozyme of T4 phage (Weaver et al. 1985). This enzyme has been proposed as an evolutionary link between the phage-type lysozymes and the g-type lysozymes. T4 lysozyme is an inverting glycosidase and it does not retain the β-configuration of the substrate in the product (Kuroki et al. 1999).

All in all, this evolutionary theory proposes the existence of a common ancestor for the three types of lysozymes. Another theory proposed is that a c-type-like common precursor diverged, giving raise to a g-type lysozyme, which in turn gave rise to a phage-type lysozyme (Grütter et al. 1983; Weaver et al. 1995). Phylogenetic studies based on a high number of sequences have revealed that bacteriophage lysozymes and g-type lysozymes are more closely related than c-types lysozymes (Hikima et al. 2001).

Like all bacteriophages, T4 is an obligate parasite of *E. coli*. This fact was fundamental for considering that it is more closely related to bacteria than to eukaryotes. However, T4 contains self-splicing introns in its genome (Gott et al. 1986; Sjoberg et al. 1986). Since these structures are typical of eukaryotes, there is evidence of genetic homologies with both eukaryotes and prokaryotes along the T4 genome (Bernstein and Bernstein 1989). Moreover, the T4 lysozyme is also able to bind chitin, a polysaccharide well known to be typically present in lower eukaryotes such as filamentous fungi (Kleppe et al. 1981).

The bacteriophage λ lysozyme shows a different action mechanism from other known lysozymes; it is not a hydrolase but a transglycosylase (Bienkowska-Szewczyk et al. 1981). This protein shows the structural features of lysozymes from the different types (Evrard et al. 1998). Glutamic acid 19 is the essential residue for catalytic activity (Jespers et al. 1992), and it has been proposed that λ lysozyme is closer to the phage lysozymes, although two features link it to eukaryotic lysozymes: the shortening of the helix connecting the two domains and the presence in the β-sheet of a segment that is very similar to the c-type lysozymes (Evrard et al. 1998; Fig. 4.3).

The sequence similarities found among the different types of lysozymes have been detected in other proteins such as the soluble lytic transglycosylase of *E. coli* (Thunnissen et al. 1994); the barley chitinases and the chitosanase of *Streptomyces* N174 (Evrard et al. 1998).

Preliminary studies of human lysozyme have revealed considerable similarity to the structure of HEWL; in fact, it may be concluded that these proteins evolved from the same gene and that they have an essentially identical mechanism of action. When a likelihood ratio test was employed to test the evolution of primate lysozyme, it was observed that the d_N/d_S ratios differed significantly among lineages, indicating that the evolution of primate lysozymes is episodic, which is incompatible with the neutral theory of evolution (Yang 1998). The lysozymes have been evolving under negative pressure inside primate lineages.

The amino acid composition of α-lactalbumin, a protein in cow's milk, is quite similar to HEWL. Based on comparison of the amino acid sequences of HEWL, human lysozyme, and α-lactalbumin, it has been postulated that a common ancestral gene would have existed for these three proteins. Duplication from the c-type generated the

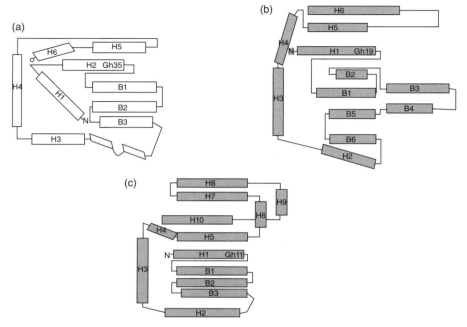

Figure 4.3. Schematic representation of three-dimensional models of lysozymes. (a) HEWL; (b) λ lysozme; (c) T4 lysozyme.

ancestral α-lactalbumin gene in the mammalian lineage, with the successive loss of lysozyme function and the gain of a novel one (Prager and Wilson 1988).

Evolutionary studies on ruminant stomach lysozymes have suggested that evolution would have occurred in an episodic manner, whereas non-stomach lysozymes would have evolved in a different way (Irwin 1995; Yu and Irwin 1996). Thus, the acquisition of an efficient stomach lysozyme was associated with the evolutionary success of ruminants. The pig lysozyme c-gene is similar in size to the lysozymes of mammalian species and this indicates that gene duplication in higher ruminants occurred after the divergence of the porcine lineage from the lineage leading to the higher ruminants (Yu and Irwin 1996), although most of the gene duplications occurred before the divergence of the cow and sheep lineages (Wen and Irwin 1999). The evolution of the coding regions of the stomach lysozyme genes in ruminants probably occurred in a concerted fashion whereas noncoding regions would have evolved in a divergent manner. Analysis of the synonymous and non-synonymous substitutions revealed that the evolution of these genes was carried out under positive selective pressure. Non-stomach lysozymes conserved the characteristic catalytic residues in vertebrate lysozymes. Tracheal and intestinal lysozymes appear to share some sequence structure properties with stomach lysozymes. Phylogenetic analyses have suggested that tracheal and intestinal lysozyme genes in cows have a more recent ancestry with stomach lysozymes than with other non-stomach lysozymes (Irwin 2004).

Regarding the ruminant stomach another consideration must be taken into account. Rumen bacteria represent an important component of the rumen biomass. The stomach lysozymes of ruminants are able to act against Gram-positive bacteria. The current bacteria in ruminant stomach have developed resistance against nisin and this resistance also confers lysozyme resistance (Dominguez-Bello et al. 2004).

As mentioned above, the i-type lysozymes are formed by lysozymes from invertebrates. The lysozymes of bivalve species present a high content in cysteine residues. Phylogenetic studies have shown that i-type lysozymes constitute a monophyletic group that has evolved from a common ancestral domain with c-type lysozymes and α-lactalbumin (Bachali et al. 2002). Invertebrate-type proteins are present in many prototosme phyla whereas the c-type is present in deuterostomes and protostomes, which suggests that both types were present in the ancestor of Bilateria, about 600 million years ago. They have been found in species of *Lepidoptera*, *Orthoptera*, and *Diptera* (Hultmark 1996).

Some insects show c-type lysozymes. When the c-type lysozyme gene family of *Anopheles gambiae* was analyzed, eight lysozymes were found. Two of these, lys c-3 and lys c-8, showed similarity with calcium-binding lysozymes. This was the first report of calcium-binding lysozymes outside mammalian and avian groups (Li et al. 2005). The analyses revealed that these proteins are involved not only in defense mechanisms but also in development.

The shrimp *Marsupenaeus japonicus* encodes a c-type lysozyme active against *Vibrio* species. This enzyme is more closely related to vertebrate c-type enzymes than to invertebrate c-type enzymes (Hikima et al. 2003).

3. GLUCANASES AND CHITINASES

Fungal cell walls contain two major highly resistant polysaccharides, namely 1,3- β-glucan and chitin, and these must be hydrolyzed by endogenous fungal lysins when normal morphogenetic processes such as cell wall elongation or yeast budding occur. When either of these endogenous lysins acts uncoordinatedly with the biosynthetic enzymes or is added from the outside, the cell wall degrades and the fungal cell explodes.

O-glycosyl hydrolases (EC 3.2.1.X) are a widespread group of enzymes that hydrolyse the glycoside bond between two monosaccharides, or between a carbohydrate and a non-carbohydrate moiety. A classification system for glycosyl hydrolases, based on sequence similarity, has led to the definition of 85 different families (Davies and Henrissat 1995; Henrissat et al. 1995; Bairoch 1999). This classification is available on the Carbohydrate-Active EnZymes (CAZY) web site (Coutinho and Henrissat 1999). Because the fold of the proteins is better conserved than their sequences, some of the families can be grouped in clans (Bourne and Henrissat 2001) that share a common ancestor. There are 88 glycosyl hydrolase families. Family 55 is formed only by filamentous fungal 1,3-β-glucanases and its components are characterized in that they display two complete β-helix domains, which are exclusive to this family (Kawai et al. 2006). Some glycoside hydrolases comprise a unique catalytic domain whereas others are modular proteins of different complexities, with two-six domains. The members of a family have the same mode of action and they can be inverting or retaining enzymes. The retaining enzyme catalyzes the hydrolysis of substrate, maintaining the configuration at the anomeric center, whereas the inverting enzymes change the configuration of the anomeric center during hydrolysis reactions (Warren 1996). There are members of the same family with different substrate specificities, which suggests a divergence adaptive phenomenon during evolution (Davies and Henrissat 1995). The catalytic domain is the site responsible for substrate hydrolysis. Another important domain is the binding domain, which can act independently of the catalytic domain. For example, starch-binding domains are present in glucoamylases, β-amylases, and α-amylases, and likewise, cellulose-binding domains are present not only in cellulases but also in some β-glucosidases (Svensson et al. 1989; Lymar et al. 1995). The classification of these enzymes is continuously reviewed and updated (Henrissat and Bairoch 1996). Such binding domains are called carbohydrate-binding modules (CBMs) and their mission is to enhance enzymatic activity by improving access to the substrate (Boraston et al. 2004). Based on the primary structure, the binding domains have been classified into 42 families and are grouped in the CAZY database.

The β-glucanases can act at the nonreducing end (exo-acting enzymes, EC 3.2.1.58) or in the middle of the polysaccharide chain (endo-acting, EC 3.2.1.39). There are two main groups, depending on the type of the active amino acid involved in their active site: either aspartic acid or glutamic acid (Ring et al. 1988; Sinnot 1990).

β-1,3-glucanases are ubiquitous in fungi and they can be exo- or intracellularly located (Mouyna et al. 2002) and β-1,6-glucanases are usually present too (Pitson et al. 1997). *Bacillus circulans* WL-12 produces at least six β-1,3-glucanases (Yahata et al. 1990), although it seems that five of them are formed from one enzyme through a proteolytic process. These enzymes are involved in functions such as nutrition, growth, and parasitism. In tobacco plants, five distinct classes and 12 β-1,3-glucanases have been found (Linthorst et al. 1991). The soybean has 12 classes of β-1,3-glucanase genes (Jin et al. 1999). The distribution of the glucanases genes in different loci must be to protect the cell against losses by unequal crossing or slippage. The majority of β-1,3-glucanase and β-1,6-glucanase sequences show a single cleavage site (Peberdy 1994). The function of this proteolytic site is probably to convert an inactive zymogene into an active enzyme and this could in turn offer protection to the producer cell against auto-degradation.

The endo-β-1,6-glucanase from the fungus *Trichoderma harzianum* shows a high relationship with family 5 of the glycoside hydrolases at the amino acid sequence level (Kim et al. 2002), and interestingly, the members of family 5 are mainly exo-β-1,3-glucanases and endo-β-1,4-glucanases.

An involvement of β-1,3-glucanases in immunoglobulin E (IgE) cross-reactivity has been suggested (Sunderasan et al. 1995). Patients allergic to tomato, potato, and banana show a reaction to Hev b 2, which is an allergenic β-1,3-glucanase from latex (Yamagi et al. 1998). The major allergen of olive pollen Ole e 9 is a β-1,3-glucanase (Huescas et al. 2001). It has been shown that β-1,3-glucanases could be involved in pollen-fruit-latex cross reactivity because it is possible to find Ig epitopes in several pollen β-1,3-glucanases of plant species belonging to the Oleaceae and Fagales (Palomares et al. 2005). Ole e 9 has been proposed as the β-1,3-glucanase allergen from the pollen of *Olea europea*. Another allergen from *Olea europea* pollen—Ole e 10—has also been found. It has been described as the major allergen of class I in humans and its analysis has allowed a new class of CBM to be described (Barral et al. 2005).

As mentioned above, β-1,3-glucanases are well characterized in bacteria, fungi, and plants. However, the distribution of β-1,3-glucanases in the animal kingdom is limited to the eggs and digestive tract of echinoderms and frogs (Sova et al. 1969). Their function in gut-located glucanases must be related to the digestion of β-glucans such as those found in higher brown algae. When the localization of β-1,3-glucanases in the eggs of three species of sea urchin (*Lytechinus variegatus, Strongylocentrotus purpuratus*, and *Arbacia punctulata*) was studied, it was found that the enzyme was located specifically in the cortical granules (Wessel et al. 1987). It was observed that the β-1,3-glucanase of *S. purpuratus* was concentrated in the spiral lamellae prior to fertilization and in the hyaline layer thereafter (Sommers and Shapiro 1989). Analysis of the amino acid sequences of β-1,3-glucanase of *S. purpuratus* revealed a similarity with bacterial enzymes (Bachman and McClay 1996), and hence horizontal gene transfer from *Bacillus* was suggested as an explanation for the presence of β-1,3-glucanase activity in sea urchins. However, the observed divergences did suggest a divergence from a common ancestor at an early stage of evolution. When the β-1,3-glucanase from the eggs of the sea urchin *Strongylocentrotus intermedius* was compared with marine and terrestrial molluscs (*Spisula sacchalinensis* and *Eulota maakii*), it was observed that *S. intermedius* glucanases were more related to marine endo-β-1,3-glucanases than to terrestrial exo-β-1,3-glucanases (Sova et al. 2003).

Endo-β-1,4-glucanases (EC 3.2.1.4) are enzymes produced in bacteria, fungi, and plants. Hydrophobic cluster analysis has shown that the catalytic core of these enzymes can be ordered in six or more families (Beguin 1990). Plant endoglucanases belong to the E family. Microbial E-type typically possesses cellulose-binding domains, whereas the plant counterparts lack these domains (Warren 1996). Plant endoglucanases are encoded by multi-gene families (Lashbrook et al. 1994) and there are plant endoglucases, such as Cel3 from *Lycopersicon esculentum*, in which the amino acid sequence diverges strongly from that of other plant or microbial E-type (Brummel et al. 1997). Its structure is in fact more similar to *Agrobacterium tumefaciens* CelC than to other plant endoglucaneses.

Chitinases or endochitinases (EC 3.2.1.14) hydrolyze the β-1,4-glycoside linkage between two *N*-acetylglucosamine residues in chitin. Chitinases are found in chitin-producing organisms as well as in plants, bacteria, and vertebrates, where they play a defensive role against pathogens (Table 4.1; Leah et al. 1991; Gohel et al. 2006).

Chitinases are classified within families 18 and 19 of the glycosyl hydrolase families. Family 18 contains chitinases from bacteria, fungi, viruses, and animals, and classes III and V chitinases from plants. Family 19 contains classes I, II, and IV plant chitinases (Henrissat and Bairoch 1993). It has been suggested that classes I and IV were derived from a common ancestral sequence that predated the divergence of dicots and monocots. It is not yet clear, however, whether the gene

TABLE 4.1. Role of Chitinases in Different Organisms

Organism	Action of Chitinase
Bacteria	Nutrition and parasitism
Fungi	Cell division, nutrition, and differentiation
Yeast	Cell division, development, and defense
Insects	Development
Protozoa	Attack
Human and vertebrates	Defense

transfer events occurred before or after the divergence of class IV and class I/II chitinases (Hamel et al. 1997)

The first report of a non-plant chitinase member of family 19 was the chitinase C1 from *Streptomyces griseus* HUT 6037, thus suggesting horizontal transfer from plants to bacteria (Ohno et al. 1996).

Families 18 and 19 do not show sequence similarity and display different 3D structures (Robertus and Monzingo 1999; Fukamizo 2000). Also, they have different enzymatic action mechanisms, since family 18 exhibits a retaining mechanism, whereas family 19 shows an inverting one (Koga et al. 1999). When exhaustive analyses of chitinases of *Trichoderma* Genera were carried out, it was found that all were members of family 18. These analyses allow family 18 chitinases to be divided into three subgroups: A, B, and C, except for one chitinase: Chi18–15 (Seidl et al. 2005). Except for this latter enzyme, ortholog genes have been found in other filamentous fungi. This suggests that their common ancestor appeared very early on in fungal evolution. Interestingly, the subgroup C members show a domain structure similar to that of *Kluyveromyces lactis* killer toxins. Chi18–15 shows orthology with chitinases from *Streptomyces* spp., *Trichoderma asperellum* and *Cordyceps bassiana*, an entomopathogen. This gene was probably acquired by horizontal transference.

Classes I and IV of family 19 chitinases have a cysteine-rich domain at their N-terminus that is involved in chitin binding. Class II lacks this domain. As mentioned above, ChiC from *Streptomyces griseus* HUT 6037 was the first non-plant member chitinase of family 19. When more species of *Streptomyces* were analyzed, the normal presence of family 19 chitinase genes, which mainly encode class IV chitinases (Watanabe et al. 1999), was observed. The high levels of sequence similarity between *Streptomyces* and plants suggest that *Streptomyces* species would have acquired the chitinases of family 19 by horizontal transfer in the recent past. Advances in molecular techniques have allowed more family 19 chitinase members to be identified in other bacterial species and even in a nematode (Kawase et al. 2004). Bacteria with family 19 genes are included in the order Actinomycetales. It is probable that an ancestor of the Streptomycineae would have acquired the chitinase gene from plants and transferred it to Actinobacteria again through horizontal gene transfer. When structural differences between the family 19 chitinases of bacteria and plants were studied, the most significant one was found to be the reduction in the substrate-binding site in *Streptomyces* chitinases, whereas the catalytic glutamic acid residues are well conserved in plants and bacteria at positions 68 and 77 (Hoell et al. 2006).

Bacterial chitinases mainly belong to family 18 and it has been proposed that they should be subdivided into several subgroups—A, B, and C—on the basis of their catalytic domain sequences (Suzuki et al. 1999). Moreover, recent studies have suggested the existence of new groups of bacterial chitinase in the bulk of rhizosphere soils (Ikeda et al. 2007). Chitinase gene sequences retrieved from diverse aquatic habitats reveal an environment-specific distribution (LeCleir et al. 2004). In some bacteria, chitinases from family 18 and family 19 are present. This is the case of *Streptomyces coelicolor* A3, which shows two genes of family 19 chitinases and 11 family 18 gene members (Saito et al. 2000; Kawase et al. 2006).

Recently, a chitinase from *Helicoverpa armigera* single nucleocapsid nucleopolyhedrovirus has been identified (Wang et al. 2004). The phylogeny of this baculoviral chitinase shows that the gene clusters exclusively with γ-proteobacteria. The authors suggested that the virus would have acquired the chitinase gene from bacteria.

Mammals express two active chitinases, chitotriosidase (CHIT1) and the prototypic chitinase acidic mammalian chitinase (AMCase). The importance of chitinases in humans is not only that they act as a defense mechanism but that they are also related to different syndromes such as Gaucher disease, lysosomal lipid storage disorders, sarcoidosis, thalassemia, and visceral Leishmaniasis (Hollak et al. 1994; Aguilera et al. 2003; Goto et al. 2003).

The human chitinolytic enzyme (CHIT1) is expressed by activated macrophages and found in lymph nodes, bone narrow, and lungs, whereas AMCases predominate in the gastrointestinal tract and the lungs (Boot et al. 2001). When phylogenetic studies comparing rodent and primate genes were performed, it was observed that the gene was well conserved in rodents and primates, and that a 24 pb duplication not present in primates was present in humans. This duplication must have originated during human evolution, which determined an enzymatically inactive protein (Gianfrancesco and

Musumeci 2004). The sequence analysis of mammalian chitinases was carried out to propose a mammalian chitinase gene family (Zhenga et al. 2005).

Deep evolutionary analysis of family 18 of the animal chitinases revealed three major phylogenetic groups: chitobiases, chitinases/chitolectins, and stabling-I interacting chitolectins, where only the first one is associated with expansion in late deuterostomes (Funkhouser and Aronson 2007). The phylogeny of this family is starting to show a birth-and-death model of evolution. Human chitinases are closely associated with human MHC (Major Histiocompatibility Complex) paralogon on chromosome 1. This suggests that late expansion in some way occurred related to an emerging interface of innate and adaptive immunity during early vertebrate history.

It has been proposed that chitinase family members in humans are important mediators in allergic diseases, including asthma (Donnelly and Barnes 2004). AMCase is induced during TH2 inflammation through an IL-13–dependent mechanism (Elias et al. 2005) and its levels are significantly increased in allergic reactions (Burton and Zaccone 2007). The AMCase shows a histidine residue at its active site (Hist187), which is essential for this activity (Bussink et al. 2008). When mycolytic activity was assayed in human chitinases, none of them showed such activity, whereas bacterial chitinases did (Sanders et al. 2007).

YKL-40 is a mammalian member of chitinase-like proteins, also called Chitinase 3-like protein. This protein shows sequence similarity to fungal and bacterial chitinases; it is able to bind chitin but not to degrade it (Renkema et al. 1998). The levels of expression of YKL-40 are high in non-pathological and pathological states, characterized by inflammation, tissue destruction, and the development of fibrosis. Diverse types of stress result in increased YKL-40 levels, which strongly supports the involvement of YKL-40 in the malignant phenotype as a cellular survival factor in an adverse microenviroment (Junker et al. 2005).

As previously described for β-1,3, glucanases with regard to their involvement in allergic processes in humans, the same is apparent for chitinase proteins. Thus, class I chitinases show a binding domain (Hev b) that bears the IgE-binding epitopes (Posch et al. 1999). Class I chitinases have been identified in chestnut, avocado, and banana as relevant allergens. Hev b 11 is a chitinase from *Hevea brasiliensis* latex that also displays IgE-binding ability (O'Riordain et al. 2003; Rihs et al. 2003). Another chitinase family-related allergen is Ziz m 1 Z. This was isolated from *Zizyphus mauritiana* that shows chitinase activity, IgE-binding capacity, and cross-reactivity with the latex allergen. Ziz m 1

Z possessed sequence similarity with the class III chitinases of family 18 (Lee et al. 2006).

Insect chitinases belong to family 18 and are highly conserved. The chitinase from the beetle *Tenebrio molitor*, however, possesses an unusual structure even though it is a member of family 18 of glycosydases (Royer et al. 2002). In *B. circulans* WL-12, the chitinase system includes at least six different chitinase molecules: chitinases A1, A2, B1, B2 C, and D, where chitinase A1 is the key enzyme of the system. Amino acid analysis of A1 reveals 33% similarity to the chitinase A of *Serratia marcenscens* in the N-terminal region. This region of the chitinase is followed by tandem repeats of 95 amino acid segments that show 70% homology and that are similar to the type III homology units of fibronectin, a plasma protein of higher eukaryotes. This homology suggests that type III homology units would have originated prior to the emergence of eukaryotes (Watanabe et al. 1990). The C-terminal domain of this enzyme is a chitin-binding domain required for specific binding to chitin, whereas the type III regions are important for the hydrolysis of chitin by the enzyme bound to chitin (Watanabe et al. 1994). Fibronectin type III-like sequences have also been identified in amylases and cellulases (Tomme et al. 1995). The chitinase ChiC from *Streptomyces lividans* has a family II cellulose-binding domain at the N-terminus (Fujii and Miyashita 1993). Another interesting chitinase is the one from *Streptomyces olivaceoviridis*. This enzyme shows a proteolytic domain at the N-terminus (Radwan et al. 1994).

When phylogenetic analysis of chitinases was carried out employing the Neighbor-Joining algorithm with Kimura corrections, a representation of both families was obtained (Fig. 4.4). However, no other glycosidase family members were found. When the phylogenetic studies were carried out for endoglucanases, members of different families were found (Figs. 4.5 and 4.6). Preliminary studies showed that exo- and endo-β-1,3-glucanases are mainly separated in two clusters: the sequences of filamentous fungi and yeast and Basidiomycotina (Martin et al. 2007). The β-1,6-glucanases showed only a common ancestor (Martin et al. 2007).

Evolution within glycanase families has occurred by domain shuffling, with subsequent modifications of the domains (Gilkes et al. 1991). The diversity between the linkers along the different families must be due to the different activities of the domains they join. It has been proposed that these enzymes have been subjected to two contradictory evolutionary pressures: the optimization of catalytic efficiency for the original substrate, on the one hand, and divergence to acquire a new specificity, on the other (Henrissat 1991).

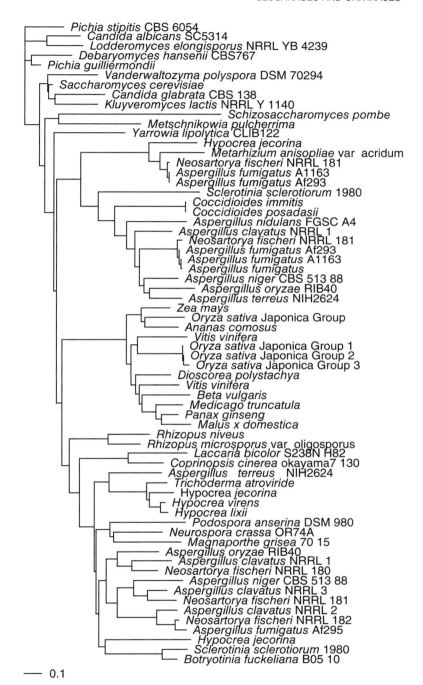

Figure 4.4. Phylogenetic tree of chitinases.

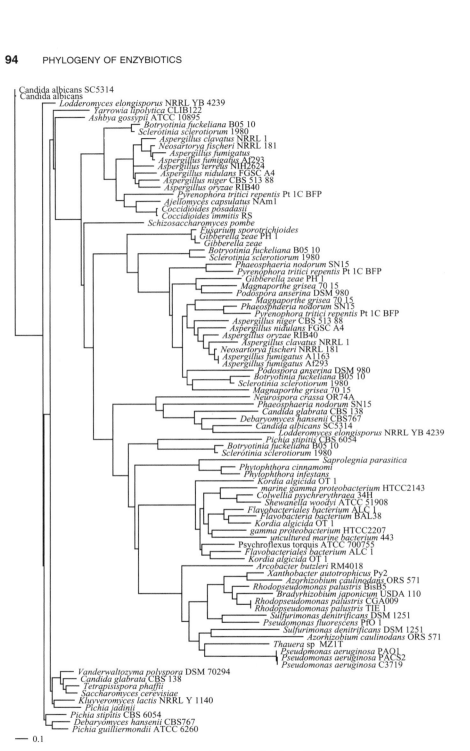

Figure 4.5. Phylogenetic tree of β-1,3-endoglucanases.

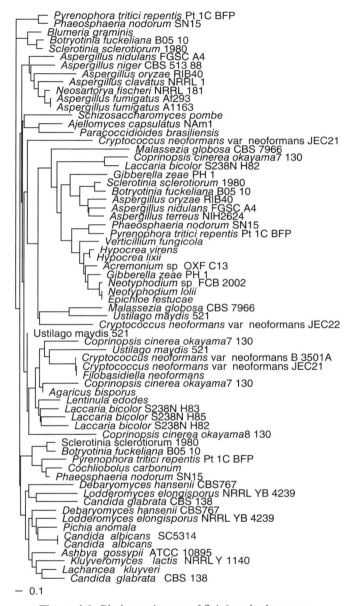

Figure 4.6. Phylogentic tree of β-1,6-endoglucanases.

Chitosanases split the β-1,4 linkage between aminoglucose residues in chitosan, which may contain some residues of N-acetylglucosamine. The structures of chitosan, chitin and peptidoglycan are very similar, both chemically and structurally. This is in turn responsible for the fact that some lysozymes, such as HEWL, can act as chitinases. However, their amino acid sequences are very different, as has been commented above. Structural analysis also reflects similarities between barley chitinases and HEWL (Holm and Sander 1994; Hart et al. 1995). When detailed comparative studies were carried out for barley chitinases, bacterial chitosanase, GEWL, T4 lysozyme, and HEWL, elements of a common core outside the catalytic site were found. Thus, it was always possible to detect a bulky hydrophobic amino acid seven residue regions upstream from the catalytic glutamic acid on the C helix. This hydrophobic residue always fits into a hydrophobic socket on the second conserved helix of the core. The β2 strand contains an invariant glycine residue, conserved in all the enzymes. There is always a hydrophobic amino acid two residues downstream from the conserved glycine (Monzingo et al. 1996). This common core had been observed previously for lysozymes (Matthews et al. 1981). All these enzymes share similar substrates and this, together with their secondary structure conserved elements, strongly suggests that lysozymes, chitinases, and chitosanases had a common ancestral precursor with an inverting mechanism of action, like the one reported for T4 lysozyme (Kuroki et al. 1995). These enzymes can be divided into the prokaryotic family and the eukaryotic family and although HEWL is slightly different from these families, it falls within the second type (Monzingo et al. 1996).

4. OTHER FORMS OF COEVOLUTION

The evolution of glycosyl hydrolases is a complicated process as it is for all kinds of life forms since not only divergence but also convergence phenomena occur simultaneously. In the case of the foregoing, whereas phages evolved their lytic enzymes, bacteria developed new peptidoglycan modifications in order to acquire resistance against infection by phages. The development of stomach lysozymes started the development of bacterial resistance in ruminant stomachs, as mentioned above.

Glycosyl hydrolases are involved in innate immune response to pathogen but the question arises as to how this is triggered. β-1,3-glucan recognition proteins (βGRP) show strong affinity towards β-1,3-glucan. When these proteins recognize polysaccharide, the innate immune response is activated. The homologies observed between the βGRP response in mammals and in insects reveal close links between both groups (Kang et al. 1998; Ochiai and Ashida 1999).

The βGRP isolated from the silkworm *Bombyx mori* shows sequence similarity to the catalytic region of bacterial glucanases and strong similarity to the peptidoglycan-binding proteins (PGRP) of *B. mori* and *Anopheles gambiae* (Ochiai and Ashida 1999). Similar results were found when the βGRP from *Plodia interpunctella* was studied. Similarity was observed not only to bacterial glucanases but also to the β-1,3-glucanases of the sea urchin *S. purpuratus* (Fabrick et al. 2003).

The βGRP of crustaceans are important components of the immune system. They show differences from the βGRP of insects in terms of molecular weight and amino acid sequences (Jimenez-Vega et al. 2002). The crayfish *Pacifastacus leniusculus* has a βGRP that is able to bind bacterial lipopolysaccharide. The structure and function of this protein is similar to that of the coelomic cytolytic factor-1 from *Eisenia foetida* and shows sequence homology with Gram-negative-binding proteins and β-1,3-glucanases (Lee et al. 2000).

The glucan-binding proteins of bacteria seem to be involved in pathogenesis because they are related to the cohesiveness of bacteria, that is, in plaque formation by oral pathogens (Landale and McCabe 1987; Banas et al. 1990).

In humans, β-1,3-glucans stimulate macrophages to release pro-inflammatory cytokines vitronectin and fibronectin being the circulating factors that recognize the β-1,3-glucan and stimulate macrophages (Vasallo et al. 2001). Thus, βGRP, PGRP, and glucanases probably share a common origin in their molecular evolution.

As has been seen, species develop defense mechanisms against β-glucan-harboring pathogens, two important ones being the production of glucanases and the production of glucan-binding proteins. However, at the same time pathogens also develop defense mechanisms against glucanases. The oomycete *Phytophthora sojae*, a soybean pathogen, produces a protein called GIP1, a member of the class of proteins known as Glucanase Inhibitor Proteins (GIPs), which specifically inhibit the endoglucanase activity of the plant host (Rose et al. 2002). GIPs have also been found in species of the Genus *Phytophthora*. Structural analysis of the GIP family has suggested that these proteins and plant endo-β-1,3-glucanases have been coevolving together for a long period of time (Damasceno et al. 2008). The inhibitor protein of endo-1,3-β-glucanases isolated from the alga *Laminaria cichorioides* blocks the biosynthesis of almost all O-glycosyl hydrolases of different species of marine fungi (Verigina et al. 2005).

Additionally, during polysaccharide hydrolysis glucanases from plants produce oligosaccharides that act as elicitors of the defense response and systemic-acquired resistance.

Recent studies have suggested the existence of inhibitors of class I plant chitinases. The rate of amino acid substitution observed in the chitinases of the Genus *Arabis* reflects the effects of positive selection aimed at modifying the capacity of the enzymes to bind either to the chitin or to the inhibitor (Stahl and Bishop 2000). In contrast to primate lysozymes and class III chitinases, the adaptive replacements of class I occur mainly at the active site. All these data suggest that plant defense proteins not involved in pathogen recognition also evolved with rapid coevolutionary interactions (Bishop et al. 2000) and that such a mechanism could be more important in overall evolution than initially believed.

ACKNOWLEDGMENTS

The authors wish to express their appreciation to the Ramon Areces Foundation from Madrid for partly furnishing the laboratory where some of the experiments reported here were conducted. Patricia Veiga-Crespo also thanks the Xunta de Galicia and the University of Santiago de Compostela for financial support from the Anxeles Alvariño Grants.

REFERENCES

Aguilera B., K. Ghauharali-van der Vlugt, M. T. J. Helmond, J. M. M. Out, W. E. Donker-Koopman, J. E. M. Groener, R. G. Boot, G. H. Renkema, G. A. van der Marel, J. H. van Boom, H. S. Overkleeft, and J. M. F. G. Aerts (2003) *Journal of Biological Chemistry* **278**, 40911–40916.

Bachali S., M. Jager, A. Hassanin, F. Schoentgen, P. Jolles, A. Fiala-Medioni, and J. S. Deutsch (2002) *Journal of Molecular Evolution* **54**, 652–664.

Bachman E. S. and D. R. McClay (1996) *Proceedings of the National Academy of Sciences of the United States of America* **93**, 6808–6813.

Bairoch A. (1999). http://www.expasy.ch/cgi-bin/lists?glycosid.txt (accessed July 2008).

Banas J. A., R. R. B. Russell, and J. J. Ferretti (1990) *Infection and Immunity* **58**, 667–673.

Barral P., C. Suarez, E. Batanero, A. Carlos, J.-D. Alche, M. I. Rodriguez-Garcia, M. Villalba, G. Rivas, and R. Rodriguez (2005) *Biochemical Journal* **390**, 77–84.

Beguin P. (1990) *Annual Reviews in Microbiology* **44**, 219–248.

Bernstein H. and C. Bernstein (1989) *Journal of Bacteriology* **171**, 2265–2270.

Bienkowska-Szewczyk K., B. Lipinska, and A. Taylor (1981) *Molecular Genetics and Genomics* **184**, 111–114.

Bishop J. G., A. M. Dean, and T. Mitchell-Olds (2000) *Proceedings of the National Academy of Sciences of the United States of America* **97**, 5322–5327.

Boot R. G., E. F. Bommaart, E. Swart, K. Ghauharali-van der vlugt, N. Bijl, C. Me, A. Place, and J. M. Aerts (2001) *Journal of Biological Chemistry* **276**, 6770–6778.

Boraston A. B., D. N. Bolam, H. J. Gilbert, and G. J. Davies (2004) *Biochemistry Journal* **382**, 769–781.

Bourne Y. and B. Henrissat (2001) *Current Opinion in Structural Biology* **11**, 593–600.

Brummel D. A., C. Catala, C. C. Lashbrook, and A. B. Bennett (1997) *Proceedings of the National Academy of Sciences of the United States of America* **94**, 4794–4799.

Bull J. J., D. W. Pfenning, and I. N. Wang (2004) *TRENDS in Ecology and Evolution* **19**, 76–82.

Burton O. T. and P. Zaccone (2007) *Trends in Immunology* **28**, 419–422.

Bussink A. P., J. Vreede, J. M. F. G. Aerts, and R. G. Boot (2008) *FEBS Letters* **582**, 931–935.

Chatterjee R. and L. Yuan (2006) *Trends in Biotechnology* **24**, 28–38.

Cheng Q., D. Nelson, S. Zhu, and V. A. Fischetii (2005) *Antimicrobial Agents Chemotherapy* **49**, 111–117.

Coutinho P. M. and B. Henrissat (1999). http://www.cazy.org/index.html (accessed July 2008).

Damasceno C. M. B., J. G. Bishop, D. R. Ripoll, J. Win, S. Kamoun, and J. K. C. Rose (2008) *Molecular Plant-Microbe Interactions* **21**, 820–830.

Davies G. and B. Henrissat (1995) *Structure* **3**, 853–859.

Dominguez-Bello M. G., M. A. Pacheco, M. C. Ruiz, F Michelangeli, M. Leippe, and M. A. de Pedro (2004) *BMC Ecology* **4**, 7.

Donnelly L. E. and P. J. Barnes (2004) *Trends in Pharmacological Sciences* **25**, 509–511.

Elias J. A., R. J. Homer, Q. Hamid, and C. G. Lee (2005) *Journal of Allergy and Clinical Immunology* **116**, 497–500.

Evrard C., J. Fastrez, and J.-P. Declercq (1998) *Journal of Molecular Biology* **276**, 151–164.

Fabrick J. A., J. E. Baker, and M. R. Kanost (2003) *Insect Biochemistry and Molecular Biology* **33**, 579–594.

Fischetti V. A. (2005) *Trends in Microbiology* **13**, 491–496.

Fujii T. and K. Miyashita (1993) *Journal of General Microbiology* **139**, 677–686.

Fukamizo T. (2000) *Current Protein & Peptide Science* **1**, 105–124.

Funkhouser J. D. and N. N. Aronson Jr. (2007) *BMC Evolutionary Biology* **7**, 96.

Gianfrancesco F. and S. Musumeci (2004) *Cytogenetic and Genome Research* **105**, 54–56.

Gilkes N. R., B. Henrissat, D. G. Kilburn, R. C. Miller Jr., and R. A. J. Warren (1991) *Microbiological Reviews* **55**, 303–315.

Gohel V., A. Singh, M. Vimal, P. Ashwini, and H. S. Chhatpar (2006) *African Journal of Biotechnology* **5**, 54–72.

Goto M., W. Fujimoto, J. Nio, T. Iwanaga, and T. Kawasaki (2003) *Archives of Oral Biology* **48**, 701–707.

Gott J. M., D. A. Shub, and M. Belfort (1986) *Cell* **47**, 81–87.

Grütter M. G., L. H. Weaver, and B. W. Matthews (1983) *Nature* **303**, 828–831.

Hamel F., R. Boivin, C. Tremblay, and G. Bellemare (1997) *Journal of Molecular Evolution* **44**, 614–624.

Hart P. J., H. D. Pflugger, A. F. Monzingo, T. Hollis, and J. D. Robertus (1995) *Journal of Molecular Biology* **248**, 402–413.

Hendrix R. E., J. G. Lawrance, G. F. Hatfull, and S. Casjens (2000) *Trends in Microbiology* **8**, 504–508.

Henrissat B. (1991) *Biochemistry Journal* **280**, 309–316.

Henrissat B. and A. Bairoch (1993) *Biochemistry Journal* **293**, 781–788.

Henrissat B. and A. Bairoch (1996) *Biochemistry Journal* **316**, 695–696.

Henrissat B., I. Callebaut, S. Fabrega, P. Lehn, J. P. Mornon, and G. Davies (1995) *Proceedings of the National Academy of Sciences of the United States of America* **92**, 7090–7094.

Hermoso J. A., B. Monterroso, A. Albert, B. Galan, O. Ahrazen, P. Garcia, M. Martinez-Ripoll, J. L. Garcia, and M. Menendez (2003) *Structure* **11**, 1239–1249.

Hibbert E. G. and P. A. Dalby (2005) *Microbiological Cell Factories* **4**, 29.

Hikima S., J.-I. Hikima, J. Rojtinnakorn, I. Hirono, and T. Aoki (2003) *Gene* **316**, 187–195.

Hikima J. I., S. Minagawa, I. Hirona, and T. Aoki (2001) *Biochimica et Biophysica Acta* **1520**, 35–44.

Hoell I. A., B. Dalhus, E. B. Heggset, S. I. Aspmo, and V. G. H. Eijsink (2006) *The FEBS Journal* **273**, 4889–4900.

Hollak C. E. M., S. van Weely, M. H. J. van Oers, and J. M. F. G. Aerts (1994) *Journal of Clinical Investigation* **93**, 1288–1292.

Holm L. and C. Sander (1994) *FEBS Letters* **340**, 129–132.

Huescas S., M. Villalba, and R. Rodriguez (2001) *Journal of Biological Chemistry* **276**, 27956–27966.

Hultmark D. (1996) Insect lysozymes. In *Lysozymes: Model Enzymes in Biochemistry and Biology*, ed. P. Jolles, 87–102. Basel, Switzerland: Birkhauser Verlag.

Iacono V. J., S. M. Zove, B. L. Grossbard, J. J. Pollock, D. H. Fine, and L. S. Greene (1985) *Infection and Immunity* **47**, 457–464.

Ibrahim H. R., T. Matsuzaki, and T. Aoki (2001) *FEBS Letters* **506**, 27–32.

Ikeda S., N. Ytow, H. Ezura, K. Minamisawa, R. Miyashita, and T. Fujimura (2007) *Microbes Environment* **22**, 71–77.

Irwin D. M. (1995) *Journal of Molecular Evolution* **41**, 299–312.

Irwin D. M. (2004) *Genome* **47**, 1082–1090.

Irwin D. M and A. C. Wilson (1989) *Journal of Biological Chemistry* **264**, 11387–11393.

Jado I., R. Lopez, E. Garcia, A. Fenoll, J. Casal, and P. Garcia (2003) *The Journal of Antimicrobial Chemotherapy* **52**, 957–973.

Jespers L., E. Sonveaux, and J. Fastrez (1992) *Journal of Molecular Biology* **228**, 529–538.

Jimenez-Vega F., R. R. Sotelo-Mundo, F. Ascencio, and F. Vargas-Albores (2002) *Fish and Shellfish Immunology* **13**, 171–181.

Jin W., H. T. Horner, R. G. Palmer, and R. C. Shoemaker (1999) *Genetics* **153**, 445–452.

Jolles P. (1996) *Lysozymes: Model Enzymes in Biochemistry and Biology.* Basel, Switzerland: Birkhäuser Verlag.

Junker N., J. S. Johansen, L. T. Hansen, E. L. Lund, and P. E. G. Kristjansen (2005) *Cancer Science* **96**, 183–190.

Kang D., G. Liu, A. Lundstrom, E. Gelius, and H. Steiner (1998) *Proceedings of the National Academy of Sciences of the United States of America* **95**, 10078–10082.

Kawai R., K. Iigarashi, and M. Samejima (2006) *Biotechnology Letters* **28**, 365–371.

Kawase T., A. Saito, T. Sato, R. Kanai, T. Fujii, N. Nikaidou, K. Miyashita, and T. Watanabe (2004) *Applied and Environmental Microbiology* **70**, 1135–1144.

Kawase T., S. Yokokawa, A. Saito, T. Fujii, N. Nikaidou, K. Miyashita, and T. Watanabe (2006) *Bioscience, Biotechnology, and Biochemistry* **70**, 988–998.

Kim D. J., J. M. Baek, P. Uribe, C. M. Kenerley, and D. R Cook (2002) *Current in Genetics* **40**, 374–384.

Kleppe G., E. Vasstrand, and H. B. Jensen (1981) *European Journal of Biochemistry* **119**, 589–593.

Koga D., M. Mitsutomi, M. Kono, and M. Matsumiya (1999) *EXS* **87**, 111–123.

Kuroki R., L. H. Weaver, and B. W. Matthews (1995) *Nature Structural Biology* **2**, 1007–1011.

Kuroki R., L. H. Weaver, and B. W. Matthews (1999) *Proceedings of the National Academy of Sciences of the United States of America* **96**, 8949–8954.

Laible N. J. and G. R. Germaine (1985) *Infection and Immunity* **48**, 720–728.

Landale E. C. and M. M. McCabe (1987) *Infection and Immunity* **55**, 3011–3016.

Lashbrook C. C., C. Gonzalez-Bosch, and A. B. Bennet (1994) *Plant Cell* **6**, 1485–1493.

Leah R., H. Tommerup, I. Svendsen, and J. Mundy (1991) *Journal of Biology Chemistry* **266**, 1564–1573.

LeCleir L. R., A. Buchan, and J. T. Hollibaugh (2004) *Applied and Environmental Microbiology* **70**, 6977–6983.

Lee M.-F., G.-Y. Hwang, Y.-H. Chen, H.-C. Lin, and C.-H. Wu (2006) *Molecular Immunology* **43**, 1144–1151.

Lee S. Y., R. Wang, and K. Soderhall (2000) *Journal of Biological Chemistry* **275**, 1337–1343.

Li B., E. Calvo, O. Marinotti, A. A. Jamesand, and S. M. Paskewitz (2005) *Gene* **360**, 131–139.

Linthorst H. J. M., L. S. Melchers, A. Meyer, J. S. C. Van Roekel, B. J. C. Cornelissen, and J. F. Bol (1991) *Proceedings of the National Academy of Sciences of the United States of America* **87**, 8756–8760.

Lopez R. and E. Garcia (2004) *FEMS Microbiology Reviews* **28**, 553–580.

Lopez R., E. Garcia, P. Garcia, and J. L. Garcia (1997) *Microbial Drug Resistance* **3**, 199.

Lymar E. S., B. Li, and V. Enganthan (1995) *Applied and Environmental Microbiology* **61**, 2976–2980.

Martin K., B. M. McDougall, S. McIlroy, J. Jayus, J. Chen, and R. J. Seviour (2007) *FEMS Microbiology Reviews* **31**, 168–192.

Masschalck B., D. Deckers, and C. W. Michiels (2002) *Journal of Food Protection* **65**, 1916–1923.

Matthews B. W., M. G. Grutter, W. F. Anderson, and S. J. Remington (1981) *Nature* **290**, 334–335.

Monzingo A. F., E. M. Marcotte, P. J. Hart, and J. D. Robertus (1996) *Nature* **3**, 133–140.

Mouyna I., J. Sarfati, P. Recco, T. Fontaine, B. Henrissat, and J. P. Latge (2002) *Medical Mycology* **40**, 455–464.

Obregon V., J. L. Garcia, E. Garcia, R. Lopez, and P. Garcia (2003) *Journal of Bacteriology* **185**, 2362–2368.

Ochiai M. and A. Ashida (1999) *Journal of Biological Chemistry* **274**, 11854–11858.

Ohno T., S. Armand, T. Hata, N. Nikaidou, B. Henrissat, M. Mitsutomi, and T. Watanabe (1996) *Journal of Bacteriology* **178**, 5065–5070.

O'Riordain G., C. Radauer, K. Hoffmann-Sommergruber, F. Adhami, C. K. Peterbauer, C. Blanco, J. Godnic-Cvar, O. Scheiner, C. Ebner, and H. Breiteneder (2003) *Clinical and Experimental Allergy* **32**, 455–462.

Palomares O., M. Villalba, J. Quiralte, F. Polo, and R. Rodriguez (2005) *Clinical and Experimental Allergy* **35**, 345–351.

Park T., D. K. Stuck, J. F. Deaton, and R. Young (2006) *Proceedings of the National Academy of Sciences of the United States of America* **52**, 19713–19718.

Peberdy J. (1994) *Trends in Biotechnology* **12**, 50–57.

Pitson S., R. Seviour, and B. McDougall (1997) *Canadian Journal of Microbiology* **43**, 432–439.

Posch A., C. H. Weeler, Z. Chen, A. Flagge, M. J. Dunn, F. Papenfuss, M. Raulf-Heimsoth, and X. Baur (1999) *Clinical and experimental allergy* **29**, 667–672.

Prager E. M. and A. C. Wilson (1988) *Journal of Molecular Evolution* **27**, 326–335.

Qasba P. K. and S. Kumar (1997) *Critical Reviews in Biochemistry and Molecular Biology* **32**, 255–306.

Radwan H. H., H. J. Plattner, U. Menge, and H. Diekmann (1994) *FEMS Microbiology Letters* **120**, 31–36.

Renkema G. H., R. G. Boot, F. L. Au, W. E. Donker-Koopmand, A. Strijland, A. O. Muijsers, M. Hrebicek, and J. M. Aerts (1998) *European Journal of Biochemistry* **251**, 504–509.

Rihs H. P., B. Dumont, P. Rozynek, M. Lundberg, R. Cremer, T. Bruning, and M. Raulf-Heimsoth (2003) *Allergy* **58**, 246–251.

Ring M., D. E. Bader, and M. Hubber (1988) *Biochemical and Biophysical Research Communications* **152**, 1050–1055.

Robertus J. D. and A. F. Monzingo (1999) *EXS* **87**, 125–135.

Rose J. K. C., K.-S. Ham, A. G. Darwill, and P. Albersheim (2002) *The Plant Cell* **14**, 1329–1345.

Royer V., S. Fraichard, and H. Bouhin (2002) *Biochemistry Journal* **366**, 921–928.

Saito A., M. Ishizaka, P. B. Francisco, T. Fujii, and K. Miyashita (2000) *Microbiology* **146**, 2937–2946.

Sanders N. N., V. G. H. Eijsink, P. S. van den Pangaart, R. J. Joost van Neerven, P. J. Simons, S. C. De Smedt, and J. Demeester (2007) *Biochemica et Biophysica Acta* **1770**, 839–846.

Schoemaker H. E., D. Mink, and M. G. Wubbolts (2003) *Science* **299**, 1694–1697.

Seidl V., B. Huemer, B. Seiboth, and C. P. Kubicek (2005) *FEBS Journal* **272**, 5923–5939.

Smith D. L., D. K. Struck, J. M. Scholtz, and R. Young. (1998) *Journal of Bacteriology* **180**, 2531–2540.

Sinnot M. L. (1990) *Chemistry Review* **90**, 1171–1202.

Sjoberg B. M., S. Hahne, C. Z. Mathews, C. K. Mathews, K. N. Rand, and M. J. Gait (1986) *EMBO Journal* **5**, 2031–2036.

Sommers C. E. and B. M. Shapiro (1989) *Development Growth and Differentiation* **31**, 1–7.

Sova V. V., L. A. Elyakova, and V. E. Vaskovsky (1969) *Comparative Biochemistry and Physiology* **32**, 465–474.

Sova V. V., N. I. Shirokova, M. I. Kusaykin, A. S. Scobun, L. A. Elyakova, and T. N Zvyagintseva (2003) *Biochemistry (Moscow)* **68**, 529–533 (Translated from Biokhimiya **68 (5)**, 650–655).

Stahl, E. A. and J. G. Bishop (2000) *Current Opinion in Plant Biology* **3**, 299–304.

Stojkovic E. A. and L. B. Rothman-Denes (2006) *Journal of Molecular Biology* **366**, 406–419.

Sunderasan E., S. Hamzah, S. Hamid, M. A. Ward, H. Y. Yeang, and M. J. Cardosa (1995) *Journal of Natural Rubber Research* **10**, 82–99.

Suzuki K., M. Taiyoji, N. Sugawara, N. Nikaidou, B. Henrissat, and T. Watanabe (1999) *Biochemistry Journal* **343**, 587–596.

Svensson B., H. Jespersen, M. R. Sierks, and E. A. MacGregor (1989) *Biochemistry Journal* **264**, 309–311.

Thunnissen A. M., A. J. Dijkstra, K. H. Kalk, H. J. Rozeboom, H. Engel, W. Keck, and B. W. Dijkstra (1994) *Nature* **367**, 750–753.

Thunnissen A. M., H. J. Rozeboom, K. H. Kalk, and B. W. Dijkstra (1995) *Biochemistry* **34**, 12729–12737.

Tomasz A.(1979) *Annual Reviews of Microbiology* **33**, 113–137.

Tomme P., R. A. J. Warren, and N. R. Gilkes (1995) *Advances in Microbial Physiology* **37**, 1–81.

Varea J., B. Monterroso, J. Saiz, C. Lopez-Zumel, J. L. Garcia, J. Laynez, P. Garcia, and M. Menendez (2004) *Journal of Biological Chemistry* **279**, 43697–43707.

Vasallo R., T. J. Kottom, J. E. Standing, and A. H. Limper (2001) *American Journal of Respiratory Cell and Molecular Biology* **25**, 203–211.

Verigina N. S., Y. V. Burtseva, S. P. Ermakova, V. V. Sova, M. V. Pivkin, and T. N. Zvyagintseva (2005) *Applied Biochemistry and Microbiology* **41**, 354–360 (Translated from *Prikladnaya Biokhimiya i Mikrobiologiya* **41 (4)**, 402–408).

Wang I. N. (2006) *Genetics* **172**, 17–26.

Wang I. N., D. E. Dykhuizen, and L. B. Slobodkin (1996) *Evolutionary Ecology* **10**, 545–598.

Wang I. N., D. L. Smith, and R. Young (2000) *Annual Reviews of Microbiology* **54**, 799–825.

Wang H., D. Wu, F. Deng, H. Peng, X. Chen, H. Lauzon, B. M. Arif, J. A. Jehle, and Z. Hu (2004) *Virus Research* **100**, 179–189.

Warren R. A. J. (1996) *Annual Reviews in Microbiology* **50**, 183–212.

Watanabe T. K., Y. Ito, T. Yamada, M. Hashimoto, S. Sekine, and H. Tanaka (1994) *Journal of Bacteriology* **176**, 4465–4472.

Watanabe T., R. Kanai, T. Kawase, T. Tanabe, M. Mitsutomi, S. Sakuda, and K. Miyashita (1999) *Microbiology* **145**, 3353–3363.

Watanabe T., K. Suzuki, W. Oyanagi, K. Ohnishis, and H. Tanaka (1990) *Journal of Biological Chemistry* **265**, 15659–15665.

Weaver L. H., M. G. Grütter, S. J. Remington, T. M. Gray, N. W. Isaacs, and B. W. Matthews (1984) *Journal of Molecular Evolution* **21**, 97–111.

Weaver L. H., M. G. Grutter, S. J. Remington, T. M. Gray, N. W. Isaacs, and B. W. Matthews (1995) *Journal of Molecular Evolution* **245**, 54–68.

Weaver L. H., D. Rennell, A. R. Poteete, and B. W. Matthews (1985) *Journal of Molecular Evolution* **184**, 189–199.

Wen Y. and D. M. Irwin (1999) *Molecular Phylogenetics and Evolution* **13**, 474–482.

Wessel G. M., M. R. Truschel, S. A. Chambers, and D. R. McClay (1987) *Gamete Research* **18**, 339–348.

Williams G. J., A. S. Nelson, and A. Berry (2004) *Cellular and molecular life sciences* **61**, 3034–3046.

Xu M., A. Arulandu, D. K. Struck, S. Swanson, J. C. Sacchettini, and R. Young (2005) *Science* **307**, 113–117.

Yamagi T., M. Sato, A. Nakamura, K. Kitagawa, A. Akasawa, and Z. Ikezawa (1998) *Journal of Allergy and Clinical Immunology* **101**, 379–385.

Yahata N., T. Watanabe, Y. Nakamura, Y. Yamamoto, S. Kamimiya, and H. Tanaka (1990) *Gene*, **86**, 113–117.

Yang Z. (1998) *Molecular Biology and Evolution* **15**, 568–573.

Young R. (1992) *Microbiology Reviews* **56**, 430–481.

Yu M. and D. M. Irwin (1996) *Molecular Phylogenetics and Evolution* **5**, 298–308.

Zhenga T., M. Rabach, N. Y. Chen, L. Rabach, X. Hu, J. A. Elias, and Z. Zhua (2005) *Gene* **357**, 37–46.

CHAPTER 5

BACTERIOPHAGE LYSINS: THE ULTIMATE ENZYBIOTIC

VINCENT A. FISCHETTI
Laboratory of Bacterial Pathogenesis, Rockefeller University, New York, NY

1. BACKGROUND

Viruses that specifically infect bacteria are called bacteriophages, or phages. After replication inside its bacterial host the phage is faced with a problem: it needs to exit the bacterium to disseminate its progeny phage. To solve this, double-stranded DNA phages have evolved a lytic system to weaken the bacterial cell wall, resulting in bacterial lysis. Phage lytic enzymes, or lysins, are highly efficient molecules that have been refined over millions of years of evolution for this very purpose. These enzymes target the integrity of the cell wall and are designed to attack one of the five major bonds in the peptidoglycan. With few exceptions (Loessner et al. 1997), lysins do not have signal sequences, so they are not translocated through the cytoplasmic membrane to attack their substrate in the peptidoglycan. This movement is tightly controlled by a second phage gene product in the lytic system, the holin (Wang et al. 2000). During phage development in the infected bacterium, lysin accumulates in the cytoplasm in anticipation of phage maturation. At a genetically specified time, holin molecules are inserted in the cytoplasmic membrane, forming patches, ultimately resulting in generalized membrane disruption (Wang et al. 2003). This allows the cytoplasmic lysin to access the peptidoglycan, thereby causing cell lysis and the release of progeny phage (Wang et al. 2000). In contrast to large DNA phages, small RNA and DNA phages use a different release strategy. They call on a phage-encoded protein to interfere with bacte-

Enzybiotics: Antibiotic Enzymes as Drugs and Therapeutics. Edited by Tomas G. Villa and Patricia Veiga-Crespo
Copyright © 2010 John Wiley & Sons, Inc.

rial host enzymes responsible for peptidoglycan biosynthesis (Young et al. 2000; Bernhardt et al. 2001), resulting in misassembled cell walls and ultimate lysis. Scientists have been aware of the lytic activity of phages for nearly a century, and while whole phages have been used to control infection (Matsuzaki et al. 2005), not until recently have lytic enzymes been exploited for bacterial control *in vivo* (Nelson et al. 2001; Schuch et al. 2002; Loeffler et al. 2003). One of the main reasons that such an approach is now even being considered is the sharp increase in antibiotic resistance among pathogenic bacteria. Current data indicate that lysins work only with Gram-positive bacteria, since they are able to make direct contact with the cell wall carbohydrates and peptidoglycan when added externally, whereas the outer membrane of Gram-negative bacteria prevents this interaction. This review outlines the remarkable potency these enzymes have in killing bacteria both *in vitro* and *in vivo*.

Most human infections (viral or bacterial) begin at a mucous membrane site (upper and lower respiratory, intestinal, urogenital, or ocular). In addition, the human mucous membranes are the reservoirs (and sometimes the only reservoirs) for many pathogenic bacteria found in the environment (i.e., pneumococci, staphylococci, streptococci), some of which are resistant to antibiotics. In most instances, it is this mucosal reservoir that is the focus of infection in the population (Coello et al. 1994; de Lencastre et al. 1999; Eiff et al. 2001). To date, except for polysporin and mupirocin ointments, which are the most widely used topically, there are no anti-infectives that are designed to control colonizing pathogenic bacteria on mucous membranes (Hudson 1994); we usually first wait for infection to occur before treating. Because of the fear of increasing the resistance problem, antibiotics are not indicated to control the carrier state of disease bacteria. It is acknowledged, however, that by reducing or eliminating this human reservoir of pathogens in the community and in controlled environments (i.e., hospitals and nursing homes), the incidence of disease will be markedly reduced (Hudson 1994; Eiff et al. 2001). Toward this goal, lysins have been developed to prevent infection by safely and specifically destroying disease bacteria on mucous membranes. For example, based on extensive animal results, enzymes specific for *Streptococcus pneumoniae* and *Streptococcus pyogenes* may be used nasally and orally to control these organisms in the community as well as in nursing homes and hospitals to prevent or markedly reduce serious infections caused by these bacteria. This has been accomplished by capitalizing on the efficiency by which phage lysins kill bacteria (Young 1992). Like antibiotics, which are used by bacteria to control the organisms around them in the environment, phage lysins are the culmination of millions of years of development by the bacteriophage in their association with bacteria. Specific lysins that are able to kill specific Gram-positive bacteria seconds after contact have now been identified and purified (Loeffler et al. 2001; Nelson et al. 2001). For example, nanogram quantities of lysin could reduce 10^7 *S. pyogenes* by >6 log seconds after enzyme addition. No known biological compounds, except chemical agents, kill bacteria this quick. Because of their highly effective activity against bacteria for the control of disease, the term "enzybiotics" was coined (Nelson et al. 2001) to describe these novel anti-infectives.

2. PHYSICAL CHARACTERISTICS

Lysins from DNA phages that infect Gram-positive bacteria are generally between 25 and 40 kDa in size except the PlyC for streptococci, which is 114 kDa. This enzyme is unique because it is composed of two separate gene products, PlyCA and PlyCB. Based on biochemical and biophysical studies, the catalytically active PlyC holoenzyme is composed of eight PlyCB subunits for each PlyCA (Nelson et al. 2006). A feature of all other Gram-positive phage lysins is their two-domain structure (Fig. 5.1; Diaz et al. 1990; Garcia et al. 1990). With no exceptions, the N-terminal domain contains the catalytic activity of the enzyme. This activity may be either an endo-β-N-acetylglucosaminidase or an N-acetylmuramidase (lysozymes), both of which act on the sugar moiety of the bacterial wall, an endopeptidase that acts on the peptide moiety, or an N-acetylmuramoyl-L-alanine amidase (or amidase), which hydrolyzes the amide bond connecting the glycan strand and peptide moieties (Young 1992; Loessner 2005). Recently, an enzyme with γ-D-glutaminyl-L-lysine endopeptidase activity has also been reported (Pritchard et al. 2007). In some cases, particularly staphylococcal lysins, two and perhaps even three different catalytic domains may be linked to a single binding domain (Navarre et al. 1999). The C-terminal cell-binding domain (termed the CBD domain), on the other hand, binds to a specific substrate (usually carbohydrate) found in the cell wall of the host bacterium (Lopez et al. 1992, 1997; Garcia et al. 1998). Efficient cleavage requires that the binding domain bind to its cell wall substrate, offering some degree of specificity to the enzyme, since these substrates are found only in enzyme-sensitive bacteria. The first complete crystal structure for the free and choline-bound states of the Cpl-1 lytic enzyme has recently been published (Hermoso et al. 2003). As suspected, the data suggest that choline

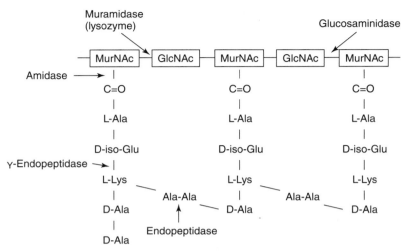

Figure 5.1. Activity of lysins on the peptidoglycan. A diagrammatic representation of the cell wall peptidoglycan showing the bonds cleaved by various lysins. Amidase, *N*-acetylmuramoyl-L-alanine amidase, γ-endopeptidase, γ-D-glutaminyl-L-lysine endopeptidase.

recognition by the choline-binding domain of Cpl-1may allow the catalytic domain to be properly oriented for efficient cleavage. An interesting feature of this lysin is its hairpin conformation, suggesting that the two domains interact with each other prior to the interaction of the binding domain with its substrate in the bacterial cell wall. Other lytic enzymes need to be crystallized to determine if this is a common feature of all lysins.

When we compared the sequences between lytic enzymes of the same enzyme class, we observed high sequence homology within the N-terminal catalytic region and very little homology within in the C-terminal cell-binding region. It seemed counterintuitive that the phage would design a lysin that was uniquely lethal for its host organism; however, as we learned more about how these enzymes function, a possible reason for this specificity became apparent (see the section below on resistance). Because of their specificity, enzymes that spilled out after cell lysis had a good chance of killing potential bacterial hosts in the vicinity of the released phage progeny. Thus, the enzymes have evolved to bind to their cell wall-binding domains at a high affinity (Loessner et al. 2002) to limit the release of free enzyme.

It seemed plausible that, due to their domain structure, different enzyme domains could be swapped, resulting in lysins with different bacterial and catalytic specificities. This was actually accomplished by the excellent detailed studies of Garcia et al. (1990; Weiss et al. 1999), in which the catalytic domains of lytic enzymes for *S. pneumoniae* phage were swapped, resulting in a new enzyme having the same binding domain for pneumococci, but able to cleave a different bond in the peptidoglycan. This capacity allows for enormous potential in creating designer enzymes with high specificity and equally high cleavage potential.

Although uncommon, introns have been associated with certain lysins. For example, 50% of *Streptococcus thermophilus* phages have been reported to have their lysin gene interrupted by a self-splicing group I intron (Foley et al. 2000). This also appears to be the case for a *Staphylococcus aureus* lytic enzyme (Flaherty et al. 2004) and perhaps the C1 lysin for group C streptococci (Nelson et al. 2003). While introns have been previously reported in phage genes, they have rarely been identified in the host genome (Fernandez-Lopez et al. 2005; Tan et al. 2005).

3. MECHANISMS OF ACTION

When examined by thin section electron microscopy, it seems obvious that lysins exert their lethal effects by forming holes in the cell wall through peptidoglycan digestion. The high internal pressure of bacterial cells (roughly 3–5 atmospheres) is controlled by the highly cross-linked cell wall. Any disruption in the wall's integrity will result in the extrusion of the peptidoglycan cytoplasmic membrane and ultimate hypotonic lysis (Fig. 5.2). Catalytically, a single-enzyme molecule should be sufficient

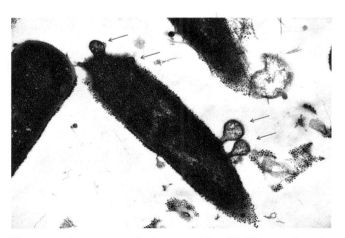

Figure 5.2. Electron microscopy of lysin-treated bacilli. Thin section electron micrograph of *B. cereus* (RSVF) (Schuch et al. 2002) after treatment with phage lytic enzyme (PlyPH; Yoong et al. 2004). High magnification of a bacillus exhibiting externalization of the cytoplasmic membrane after treatment with enzyme for 1 min (arrows).

to cleave an adequate number of bonds to kill an organism; however, it is uncertain at this time whether this theoretical limit is possible. The reason comes from the work of Loessner et al. (2002), showing that a listeria phage enzyme had a binding affinity approaching that of an IgG molecule for its substrate, suggesting that phage enzymes, like cellulases (Jervis et al. 1997), are one-use enzymes, likely requiring several molecules attacking a local region to sufficiently weaken the cell wall.

4. LYSIN EFFICACY

In general lysins kill only the species (or subspecies) of bacteria from which they were produced. For instance, enzymes produced from streptococcal phage kill certain streptococci, and enzymes produced by pneumococcal phage kill pneumococci (Loeffler et al. 2001). Specifically, a lysin from a group C streptococcal phage (PlyC) will kill group C streptococci, as well as groups A and E streptococci, the bovine pathogen *Streptococcus uberis*, and the horse pathogen *Streptococcus equi*, but will have essentially no effect on streptococci normally found in the oral cavity of humans and other Gram-positive bacteria. Similar results are seen with a pneumococcal-specific lysin; however, in this case, the enzyme was also tested against strains of penicillin-resistant pneumococci and the killing efficiency was the same. Unlike antibiotics, which are usually broad-spectrum and kill many different bacteria found in the human body (some of which are beneficial), lysins that kill only the disease organism with little to no effect on the normal human bacterial flora may be identified. In some cases, however, phage enzymes may be identified with broad lytic activity. For example, an enterococcal phage lysin has recently been reported to kill not only enterococci but a number of other Gram-positive pathogens such as *S. pyogenes*, group B streptococci, and *Staphylococcus aureus*, making it one of the broadest acting lysins identified (Yoong et al. 2004). However, its activity for these other pathogens was lower than for enterococci.

A significant lysin with respect to infection control is one directed to *S. aureus* (Sonstein et al. 1971; Clyne et al. 1992; O'Flaherty et al. 2005; Rashel et al. 2007; Sass and Bierbaum 2007). However, in most cases these enzymes show low activity or are difficult to produce in large quantities. In one recent publication (Rashel et al. 2007), a staphylococcal enzyme that could be easily produced recombinantly and had a significant lethal effect on methicillin-resistant *S. aureus* (MRSA) both *in vitro* and in a mouse model was described. In the animal experiments the authors show that the enzymes may be used to decolonize staphylococci from the nose of the mice as well as protect the animals from an intraperitoneal challenge with MRSA. However, in the latter experiments, the best protection was observed if the lysin was added up to 30 min after the MRSA.

5. ANTIBIOTIC AND LYSIN SYNERGY

Several lysins have been identified from pneumococcal bacteriophages, which are classified into two groups: amidases and lysozymes. Exposure of pneumococci to either of these enzymes leads to efficient lysis. Both enzymes have very different N-terminal catalytic domains but share a similar C-terminal choline cell-binding domain. These enzymes were tested to determine whether their simultaneous use is competitive or synergistic (Loeffler and Fischetti 2003).

To accomplish this, three different analytical methods were used to determine synergy: time-kill in liquid, disk diffusion, and checkerboard broth microdilution analysis. All three are standard methods used in the antibiotic industry to determine synergy (Eliopoulos and Moellering 1991). In all three assays, the results revealed a clear synergistic effect in the efficiency of killing when both enzymes were used (Loeffler and Fischetti 2003). *In vivo*, the combination of two lysins with different peptidoglycan specificities was found to be more effective in protecting against disease than each of the single enzymes (Jado et al. 2003; Loeffler and Fischetti 2003). Thus, in addition to more effective killing, the application of two different lysins may significantly retard the emergence of enzyme-resistant mutants.

When the pneumococcal lysin Cpl-1 was used in combination with certain antibiotics, a similar synergistic effect was seen. Cpl-1 and gentamicin were found to be increasingly synergistic in killing pneumococci with a decreasing penicillin minimum inhibitory dose (MIC), while Cpl-1 and penicillin showed synergy against an extremely penicillin-resistant strain (Djurkovic et al. 2005). Synergy was also observed with a staphylococcal-specific enzyme and glycopeptide antibiotics (Rashel et al. 2007). Thus, the right combination of enzyme and antibiotic could help in the control of antibiotic-resistant bacteria as well as reinstate the use of certain antibiotics for which resistance has been established.

6. ANTIBODIES TO LYSIN

A potential concern in the use of lysins is the development of neutralizing antibodies that could reduce the *in vivo* activity of enzyme during treatment. Unlike antibiotics, which are small molecules that are not generally immunogenic, enzymes are proteins that stimulate an immune response, when delivered mucosally or systemically, which could interfere with the lysin's activity. To address this, rabbit hyperimmune serum raised against the pneumococcal-specific enzyme Cpl-1 was assayed for its effect on lytic activity (Loeffler et al. 2003). It was found that highly immune serum slows but does not block the lytic activity of Cpl-1. When similar *in vitro* experiments were performed with antibodies directed to an anthrax- and an *S. pyogenes*-specific enzyme, similar results were obtained (unpublished data). These results were also verified with a staphylococcal-specific lysin (Rashel et al. 2007).

To test the relevance of this *in vivo*, mice that received three intravenous (IV) doses of the Cpl-1 enzyme had tested positive for IgG against Cpl-1 in five of six cases with low but measurable titers of about 1:10. These vaccinated and naive control mice were then challenged intravenously with pneumococci and then treated by the same route with 200 µg Cpl-1 after 10 h. Within a minute, the treatment reduced the bacteremic titer of Cpl-1-immunized mice to the same degree as the naive mice, supporting the *in vitro* data that antibodies to lysins have little to no neutralizing effect. A similar experiment by Rashel et al. (2007) with a staphylococcal enzyme showed the same result and that the animals injected with lysin multiple times exhibited no adverse events.

This unexpected effect may be partially explained if the binding affinity of the enzyme for its substrate in the bacterial cell wall is higher than the antibody's affinity for the enzyme. This is supported by the results of Loessner et al. (2002), showing that the cell wall-binding domain of a listeria-specific phage enzyme binds to its wall substrate at the affinity of an IgG molecule (nanomolar affinities). However, while this may explain the inability of the antibody to neutralize the binding domain, it does not explain why antibodies to the catalytic domain do not neutralize. Nevertheless, these results are encouraging since they suggest that such enzymes may be used repeatedly in certain situations to control colonizing bacteria on mucosal surfaces in susceptible populations, such as those in hospitals, day-care centers, and nursing homes, or in blood to eliminate antibiotic-resistant bacteria in cases of septicemia and bacteremia.

7. ANIMAL MODELS OF INFECTION

Animal models of mucosal colonization were used to test the capacity of lysins to kill organisms on these surfaces, perhaps the most important use for these enzymes. An oral colonization model was developed for *S. pyogenes* (Nelson et al. 2001), a nasal model for pneumococci (Loeffler et al. 2001), and a vaginal model for group B streptococci (Cheng et al. 2005). In all three cases, when the animals were colonized with their respective bacteria and treated with a single dose of lysin, specific for the colonizing organism, these organisms were reduced by several logs (and in some cases below the detection limit of the assay) when tested again 2 to 4h after lysin treatment. These results lend support to the idea that such enzymes may be used in specific high-risk populations to control the reservoir of pathogenic bacteria and thus control disease.

Similar to other proteins delivered intravenously to animals and humans, lysins have a short half-life ($T^{\frac{1}{2}}$ = 15–20 min; Loeffler et al. 2003). However, the action of lysins for bacteria is so rapid that this may be sufficient time to observe a therapeutic effect (Jado et al. 2003; Loeffler et al. 2003). Mice intravenously infected with type 14 *S. pneumoniae* and treated 1h later with a single bolus of 2.0 mg of Cpl-1 survived through the 48-h endpoint, whereas the median survival time of buffer-treated mice was only 25h, and only 20% survival at 48h (Fig. 5.3). Blood and organ cultures of the euthanized surviving mice showed that only one Cpl-1-treated animal was totally free of infection at 48h, suggesting that multiple enzyme doses or a constant infusion of enzyme would be required to eliminate the organisms completely in this application. Similar results were obtained when animals were infected and treated intraperitoneally with lysin (Jado et al. 2003; Rashel et al. 2007).

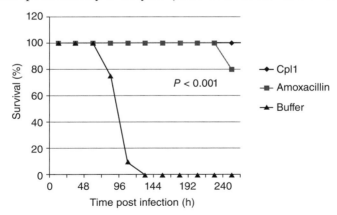

Figure 5.3. Survival of pneumococcal pneumonia after intraperitoneal treatment with Cpl-1 or Amoxacillin. Mice pretreated nasally with 10^8 pneumococci developed classic pneumonia in 24h. Beginning 24h following infection, mice were treated by intraperitoneal injections of Cpl-1, amoxicillin or buffer given every 12h for 72h (Witzenrath et al. 2008).

Because of lysin's short half-life, it may be necessary to modify the lysins with polyethylene glycol or the Fc region of IgG to extend the residence time *in vivo* to several hours (Walsh et al. 2003). In recent studies, phage lysins have also been shown to be successful in the treatment of meningitis by adding the lysin directly to the brain intrathecally (Grandgirard et al. 2008) and in the treatment of endocarditis by delivering the lysin intravenously by constant IV infusion (Entenza et al. 2005). Both these applications would also benefit from modified long-acting lysins.

The crucial challenge for lysins is to determine whether they are able to cure an established infection. To approach this, a mouse pneumonia model was developed in which mice were transnasally infected with pneumococci and treated with Cpl-1 by repeated intraperitoneal injections after infection was established (Witzenrath et al. 2008). From a variety of clinical measurements, as well as morphological changes in the lungs, it was shown that at 24 h mice suffered from severe pneumonia. When treatment was initiated at 24 h and every 12 h thereafter, 100% of the mice survived otherwise fatal pneumonia and showed rapid recovery. Cpl-1 dramatically reduced pulmonary bacterial counts and prevented bacteremia.

Using lysins systemically to kill bacteria could result in an increase in cytokine production as a result of bacterial debris being released. In one study to address this issue (Entenza et al. 2005), untreated pneumococcal endocarditis induced the release of interleukin-1α (IL-1α), IL-1β, IL-6, IL-10, gamma interferon, and tumor necrosis factor, but not IL-2, IL-4, or granulocyte-macrophage colony-stimulating factor. The use of Cpl-1 resulted in an increase in the release of these cytokines. However, in a mouse model of pneumonia, Witzenrath et al. (2008) showed that transnasal infection with pneumococci caused an increase of pro-inflammatory cytokines and chemokines within 36 h. However, treatment with Cpl-1 was associated with reduced synthesis of IL-1β, IL-6, and the chemokines KC, Mip-1α, and MCP-1, as well as G-CSF in the lungs. In addition, decreases were observed in plasma concentrations of IL-6, KC, and Mip-1α, and MCP-1, G-CSF, and IFNγ concentrations. Pulmonary and systemic IL-10 synthesis was found only in septic animals 60 h after infection, and was completely prevented by Cpl-1 treatment. The reason for this difference has not been determined as yet; however, it may depend on the amount of lysin used for the treatments. It is possible that in the former study, in which the animals were treated with a constant IV infusion of lysin, the high concentration of the enzyme resulted in the fragmentation of the bacterial cell wall, while in the second study, enzyme delivered in 12-h intervals resulted in producing holes in the bacterial cell wall without forming wall pieces, which are clearly more inflammatory (Tuomanen et al. 1985; Kengatharan et al. 1998).

Since lysins are proteins, it is unlikely that they are able to enter cells. However, there are no publications to date that address the issue of the effects of lysins on intracellular bacteria.

8. SECONDARY BACTERIAL INFECTIONS

Secondary bacterial infections following upper respiratory viral infections, such as influenza, are a major cause of morbidity and mortality (Brundage and Shanks 2008). The organisms responsible for most of these complications are *S. aureus* and *S. pneumoniae*. Furthermore, otitis media due to *S. pneumoniae* is a leading cause of morbidity and health-care expenditures worldwide; it also increases after an upper respiratory viral infection (McCullers 2006). The elimination of or reduction in the bacterial burden by these organisms will significantly reduce or eliminate these secondary infections. However, except for mupirocin and polysporin ointments, for which resistance is being developed, there is no effective way to eliminate these organisms from the upper respiratory mucosa.

In a mouse model of otitis media, using a noninvasive bioluminescent imaging technique, 80% of mice colonized nasally with *S. pneumoniae* naturally develop otitis media upon infection with influenza virus. Treatment of these mice with Cpl-1 lysin before influenza challenge was 100% effective at preventing the development of otitis media (McCullers et al. 2007). Thus, treatment of high-risk individuals with lysin during influenza season to decolonize them from pneumococci and staphylococci could prove effective in reducing secondary infections by these bacteria.

9. ANTHRAX-SPECIFIC LYSIN

Because lysins are able to kill pathogenic bacteria rapidly, they may be a valuable tool in controlling biowarfare bacteria. To determine the feasibility of this approach, a lytic enzyme was identified from the gamma phage, a lytic phage that is highly specific for *Bacillus anthracis* (Watanabe et al. 1975). The gamma lysin, termed PlyG, was purified to homogeneity by a two-step chromatography procedure and tested for its lethal action on gamma phage-sensitive bacilli (Schuch et al. 2002). Three seconds after contact, as little as 100 units (about 100 μg/mL) of PlyG mediated a 5000-fold decrease in viable counts of a suspension of ~10^{77} bacilli. When the enzyme was then tested against 10 *B. anthracis* strains from different clonal types isolated worldwide, all could be killed. In addition, the PlyG lysin was also lethal for five mutant *B. anthracis* strains lacking either capsule or toxin plasmids.

Based on physiological characteristics, sensitivity to gamma phage, and mouse lethality, a strain of *Bacillus cereus* that is closely related to *B. anthracis* was identified. *In vivo* experiments revealed that when 10^7 of these organisms were administered intraperitoneally to 10 mice, all except one died of a rapidly fatal septicemia within 4 h (10% survival). When a second set of 18 mice was also challenged intraperitoneally with these bacilli, but given a single 100 μg dose of PlyG 15 min later by the same route, only five animals died (72% survival), and in three of these animals death was delayed >24 h. We anticipate that based on the half-life of lysins (see above), higher doses, multiple doses, or constant IV infusion of lysin will result in increased survival.

Because the treatment window for individuals exposed to anthrax is about 48 h, PlyG may be used intravenously in post-exposure individuals to increase the treatment window beyond the 48-h period. This would allow more time to test the infecting bacillus for its antibiotic resistance spectrum before treatment.

10. BACTERIAL RESISTANCE TO LYSINS

Exposure of bacteria grown on agar plates to low concentrations of lysin did not lead to the recovery of resistant strains even after more than 40 cycles. Organisms in colonies isolated at the periphery of a clear lytic zone created by a 10-μL drop of dilute lysin on a lawn of bacteria always resulted in enzyme-sensitive bacteria. Enzyme-resistant bacteria could also not be identified after >10 cycles of bacterial exposure to low concentrations of lysin (from 5 to 20 units) in liquid culture (Loeffler et al. 2001; Schuch et al. 2002). These results may be explained, for example, by the fact that the cell wall receptor for the pneumococcal lysin is choline (Garcia et al. 1983), a molecule that is essential for pneumococcal viability. While not yet proven, it is possible that during a phage's association with bacteria over the millennia, to avoid becoming trapped inside the host, the binding domain of their lytic enzymes has evolved to target a unique and essential molecule in the cell wall, making resistance to these enzymes a rare event. Since through evolution the phage has performed the "high throughput" analysis to identify the "Achilles heel" of these bacteria, we may take advantage of this by identifying the pathway for the synthesis of the lytic enzyme's cell wall receptor and identify inhibitors for this pathway. This would theoretically result in new antimicrobials that would be difficult to become resistant against.

11. CONCLUDING REMARKS

Lysins are a new reagent to control bacterial pathogens, particularly those found on the human mucosal surface. For the first time we may be able to specifically kill pathogens on mucous membranes without affecting the surrounding normal flora, thus reducing a significant pathogen reservoir in the population. Since this capability has not been previously available, its acceptance may not be immediate. Nevertheless, like vaccines, we should be striving to developing methods to prevent rather than treat infection. Whenever there is a need to kill bacteria, and contact can be made with the organism, lysins may be freely utilized. Such enzymes will be of direct benefit in environments where antibiotic-resistant Gram-positive pathogens are a serious problem, such as hospitals, day-care centers, and nursing homes. The lysins isolated thus far are remarkably heat-stable (up to 60 °C) and are relatively easy to produce in a purified state and in large quantities, making them amenable to these applications. The challenge for the future is to use this basic strategy and improve upon it, as was the case for second- and third-generation antibiotics. Protein engineering, domain swapping, and gene shuffling all could lead to better lytic enzymes to control bacterial pathogens in a variety of environments. Since it is estimated that there are 10^{31} phages on Earth, the potential to identify new lytic enzymes as well as those that kill Gram-negative bacteria is enormous. Perhaps some day phage lytic enzymes will be an essential component in our armamentarium against pathogenic bacteria.

ACKNOWLEDGMENTS

I acknowledge the members of my laboratory who are responsible for much of the phage lysin work: Qi Chang, Mattias Collin, Anu Daniel, Sherry Kan, Jutta Loeffler, Daniel Nelson, Jonathan Schmitz, Raymond Schuch, and Pauline Yoong, with the excellent technical assistance of

Peter Chahales, Adam Pelzek, Rachel Shively, Mary Windels, and Shiwei Zhu. I am indebted to my collaborators Stephen Leib, Jon McCullers, Philippe Moreillon, and Martin Witzenrath for their excellent work with the lysins in their model systems. I also wish to thank Abraham Turetsky at the Aberdeen Proving Grounds and Leonard Mayer of the Centers for Disease Control for testing the gamma lysin against authentic B. anthracis strains and Richard Lyons and Julie Lovchik for the animal protection studies with anthrax. Supported by DARPA and USPHS Grants AI057472 and AI11822.

REFERENCES

Bernhardt T. G., I. N. Wang, D. K. Struck, and R. Young (2001) *Science* **292**, 2326–2329.

Brundage J. F. and G. D. Shanks (2008) *Emerging Infectious Diseases* **14**, 1193–1199.

Cheng Q., D. Nelson, S. Zhu, and V. A. Fischetti (2005) *Antimicrobial Agents and Chemotherapy* **49**, 111–117.

Clyne M., T. H. Birkbeck, and J. P. Arbuthnott (1992) *Journal of General Microbiology* **138**, 923–930.

Coello R., J. Jimenez, M. Garcia, P. Arroyo, D. Minguez, C. Fernandez, F. Cruzet, and C. Gaspar (1994) *European Journal of Clinical Microbiology Infectious Diseases* **13**, 74–81.

de Lencastre H., K. G. Kristinsson, A. Brito-Avo, I. S. Sanches, R. Sa-Leao, J. Saldanha, E. Sigvaldadottir, S. Karlsson, D. Oliveira, R. Mato, M. A. de Sousa, and A. Tomasz (1999) *Microbial Drug Resistance* **5**, 19–29.

Diaz E., R. Lopez, and J. L. Garcia (1990) *Proceedings of the National Academy of Sciences USA* **87**, 8125–8129.

Djurkovic S., J. M. Loeffler, and V. A. Fischetti (2005) *Antimicrobial Agents and Chemotherapy* **49**, 225–1228.

Eiff C.V., K. Becker, K. Machka, H. Stammer, and G. Peters (2001) *New England Journal of Medicine* **344**, 11–16.

Eliopoulos G. and R. Moellering (1991) Antimicrobial combinations. In *Antibiotics in Laboratory Medicine*, ed. V. Lorian. Baltimore, MD: Williams and Wilkins.

Entenza J. M., J. M. Loeffler, D. Grandgirard, V. A. Fischetti, and P. Moreillon (2005) *Antimicrobial Agents and Chemotherapy* **49**, 4789–4792.

Fernandez-Lopez M., E. Munoz-Adelantado, M. Gillis, A. Williams, and N. Toro (2005) *Molecular Biology and Evolution* **22**, 1518–1528.

Flaherty S. O., A. Coffey, W. Meaney, G. F. Fitzgerald, and R. P. Ross (2004) *Journal of Bacteriology* **186**, 2862–2871.

Foley S., A. Bruttin, and H. Brussow (2000) *Journal of Virology* **74**, 611–618.

Garcia E., J. L. Garcia, A. Arraras, J. M. Sanchez-Puelles, and R. Lopez (1998) *Proceedings of the National Academy of Sciences USA* **85**, 914–918.

Garcia P., E. Garcia, C. Ronda, A. Tomasz, and R. Lopez (1983) *Current Microbiology* **8**, 137–140.

Garcia P., J. L. Garcia, E. Garcia, J. M. Sanchez-Puelles, and R. Lopez (1990) *Gene* **86**, 81–88.

Grandgirard D., J. M. Loeffler, V. A. Fischetti, and S. L. Leib (2008) *Journal of Infectious Diseases* **197**, 1519–1522.

Hermoso J. A., B. Monterroso, A. Albert, B. Galan, O. Ahrazem, P. Garcia, M. Martinez-Ripoli, J. L. Garcia, and M. Menendez (2003) *Structure* **11**, 1239–1249.

Hudson I. (1994) *Journal of Hospital Infectious* **28**, 235.

Jado I., R. Lopez, E. Garcia, A. Fenoll, J. Casal, and P. Garcia (2003) *Journal of Antimicrobial Chemotherapy* **52**.

Jervis E. J., C. A. Haynes, and D. G. Kilburn (1997) *Journal of Biological Chemistry* **272**, 24016–24023.

Kengatharan K. M., K. S. De, C. Robson, S. J. Foster, and C. Thiemermann (1998) *Journal of Experimental Medicine* **188**, 305–315.

Loeffler J. M., S. Djurkovic, and V. A. Fischetti (2003) *Infection and Immunity* **71**, 6199–6204.

Loeffler J. M. and V. A. Fischetti (2003) *Antimicrobial Agents and Chemotherapy* **47**, 375–377.

Loeffler J. M., D. Nelson, V. A. Fischetti (2001) *Science* **294**, 2170–2172.

Loessner M. J. (2005) *Current Opinion in Microbiology* **8**, 480–487.

Loessner M. J., K. Kramer, F. Ebel, and S. Scherer (2002) *Molecular Microbiology* **44**, 335–349.

Loessner M. J., S. K. Maier, H. Daubek-Puza, G. Wendlinger, and S. Scherer (1997) *Journal of Bacteriology* **179**, 2845–2851.

Lopez R., E. Garcia, P. Garcia, and J. L. Garcia (1997) *Microbiology Drugs Resistance* **3**, 199–211.

Lopez R., J. L. Garcia, E. Garcia, C. Ronda, and P. Garcia (1992) *FEMS Microbiology Letters* **79**, 439–447.

Matsuzaki S., M. Rashel, J. Uchiyama, S. Sakurai, T. Ujihara, M. Kuroda, M. Ikeuchi, T. Tani, M. Fujieda, H. Wakiguchi, and S. Imai (2005) *Journal of Infection and Chemotherapy* **11**, 211–219.

McCullers J. A. (2006) *Clinical Microbiology Reviews* **19**, 571–582.

McCullers J. A., A. Karlstrom, A. R. Iverson, J. M. Loeffler, and V. A. Fischetti (2007) *PLoS.Pathogenesis* **3**, e28.

Navarre W. W., H. Ton-That, K. F. Faull, and O. Schneewind (1999) *The Journal of Biological Chemistry* **274**, 15847–15856.

Nelson D., P. Chahalis, S. Zhu, and V. A. Fischetti (2006) *Proceedings of the National Academy of Sciences USA* **103**, 10765–10770.

Nelson D., L. Loomis, and V. A. Fischetti (2001) *Proceedings of the National Academy of Sciences USA* **98**, 4107–4112.

Nelson D., R. Schuch, S. Zhu, D. M. Tscherne, and V. A. Fischetti (2003) *Journal of Bacteriology* **185**, 3325–3332.

O'Flaherty S., A. Coffey, W. Meaney, G. F. Fitzgerald, and R. P. Ross (2005) *Journal of Bacteriology* **187**, 7161–7164.

Pritchard D. G., S. Dong, M. C. Kirk, R. T. Cartee, and J. R. Baker (2007) *Applied and Environmental Microbiology* **73**, 7150–7154.

Rashel M., J. Uchiyama, T. Ujihara, Y. Uehara, S. Kuramoto, S. Sugihara, K. Yagyu, A. Muraoka, M. Sugai, K. Hiramatsu, K. Honke, and S. Matsuzaki (2007) *The Journal of Infectious Diseases* **196**, 1237–1247.

Sass P. and G. Bierbaum (2007) *Applied and Environmental Microbiology* **73**, 347–352.

Schuch R., D. Nelson, and V. A. Fischetti (2002) *Science* **418**, 884–889.

Sonstein S. A., J. M. Hammel, and A. Bondi (1971) *Journal of Bacteriology* **107**, 499–504.

Tan K., G. Ong, and K. Song (2005) *Journal of Bacteriology* **187**, 567–575.

Tuomanen E., H. Liu, B. Hengstler, O. Zak, and A. Tomasz (1985) *The Journal of Infectious Diseases* **151**, 859–868.

Walsh S., A. Shah, and J. Mond (2003) *Antimicrobial Agents and Chemotherapy* **47**, 554–558.

Wang I.-N., J. Deaton, and R. Young (2003) *Journal of Bacteriology* **185**, 779–787.

Wang I.-N., D. L. Smith, and R. Young (2000) *Annuals Reviews in Microbiology* **54**, 799–825.

Watanabe T., A. Morimoto, and T. Shiomi (1975) *Canadian Journal of Microbiology* **21**, 1889–1892.

Weiss K., M. Laverdiere, M. Lovgren, J. Delorme, L. Poirier, and C. Beliveau (1999) *American Journal of Epidemiology* **149**, 863–868.

Witzenrath M., B. Schmeck, J. M. Doehn, T. Tschernig, J. Zahlten, J. M. Loeffler, M. Zemlin, H. Muller, B. Gutbier, H. Schutte, S. Hippenstiel, V. A. Fischetti, N. Suttorp, and S. Rosseau (2008) *Critical Care Medicine* **37**, 642–649.

Yoong P., D. Nelson, R. Schuch, and V. A. Fischetti (2004) *Journal of Bacteriology* **186**, 4808–4812.

Young R. (1992) *Microbiological Reviews* **56**, 430–481.

Young R., I.-N. Wang, and W. D. Roof (2000) *Trends in Microbiology* **8**, 120–128.

CHAPTER 6

BACTERIOPHAGE HOLINS AND THEIR MEMBRANE-DISRUPTING ACTIVITY

MARÍA GASSET

Instituto de Química-Física "Rocasolano," Consejo Superior de Investigaciones Científicas, Madrid, Spain

1. INTRODUCTION

One of the primary mechanisms used by nature to sustain life is the generation of physical barriers. These physical barriers are mainly provided by the assemblies known as biological membranes that function as both active and passive shields, isolating compartments and controlling their chemical communication. Consequently, the disruption of these barriers by allowing them to be permeable is a general route of cell life interference, which can have physiological or pathological consequences.

Despite the molecular complexity of the biological membranes, the lipid bilayer backbone is the permeability guardian and therefore the universal target of some proteins that act as perfect permeability weapons. These proteins have been tailored for providing a physiological activity, but with controlled use, they can become valuable biotechnological tools. Of the distinct fundamental processes occurring with massive membrane permeation, or lysis, the physiological event accompanying the release of the phage progeny from the host can be redesigned and used as an intelligent weapon to combat the hosts. Such a task requires understanding in molecular terms the elementary steps involved in the overall process.

To successfully escape from the bacterial cells at the end of the vegetative cycle, bacteriophages must overcome two barriers: the cyto-

Enzybiotics: Antibiotic Enzymes as Drugs and Therapeutics. Edited by Tomas G. Villa and Patricia Veiga-Crespo
Copyright © 2010 John Wiley & Sons, Inc.

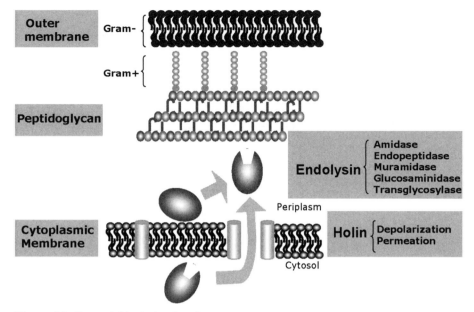

Figure 6.1. Bacterial isolation barriers as targets for the targets of the two-component phage lytic system. Bacterial cell cytosol is isolated from the environment by the cytoplasmic membrane and the peptidoglycan mesh in Gram-positive (+) cells. Gram-negative (−) cells differ in an additional barrier provided by the outer membrane.

plasmic membrane and the peptidoglycan mesh. As a function of the strategy used, phages are classified into two categories: filamentous and lytic (Young et al. 2000). Filamentous phages continuously extrude from their host without lysis. On the contrary, lytic phages compromise cell integrity. For doing so, some lytic phages (small phages containing ssDNA or ssRNA genomes) encode a single lytic factor that induces bacteriolysis by inhibition of the cell wall synthesis or other mechanism (Young et al. 2000). However, most lytic phages disrupt the host using a "two-component" cell lysis system that involves a holin and an endolysin (Fig. 6.1) (Young 1992, 2002; Young et al. 2000).

Of the two-component cell lysis system, the endolysin proteins carry the enzymatic activity degrading the peptidoglycan mesh acting on glycosidic bonds (glycosidases, transglycosidases) or on peptide bonds (amidases, endopeptidases). On the other hand, holins are small hydrophobic proteins that integrate into the cytoplasmic membrane and disrupt its isolation properties (Young 1992; Young and Bläsi 1995; Wang et al. 2000). The holin-induced membrane perturbation includes its depolarization in combination or not of its piercing that in the latter case allows either the activation by release of the endolysins or the access of the cytosolic folded endolysins to their substrate (Young et al. 2000; Xu et al. 2004, 2005; Srividhya and Krishnaswamy 2007).

Holins differ from most membrane-disrupting proteins and peptides both structurally and mechanistically. These proteins exist singularly under a membrane-bound state, and their action is irreversibly lethal. It is precisely the lethality of their function that converts them in promising candidates for pharmacological exploitation.

2. CLASSIFICATION OF HOLINS

Holins represent a structurally diverse family of proteins consisting of a single polypeptide chain of length varying from about 50 to 150 amino acid residues (Young and Bläsi 1995; Wang et al. 2000). Holins take their name from their capacity to induce large lesions in the bacterial cytoplasmic membranes that were imaginarily conceived as holes. Holins share no sequence homology but have some resemblance in terms of their polypeptide chain architecture (Fig. 6.2). Essentially, holin chains consist of a tandem array of segments with alternate hydrophilicity and hydrophobicity (Wang et al. 2000; Young 2002). In general, the holin chains start at a polar N-terminus that is followed by two or three hydrophobic regions spaced by stretches rich in polar residues and end at a long C-terminal tail rich in charged residues. For instance, the Ejh chain, the holin of EJL bacteriophage (Fig. 6.2b), is featured by an N-terminal segment 11 amino acids long with basic residues, followed by a stretch of 20 apolar amino acids. This segment is linked through a sequence of 10 polar amino acids to the following hydrophobic segment (residues 45–62), which continues into a 23 amino acid polar tail. For S^λ holin, the chain contains an additional hydrophobic segment in tandem with a stretch of polar residues.

The alternation of polar and apolar segments in the holin chains determines their existence as transmembrane proteins. Sequence analysis using theoretical algorithms shows that the hydrophobic segments exhibit both the hydrophobicity and length requirements to be transmembrane helices (TM). Consequently, these segments are often referred to as TMi (where i numbers the distance from the N-terminus, with 1 being the closest) and holins visualized as a membrane protein spanning the membrane as many times as the number of predicted TM segments.

The basic and simple chain construction has allowed the classification of holins into three groups according to the number of

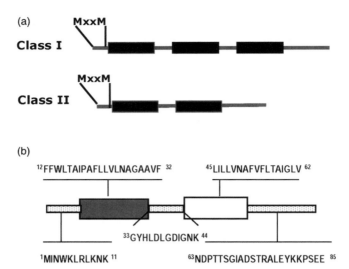

Figure 6.2. Holin protein family: architecture. a: Organization of holin chains. Analysis of the chains with theoretical algorithms permits the identification of a variable number of hydrophobic segments with the length enough to cross the bilayer as transmembrane α-helices. Attending to the number of segments, the members of the holin family can be grouped as class I, class II, and a third group that gathers the remaining chains falling out of the previous. The holin chains are formed by sequential arrays of polar (gray thin line) and hydrophobic (black boxes) stretches of amino acid residues. For membrane-integrated proteins, the different polar and charged regions constitute solvent-exposed loops, whereas the hydrophobic segments exhibit a length enough to bypass the lipid bilayer in a helical fold. Most, but not all, holins also contain a dual methionine motif at the N-terminus that dictates the synthesis of two chains, a long one and a short one. This motif is indicated as MxxM, x being a variable number of amino acid residues, some bearing positively charged side chains. b: Structural organization of Ejh, the holin of EJ-1 phage, and its use as template for solid-phase synthetic approaches. Full membrane-perturbing activity of Ejh mapped at the 1–32 region.

hydrophobic segments (Young and Bläsi 1995; Wang et al. 2000). Holins belonging to class I contain three predictable TMs and their membrane integration dictates a final topology featured by the projection of the N-terminus to the periplasm (extracellularly). This class is represented by the S^{λ} holin (Table 6.1). Class II groups the holins with only two hydrophobic segments and a topology featured by cytoplasmic N- and C-termini. The members of this class, represented by the Ejh and S^{21} holins, are made from chains shorter than those of class I (Table 6.1). Finally, class III comprises the members that diverge from the previous classes and that contain either one or four transmembrane regions (Table 6.1). It must be stressed that this classification is based

TABLE 6.1. Holin Proteins and Their Features

Class	Holin Family	Holin Gene	Accession Number	Chain Length	M-M	Phage
I	λ	*pl1*	AF157835	94	>10	APSE-1
		13	X67137	108	2	ES18
		S	AF069308	107	1	HK22
		S	AF069529	106	1	HK97
		S	J02459	107	1	λ
		13	M10997	108	2	P22
		HI1416	U32821	118	3	φflu
		orf1	AJ133022	111	>10	ORF1-Xnem
I	P1	*lydA*	X87674	109	>10	P1
I	P2	*O*	U32222	98	>10	186
		Y	AF063097	93	>10	P2
I	φCTX9	*orf9*	AB008550	117	>10	φCTX
I	φCTX10	*orf10*	AB008550	90	1	φCTX
I	PRD1	*m*	M69077	90	>10	PRD1
I	PS3	*13*	AJ011579	105	3	PS3
I	A118	*hol2438*	X89234	95	2	2438
		hol118	X85008	96	2	A118
		hol500	X85009	96	2	A500
		hol	X90511	142	>10	φg1e
		orf117	AF047001	117	>10	fOg44
I	c2	*orf37*	L33769	97	0	bIL67
		l17	L48605	96	>10	c2
		orf2	L37090	96	>10	P001
I	Cp-1	*cph1*	Z47794	134	4	Cp-1
I	Phi-29	*14*	X99260	132	1 or 2	B103
		14	X04962	131	5	φ29
		lysis	M11813	131	5	PZA
I	sk1	*l21*	AF009630	117	3	bIL170
		orf19	AF011378	117	3	sk1
		orf2	M90423	117	3	US3
II	21	*S*	M65239	71	2	21
		S	AF125520	71	2	933W
		S	U82598	71	2	DLP-12
		S	AF034975	68	0	H-19B
		S	J02580	70	2	PA-2
		S	AE000252	96	2	QIN
		S	AP000363	96	2	VT2-Sa
		hol	U24159	78	>10	HP1
		orf6	U28154	74	3	orf6-Hsom
II	Hol-Paeru	*hol*	AB030825	117	9	hol-Paeru
II	N4	*orf63*			>10	N4
II	N15	*53*	AF064539	101	1	N15
II	NucE	*nucE*	U11698	89	>10	nucE-Smar
		regA	U31763	88	>10	regA-Smar
		13	AJ011581	67	2	PS119
		13	AJ011580	67	2	PS34

TABLE 6.1. *Continued*

Class	Holin Family	Holin Gene	Accession Number	Chain Length	M-M	Phage
II	T7	*lys*	M14784	67	>10	T3
		17.5	V01146	67	>10	T7
II	80α	?		92	2	393-A2
		holin	U72397	145	>10	80α
		orf24	AF085222	80	>10	DT1
		ejh	S43512	85	>10	EJ-1
		orf3	L34781	145	>10	φ11
		lysA	U04309	88	>10	φΛX3
		lyt50	U88974	80	>10	φO1205
		orf87	AF057033	87	6	Sfi11
		orf87	AF115102	87	6	Sfi19
		orf87	AF115103	87	6	Sfi21
		S	L31364	88	>10	Tuc2009
		holTW	Y07739	185	>10	Twort
II	BK5-T	*orf95*	L44593	95	>10	BK5-T
		xhlB	L25924	87	>10	PBSX
		orf24	AB009866	100	>10	φPVL
		bhlB	AF021803	88	>10	SPβ
		26	X97918	82	1	SPP1
		xpaF2	D13377	87	>10	xpaF2-Blich
		xpaG2	D49712	89	0	xpaG2-Blich
		xpaL2	M63942	87	>10	xpaL2-Blich
II	Dp-1	*dph*	Z93946	74	>10	Dp-1
II	L5	*11*	AF022214	141	>10	D29
		11	Z18946	131	>10	L5
II	LL-H	*hol*	M96254	107	>10	LL-H
			M60167			mv1
		lysB	Z26590	124	>10	mv4
II	φadh	*hol*	Z97974	114	>10	φadh
II	φC31	*orf1*	X91149	78	>10	φC31
II	r1t	*orf48*	U38906	75	6	r1t
U (1)	φ6	*P10*	M17462	42	>10	φ6
U (1)	φ6	*P10*	AF125681	42	>10	φ7
U (1)	T4	*t*	M16812	218	>10	K3
		t	AF158101	218	>10	T4
U (P)		*t*	X05676		>10	M1
		t	X05675		>10	Ox2
U (1)	187	*hol187*	Y07740	57	1	187
U (1)	10MC	*P98*	AF049087	98	9	10MC
U (0)	A511	*hol511*	X85010	126	>10	A511
U (1)	PBSX	*xhlA*	L25924	89	>10	PBSX
U (0)		*24.1*	X97918	91	2	SPP1
U (1)		*xpaF1*	D13377	89	>10	xpaF1-Blich
		xpaG1	D49712	89	>10	xpaG1-Blich
		xpaL1	M63942	89	>10	xpaL1-Blich

TABLE 6.1. *Continued*

Class	Holin Family	Holin Gene	Accession Number	Chain Length	M-M	Phage
U (1)	SP-beta	*bhlA*	AF021803	70	1	SPβ
U (4)	LrgA	*ipa-23r*	X73124	128	>10	ipa-23r-Bsub
		ysbA	Z75208	146	>10	ysbA-Bsub
		yohJ	U00007	132	>10	yohJ-Ecoli
		HI1297	L42023	140	>10	HI1297-Hinf
		PAB0239	AJ248284	109	>10	PAB0239-Paby
		PH1801	AP000007	109	>10	PH1801-Phor
		lrgA	U52961	147	>10	lrgA-Saur

Holin proteins are given as families and are classified according to the number of hydrophobic segments: classes I, II, and III. Class III is referred to as U or undetermined. The coding gene (holin gene) and its source (phage) are included together with the accession number for sequence retrieval. The chain length accounts for the predicted number of amino acid residues translated from the coding region. The number of amino acids spacing the two N-terminal methionines is indicated by M-M. References from this table can be retrieved from each accession number and from the Wang et al. (2000) supplementary material.

on the theoretical analysis of the sequence and that only in a small number of cases has the topology been experimentally determined.

In addition to the previous classification, holins can also be grouped into two classes attending to the presence or absence of two methionine residues at the N-terminal polar region. Most holins, but not all, contain two spaced methionine residues at their polar N-terminus, which indicate the possibility of a dual translation start site (Young and Bläsi 1995; Wang et al. 2000). This motif determines that each chain, in fact, duplicates into two differing in length: a long chain starting at the first methionine and a short chain produced from the second methionine. Moreover, these chains are functionally related (Bläsi and Young 1996; Graschopf and Bläsi 1999a). The short chain generally carries the membrane-damaging activity (effector), whereas the long chain functions as a delay or inhibitor factor (Bläsi et al. 1990; Bläsi and Young 1996; Takác et al. 2005). The long chain is often referred to as anti-holin and differs from the active form in the basic character of its N-terminus and in the topological architecture (see below). This anti-holin chain is essential not for the lysis process but for its control, and lytic systems lacking it work faster. As an example, S^λ holin, which belongs to the class I of holins, contains two spaced methionines at the chain N-terminus. These methionines dictate the existence of S^λ as two chains, $S^\lambda 107$ (long form) and $S^\lambda 105$ (short form). The ratio between the long and the short is about 1:2 and is determined by an RNA stem-loop structure in the ribosomal binding site (Graschopf and Bläsi 1999a;

Wang et al. 2000). Of these two chains, $S^\lambda 105$ carries the membrane-damaging activity (effector), whereas $S^\lambda 107$ functions as a delay or inhibitor factor (Bläsi et al. 1990; Bläsi and Young 1996; Takác et al. 2005).

The third criterion for holin classification regards the size of the induced membrane damage (see below). The holins that are tailored to allow the passage of a fully folded endolysin from the cytosol to the periplasm cause a two-step membrane damage feature by a first depolarization followed by a second piercing activity. On the contrary, those holins referred to as pinholins designed to activate the periplasmic located latent endolysins cause only a membrane depolarizing lesion. Examples of these types are S^λ as holin and S^{21} as pinholin.

3. HOLINS AS MEMBRANE-INTEGRATED PROTEINS

The structural characterization of holins has been largely impeded by the lack of tools allowing the production and purification of their chains to the degree and quantity needed for biophysical studies. Their membrane integration and disruption features dictate a major bottleneck in their overproduction, since the amount of production will be parallel with the death of the producing organism. On the other hand, the lack of tools for their easy and specific detection also conditioned tedious purification protocols. Knowledge of them is still highly restricted to theoretical approaches (Young and Bläsi 1995; Wang et al. 2000) and biochemical studies for topology determination (Zagotta and Wilson 1990; Bläsi et al. 1999; Gründling et al. 2000a, b). Only recently and in a few cases, a more biophysical characterization of a His-tagged form of the $S^\lambda 105$ chain and of the EJh-templated peptides has been performed (Díaz et al. 1996; Smith et al. 1998; Haro et al. 2003; Ambroggio et al. 2007; Saava et al. 2008). The new approaches open new venues for structural studies.

3.1. Sequence Analysis Defines Holins as Transmembrane Proteins and Limits the Number of Transbilayer Segments

As mentioned before, the theoretical analysis of holin sequences has provided the essential signatures for their consideration as membrane-spanning proteins (Young and Bläsi 1995). Such studies have also offered the upper and lower limits for the number of times the polypeptide chain could hypothetically cross the bilayer and the sequence of the transmembrane segments. These patterns were obtained from the agreement between the thicknesses of the lipid bilayer together and the length of the hydrophobic segments assuming them folded as trans-membrane α-helices (Young and Bläsi 1995).

Importantly, as opposed to sequences integrating from aqueous media to lipid bilayers, and causing the membrane permeation or the generation of a transmembrane pore, the holin transmembrane regions often lack any amphipathic pattern, charge distribution, or association motif that could easily explain the construction of a polar cavity through the lipid bilayer and support their damaging action. Some of these segments indeed contain charged or polar residues, but they do not shape any conserved motif along the different sequences that could explain the function along the protein family.

3.2. Biochemical and Genetic Studies Set the Topology Bounds

Holin chain topology is, in general terms, determined by the relative location of the charged N- and C-terminal segments. The study of their location was first addressed using biochemical approaches. Regardless of the class and type of holin, all chains feature a highly charged C-terminal tail. Gene fusion experiments together with protease protection assays support the cytoplasmic location for the highly charged C-terminus (Bläsi et al. 1999; Park et al. 2007). This location is conserved for most holins, and in only one, the T4 holin, has this segment been found at the periplasm (Tran et al. 2005). It must be stressed that T4 holin is indeed an unusual member. Its chain is twice as long as the conventional holins, and the theoretical analysis yields a single trans-membrane segment together with a very long C-terminal tail (Tran et al. 2005).

Regarding the N-terminus, biochemical studies coupling extensive cysteine-scanning mutagenesis with chemical modification with thiol-specific reagents showed that the active class I holin $S^\lambda 105$ crosses the bilayer three times. Similar experiments showed that the inhibitory action of $S^\lambda 107$ became impaired on the adoption of the 3-TM topology, suggesting both a dual location in the N-terminus and the relation of this location with function (see below). Regarding the class II holins, topological studies have been mainly focused on S^{21} (Fig. 6.3). Using similar strategies as before, the N- and C-terminal tails of S^{21} have been located to the cytoplasm in the short and long nascent forms (Park et al. 2006). However, in this case, in the short form ($S^{21}68$), the nascent topology evolves so the N-terminus translocates to the periplasms together with the extrusion of TM1 (Park et al. 2006).

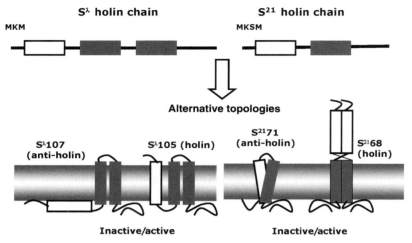

Figure 6.3. Topology flexibility of holin chains and the structural basis of their action. Holins displayed a topological flexibility at the N-terminus that is encoded in its sequence. The chain organization of the S^λ and the S^{21} holins, each coding for an inhibitor (anti-holin) and an effector (holin), is depicted at the top of the figure. The polypeptide chains are represented by a thick line in which the hydrophobic segments are denoted by rectangles of different colors according to their relative position in the sequence. The N-terminal sequences (MKM- and MKSM-) are denoted at the top of the left side of the chains. The chains integrate into the membrane generating two alternate topologies. Holin nascent chains and anti-holin chains are featured by an arrangement that projects both N- and C-terminal tails to the cytoplasm. The N-terminus interacts with the membrane surface through electrostatic binding. Holin activation involves a conformational change consisting in the movement of transmembrane segments through the bilayer so the N-terminus ends projected to the periplasm. For S^λ, activation involves the acquisition of a third trasmembrane segment, whereas for S^{21}, activation causes the extrusion from the membrane of the N-terminal transmembrane segment.

Thus, holins are unique membrane multispanning proteins exhibiting flexible topology at their N-termini. This peculiar flexibility is coded by the different polar N-termini allowing or impeding the entry and exit of chain segments in the membrane.

3.3. Probing Holin Structure Using Synthetic Approaches

The organization of the holin chain as tandem arrays of segments exhibiting well-defined polarity patterns allowed its use as template for synthetic approaches. For doing so, the shortest sequence with verified holin activity, Ejh, was taken as a template for the solid-phase synthesis of a collection of peptides encompassing the different chain regions either independently or in tandem (Diaz et al. 1996; Haro et al. 2003).

The use of a combination of both far-UV CD and ATR-FTIR, spectroscopic techniques sensitive to the polypeptide conformation, permitted the determination of the preferential secondary structure of the different segments in their lipid-free and lipid-bound state (Haro et al. 2003). The obtained secondary structures for each of the considered segments were similar if not identical to those predicted theoretically. The only disparity was found at the N-terminus, that containing a (i, i + 1) pattern of polar/apolar residues allowed the detection of β-sheet-based assemblies. In addition to the basic secondary structure preference determination, the capacity of polarized ATR-FTIR to inform about the relative order of peptide backbones with respect to the bilayer (perpendicular, parallel, tilted with respect to the bilayer normal) allowed the confirmation of the transmembrane character of the hydrophobic segments.

Parallel experiments for function screening using an assay based in the detection of the leakage of liposome-entrapped fluorescent markers allowed the ascription of the membrane-perturbing activity to the 1-32 sequence region (the N-terminal polar region together with the TM1). Such region was then exploited for the study using fluorescence emission spectroscopy of the self-assembly properties. Analysis of the fluorescent emission properties of the covalently labeled peptide showed that under active conformation, this mini holin exists as an in-membrane oligomer (Haro et al. 2003).

3.4. Tagging the Holin Chain Facilitates Structural Study

In 1998, Smith et al. pioneered the production and purification of an active holin, the $S^{\lambda}105$ chain together with lysis-defective mutants. This breakthrough allowed the first conformational characterization of a holin in a membrane-mimicking environment, such as that provided by detergent micelles. In this membrane-like environment, the holin chain adopted a structure that featured about 40% of the helical content in β-octylglucoside. Interestingly, the helical content was highly dependent on the detergent used, increasing up to 60%. The helical variability was interpreted in terms of the number of residues involved in transmembrane topology; thus, 40% would have a 2-TM conformation, whereas 60% were assigned to the topology based on a 3-TM. This helicity range was not observed in the corresponding lysis-defective $S^{\lambda}105$ mutants and was therefore interpreted in terms of a putative structure–function relation that correlated the defect in helicity with the extrusion from the membrane of one of the hydrophobic segments and the lytic activity impairment.

Regarding the $S^\lambda105$ association state, the use of reconstitution approaches with varying detergents together with gel filtration and cryo-electron microscopy experiments revealed that the protein exists as a complex oligomer (Savva et al. 2008). These oligomers exhibited different sizes and shapes, both parameters being determined by the detergent used to mimic the membrane environment (Saava et al. 2008). Importantly, the $S^\lambda105$ capacity to form high-molecular-weight oligomers is directly correlated with its lytic function, since its lysis-defective mutant $S^\lambda105$ A52V forms dimmers, but the dimers do not evolve toward any higher degree of association. The polymerization impediment observed in the lysis-defective mutant dimer is an intrinsic property independent of the structure of the detergent and suggests that the dimer polymerization is the crucial functional step.

The differences in the secondary structure and in the oligomerization capacity of $S^\lambda105$ and of its nonfunctional mutants agree with the predictions derived from genetic and biochemical experiments. With all these data, the structure of a holin chain as $S^\lambda105$ can be described by a family of conformations separated by regulatory steps. Nascent $S^\lambda105$ inserts into the membrane yielding a low-helical-content dimer. This low-helical-content dimer lacks activity and therefore constitutes an inactive dimer. This state can be generalized to other inactive forms, such as the structural state of $S^\lambda107$, of the lysis-defective mutants, and probably also of the $S^\lambda107$–$S^\lambda105$ heterodimers. For $S^\lambda105$, the inactive dimer converts into a high-helical-content dimer that can be considered as a latent state. Essentially, this conversion involves a topological transition from a state featured by 2-TM to another characterized by 3-TM. Once formed, the high-helical-content dimer with a 3-TM topology in each of the monomers displays its in-membrane oligomerization capacity in the absence of major structural rearrangements.

For the active $S^\lambda105$ chains, two major assemblies have been found: ring dimers and stacked filaments (Savva et al. 2008). The ring dimer can be described as a hyperboloid of 23-nm outer diameter, 8.5-nm inner diameter, and 4-nm height. These dimensions agree with a basic scaffold of a variable number of protomers (from 18 to 20), each protomer being an oligomer of $S^\lambda105$ chains. Regarding the stacked filaments, these assemblies are rod-shaped structures of variable length and a thickness of about 20 nm per filament. The existence of such variability in assembling mode probably simply denotes, apart from the functional implications and speculations, the high tendency of the holin chain to self-associate in membrane-mimetic environments.

Thus, despite the few holins studied, the available biophysical data support that holins are indeed classical small membrane multispanning proteins tailored to form membrane-damaging polymers. Polymerization capacity is determined by a dimer that has two alternate conformations differing in the relative location of the N-terminal tail with respect to the lipid bilayer.

4. HOLINS AND ANTI-HOLINS DIFFER IN TOPOLOGY: THE ENTRIES AND EXITS OF PEPTIDE SEGMENTS INTO THE LIPID BILAYER

As mentioned, most but not all of the genes coding for holins encode two proteins differing uniquely in the N-terminus. The discovery of these two proteins, one behaving as an anti-holin and the other bearing the true holin function, was for a long time a puzzling issue affecting both mechanistic and regulatory aspects. However, the extensive work performed in the two paradigmatic holins, S^λ and S^{21}, has allowed the establishment of the molecular basis for such a chain duality as well as shed light on its impact in function and regulation. At this moment, it is worth remembering that holins are the gatekeepers for the phage progeny release, which means that holin action is best coupled to phage progeny production and therefore requires the existence of a tightly and properly regulated timing for the trigger of action. A fast lysis does not ensure the optimal production yield of phage progeny and therefore reduces the efficiency of the process. On the other hand, a retarded lysis is meaningless in terms of the efficiency of phage production and is therefore nonoperative in cost terms. One of the processes in charge of such regulation is provided by the relation between the anti-holin and the holin. The elegant studies aimed at understanding the meaning of holin and anti-holin partners, their missions, and relations have resulted in the description at the molecular level of the holin mechanism of action.

4.1. S^λ Anti-holin and Holin: The Entry of TM1 into the Membrane Switches on the Action

$S^\lambda 107$ and $S^\lambda 105$, the anti-holin and holin chains synthesized from the S^λ holin gene, differ only in an MK dipeptide at the N-terminus (Bläsi and Young 1996). Despite this simple difference, their chains adopt different topologies in the membrane (Graschopf and Bläsi 1999b; Gründling et al. 2000a). The $S^\lambda 105$ chain folds, passing across the bilayer three times using the predicted TM1, TM2, and TM3 regions as transmembrane α-helices. On the contrary, the long chain, the anti-holin

$S^{\lambda}107$, spans the bilayer only twice using exclusively the segments TM2 and TM3 (Fig. 6.3).

The differences in topology of the anti-holin and holin chains support a mode of action in which the lysis trigger is given by the relative disposition of the TM1 segment. In this sense, the N-terminus bearing the MK dipeptide impedes the insertion into the bilayer of TM1 that remains adhered onto the lipid surface. The loss of such dipeptide allows the insertion of TM1 and the access of the chain N-terminus to the periplasm. In other words, the charge provided by a single K residue is enough to stall and impede the insertion of the transmembrane segment.

4.2. S^{21} Anti-holin and Holin: The Exit of TM1 from the Membrane Switches on the Function

Contrary to the S^{λ} holin, whose function is designed to allow the passage from the cytoplasm to the periplasm of a folded endolysin, the S^{21} holin function is tailored to provide the membrane lesions needed for the release from the membrane of a signal anchor release containing endolysin, a membrane depolarization event. Membrane depolarization requires not a major bilayer disruption but the exchange of ions.

As in the case of the S^{λ} holin, S^{21} exists as two chains of 71 and 68 residues differing in the N-terminus MKS tripeptide sequence. Sequence analysis predicts the existence of two transmembrane domains and an α-helical hairpin fold with both N- and C-termini projected to the cytoplasm (intracellular). In this case, the N-terminal domain is dispensable for the membrane-perturbing action, and the lethal form projects the TM1 to the periplasm (Park et al. 2006).

For S^{21}, the loss of the N-terminal MKS sequence on changing from the anti-holin $S^{21}71$ to the holin $S^{21}68$ dictates the green light for the bilayer exiting of TM1. This membrane extrusion involves a complex tertiary rearrangement in the molecule from a state in which TM1 is interacting with its sequential TM2 (TM1–TM2) to a state in which a dimer is formed through homologous intermolecular interaction (TM1–TM1, TM2–TM2).

4.3. Sequence Unrelated Holins and Anti-holins: The T4 Case

T4 holin is an unusual member of the holin family, diverging at both the structural level and the mode of inhibition. Its chain is made by 263 residues, 163 of which form part of the polar C-terminal tail. This chain exhibits a cytoplasmic N-terminus followed by a single transmembrane region that is followed by the periplasmic large C-terminus (Tran et al. 2005). In addition to this unique topology and size, its function is subjected to a mode of regulation not seen in the other holins. T4 holin is inhibited by a sequence unrelated anti-holin, the protein RI, through a mechanism involving a specific and direct binding and that is under environmental signal regulation (Ramanculov and Young 2001a, b, c; Tran et al. 2005). The increasing number of holins potentially belonging to this group would allow in the future a more general molecular description of this mode of action.

4.4. Inhibiting the Inhibitor: Triggers for the Transbilayer Movement of Peptide Segments

As judged from the available data on the model holins S^λ and S^{21}, the anti-holin action basically results from their capacity to retain an inactive topology and to retard the achievement of the proper conformation to the effector holin. This bimodal action provides the basis for function inhibition or, in other words, the code for nonfunctionality.

The studies using holins with length and charge modifications at the N-terminal tail have shown that the cytoplasmic retention of such a segment in the anti-holin chain is due to its adherence to the membrane surface. The positive charge present in the N-terminal tail causes its binding as a staple to the negatively charged cytoplasmic side of the inner membrane through simple electrostatic interaction. Thus, any modification of the inner cytoplasmic membrane surface will be sensed at the anti-holin function. One way to modify membrane properties is to use uncouplers (generally 2,4-dinitrophenol [DNP]) or inhibitors (potassium cyanide [KCN]) of the oxidative phosphorylation (Garrett and Young 1982). The chemical-induced membrane energy poisoning occurs with a depolarization event that inhibits the anti-holin action and confers to the anti-holin the lysis activity (Garret and Young 1982; Bläsi et al. 1990).

Provided that both anti-holin and holin partners are synthesized together, the only advantage of their simultaneous production would be their engagement into a complex of varying stoichiometry. The evidence for such an interaction came from the experiments of Bläsi et al. (1990), who, isolating the expression of each of the forms and using a co-expression strategy, showed that the anti-holin $S^\lambda 107$ indeed inhibited in *trans* the lytic action of $S^\lambda 105$. Such interaction, or the outcome of such interaction, is highly dependent on the dose of production of each of the reagents. In this sense, while under *in vitro* conditions, the dose of synthesis of each of the chains is determined by structural elements in the coding mRNA; the relative synthesis in cellular context displays an additional regulatory level. Results employing the T7 expression system suggest that, in fact, $S^\lambda 107$ production predominates early after induction of S gene expression, and then $S^\lambda 105$ synthesis dominates after 30–40 min of synthesis (Wang et al. 2000). Taken together, this strongly supports the existence of a complex in-membrane oligomerization process regulated at the level of heterogeneity (homo-oligomers and hetero-oligomers).

5. HOLINS AND PINHOLINS: THE SIZE OF THE MEMBRANE LESIONS MATTERS

The mechanism by which holins damage membranes has been a matter of mystery regarding the molecular aspects. The production of an anti-S^λ specific antibody and its use for monitoring and characterizing S^λ holin during the bacterial lysis provided the first evidence supporting the holin-induced membrane lesion as a progressive process likely dependent both on the protein dose and on a protein oligomerization event (Zagotta and Wilson 1990). Thus, the immunoblot analysis of cell extracts from induced bacteriophage lambda lysogens probed with anti-S^λ antibody demonstrated that the S^λ holin begins to appear 10 min after phage induction, that it is localized to the inner membrane at all times during the lytic cycle, and that $100–1000\,S^\lambda$ molecules per cell are present at the time of the phage-induced lysis. Coupling cross-linking approaches with the immunotools allowed the visualization of a ladder of bands at sizes compatible with the presence of homopolymers (dimers, trimers, tetramers, and hexamers) of S^λ holin at the inner membrane, some of which exhibited resistance to using SDS as a detergent solubilization.

Moreover, the function preservation on exchanging holins between lysis cassettes and their versatility toward hosts demonstrated their intrinsic activity (Wang et al. 2000). In this sense, the flexibility of holin function within the two-component lysis system made unlikely the possible interaction between the holin and the endolysin components. Another important factor is that holin activity can be considered as universal regarding the target host as judged from the absence of dependence on host factors other than the membrane bilayer (Wang et al. 2000). In this sense, if the function of holins depended on any protein component as an accessory molecule involved in its insertion, folding, assembly, or activator, the efficiency of its lytic action would have been dramatically dictated by the host under study.

Given the membrane integration nature of holins, the reduced size of their chains, and the previous functional properties, their action was soon postulated to involve the acquisition of a critical quaternary structure through an in-membrane assembly process and the rupture of the lipid bilayer continuity. However, even under this simple view, the type of lesion was difficult to bring into line with the conventional models for proteinaceous pores and channels. In fact, the size tolerance and the irreversibility of the holin-provoked lesions are better described as a local and very effective destruction than as the result of the assembly of a true channel.

5.1. Evaluating the Membrane Lesion: Genetic and Biophysical Analysis

The studies aimed at determining the molecular basis of the function of holins using *in vitro* reconstitution approaches (holin and target membranes) have been impeded for a long time by the failure in protein production in terms of both quantity and purity. Even some tightly regulated inducible systems widely used for protein production resulted in very low quantities of protein that turned out to be very difficult to purify using conventional protein and peptide chemistry methods.

Our knowledge of holin function was originally provided by genetic approaches (Wang et al. 2000). The simplicity of both holin genes and of the functional assay of holins offered an adequate scenario for fast and efficient advances (Wang et al. 2002). The construction of a large collection of both point and length shortening mutations along the S^λ sequence clarified the role of the different domains and key residues (Raab et al. 1986, 1988). Most inactivating mutations (impairing lethality) of S^λ map to the sequence segments named TM1, TM2, and the loop connecting them. Also, the truncation of the N-terminus (from $S^\lambda 107$ to $S^\lambda 105$) indicated the inhibitor action of $S^\lambda 107$ and the effector function of $S^\lambda 105$ (Steiner and Bläsi 1993). Similarly, the highly charged C-terminal tail of S^λ was found to be nonessential for function but to play a regulatory role such as in lysis timing (Johnson-Boaz et al. 1994; Bläsi et al. 1999). Taken together, these results suggested holin as the result of a basic helical hairpin TM1 loop–TM2 scaffold with regulatory elements at the N- and C-terminal tails.

Once the over-expression and purification of a holin in the form of His-tagged $S^\lambda 105$ succeeded, its function in terms of the capacity to produce vesicle leakage could be analyzed (Smith et al. 1998). Assays using liposome-entrapped calcein showed the spillage of this fluorescent probe upon addition of the holin chain (Smith et al. 1998). Importantly, under similar conditions, the lysis-defective His-tagged

$S^\lambda105$ A52V mutant did not produce any calcein leakage. These results indicate that holin activity indeed involves the permeation of the lipid bilayer with no other requirement than the lipid assembly itself (Smith et al. 1998). It must be noted that the kinetics of the permeation process, which contains the mechanistic information, differed notably from that of membrane-acting soluble toxins, supporting as expected notable differences in the elementary steps of the overall process (see Chapter 8 for a detailed description). However, the detail analysis of such elementary steps is precluded in the case of holins given their intrinsic insolubility in aqueous media.

The membrane lesion required for liposome-entrapped calcein spillage is about an order of magnitude smaller than that required to allow the free passage of a folded protein of a size comparable to that of a typical endolysin. Next, pore size was analyzed using synthetic peptides containing the holin active domains (Fig. 6.4). Quantitative comparison of the differential leakage of fluorescent compounds of varying sizes

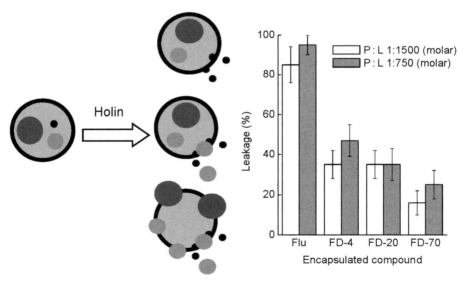

Figure 6.4. Biophysical approaches for sizing the holin-induced membrane damages. Holin activity consists in a membrane permeation event that can be followed studying the release of liposome content. To size the lesion induced by holins, the liposomes can be loaded with a set of fluorescent-labeled compounds differing in their size and quantify the extent of escape on holin addition (Ambroggio et al. 2007). The extent of lysis is determined as percentage of compound leakage, referenced to the maximal escape, which is determined as that produced by a conventional detergent as Triton X-100. P:L refers to protein to lipid molar ratio.

showed that at high lipid-to-peptide ratio (750 and 1500), the peptides indeed break the barrier into small substrates and also allow the passage of a 20-kDa dextrane (Haro et al. 2003). Such drastic membrane permeation could be explained by the formation of transmembrane holes with either fixed or variable internal diameter but big enough to allow the escape of a large polymer.

To shed light on the structural aspects of the lesion, the previous membrane leakage experiments were accompanied by studies using atomic force microscopy (AFM). In this case, the active holin peptides were added to supported lipid bilayers of varying composition prepared onto freshly cleaved mica surface and the topography obtained by AFM. As depicted in Figure 6.5, the active holin peptides indeed pierced the lipid bilayer, inducing the formation of holes. Interestingly, these holes displayed a notable internal diameter variation compatible with an intrinsic polydispersity. Simple geometric considerations of the lesion diameter together with the quantitative estimation of the

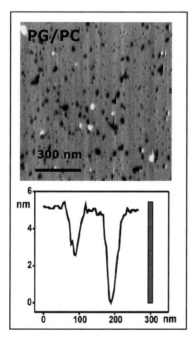
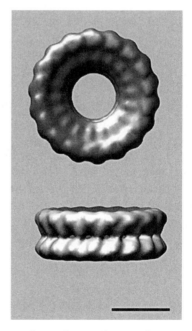

Figure 6.5. Visualization the holin assemblies and membrane damage in supported bilayers. Membrane-induced lesions were visualized by atomic force microscopy (AFM) using supported bilayers made of phosphatidylglycenol (PG) and phosphatidylcholine (PC). In this case, the holin moiety consists in the 1-32 N-terminal segment of Ejh. At the right, ring oligomers formed by S^λ holin (Saava et al. 2008) are displayed.

in-membrane peptide concentration yielded a model in which the walls of the holes would be recovered by both peptide scaffolds and lipids (Haro et al. 2003). This mixed upholstering of the hole walls derived from geometrical considerations gives some indication that the integrated peptide, through its oligomerization, is capable of causing the membrane to tear up and to enlarge the original holes.

5.2. Active and Inactive Oligomers: Scaffolding the Lesion

Even without knowledge of holin structure, genetic studies showed that holins accumulate in the membrane until the damage is produced. In other words, a holin must have two functionally distinct forms: a pre-hole state and a hole state. It was also clear from genetic experiments that the transition from the pre-hole state to the hole state is opposed by the energized membranes, since the opening of the first bilayer pass dictates the collapse of the membrane potential and the amplification of the signal.

In-membrane holin accumulation and the transition between states clearly indicated the formation of some type of oligomers in the membrane. The capacity to interact in the membrane was first found for S^λ using cysteine-scanning mutagenesis and oxidation experiments (Gründling et al. 2000b). Such a dimerization process involved intermolecular disulfide bonds between TM2 and could take place between holin–anti-holin partners and also between lytic defective mutants, suggesting that dimers could account for the minimal building block. Cross-linking experiments on holin-containing membranes showed for S^λ a dispersed degree of oligomerization, whereas for the lysis-defective mutant, such association was impeded (Zagotta and Wilson 1990). Thus, the oligomerization from the holin dimer drives function and the anti-holin–holin heterodimer formation impairs activity. Similar to cross-linking methods, fluorescence quenching/dequenching experiments using fluorescently labeled synthetic mini holins showed an in-membrane oligomerization process (Haro et al. 2003). In this case, the polydispersity of the oligomers detected simultaneously by other techniques impeded a better definition of the peptide self-assemblies through the systematic analysis of the quenching variations.

As mentioned previously, probably the key piece of knowledge has been obtained recently upon analysis of the different assemblies formed by $S^\lambda 105$ in different detergents (Savva et al. 2008). The capacity of the holin chain to generate high-molecular-size aggregates with different shapes, ring dimmers, and filaments agrees with its outstanding capacity for self-association. Thus, despite the continued search for the hole definition in terms of a defined architecture, the possibility of oligomers of varying size and shape displaying full functionality will remain.

5.3. Pinholins: A Simple Membrane Depolarization Is Enough

Although holins function to produce a membrane lesion to allow the passage of a medium-sized globular protein, in some cases the holins function simply to activate the endolysin (Park et al. 2007). Such activation consists in the release from the cytoplasmic membrane into the periplasm of the endolysin molecule which is synthesized as translocated but in a membrane-tethered inactive form (Xu et al. 2004, 2005; Park et al. 2007). These holins are thus designated as pinholins for differentiating purposes and the corresponding endolysins referred to as signal-anchor release (SAR) endolysins.

At physiological levels of expression, the pinholins as S^{21} are lethal and can mediate host lysis when co-expressed with cognate or noncognate SAR endolysins but not with cytoplasmic endolysins (Park et al. 2007). The lysis capacity limit has been shown to be related to the restriction of the membrane permeation process; pinholins depolarize membranes, making them permeable to small ions but not to macromolecules (Park et al. 2007).

5.4. Holins in Action: Sculpturing the Membrane Lesion

The recent structural characterization of the S^{λ} holin chain in detergents has provided a solid argument for building models focused on the very high tendency for the formation of large two-dimensional aggregates in the membrane as the basis for the massive membrane damage. Together with this main property, the experimental evidence supporting the occurrence of two types of aggregates differing in size and membrane effects has provided more detail to the models. These biochemical features, together with physical considerations, have resulted in a universal model in which the initial dose of holin synthesis conditions a group of aggregates, those featured by a small size, that may provoke initial changes in the permeability allowing the passage of small substances. At a higher dose of membrane-incorporated protein, which follows in time, the aggregates will increase in size, causing the collapse of the membrane and the appearance of true holes (Krupovic and Bamford 2008). Thus, it seems plausible that the holin-induced lesion would be of physical origin. Overdosing the membranes with holins may cause their segregation into holin-rich membrane domains or rafts, which, if increased in size, may perturb the physical properties of the membrane, causing the formation of cracking faults. This simple model explains all biochemical features of holin function and also provides an explanation for the impediments of hole visualizations.

6. HOLINS IN MEDICINE: UNCONTROLLED DAMAGE WITH NO LYSIS

In general, antibiotics exert their action either by killing the bacteria (bactericidal antibiotics) or by inhibiting the growth of the bacteria (bacteriostatic antibiotics). Although bactericidal agents have been preferred, bacteriostatic agents have also been beneficial since the normal host defenses, often shifting the balance of destruction over replication, are more efficient on slower-growing bacterial populations. In this sense, holins can be envisaged as potential lethal weapons displaying a spectrum activity that ranges from bacteriostatic (bacterial host inactivation) to bactericidal (bacterial pathogen killing). From a technological point of view, the correct exploitation of a system as a tool requires the knowledge of its mechanism of function, for optimization of the preferred action and for the reduction if not the elimination of the putative undesired side reactions.

6.1. Strategies for Functional Delivery of Holins

The single application of holins as a simple therapeutic protein externally added is precluded by several factors. First, holins are membrane proteins; as such, they intrinsically display low solubility in aqueous media, which is a determining requirement for membrane-mimetic carriers for solubility and transport. Second, holins are not true enzymes and their function is strictly dosage dependent, requiring to achieve a high local concentration at the desired site. Third, holins are tailored to exert their microbicidal action intracellularly and therefore their external addition might be a priori irrelevant or even a source of side effects. Thus, it follows that the exploitation of holins following the conventional approach for therapeutical proteins will require the parallel development of a carrier capable of solubilizing the chain, performing the correct transport, producing highly local concentration, and impeding side effects. Even assuming a certain degree of success in the achievement of these elementary steps, the limitation will definitively be overcome with a significant increase in the costs of production that will eventually hamper its use. Alternatively, the use of holins as part of bacteriophages looks like a very promising tool in the foreseeable future.

6.2. Biotechnological Added Value of Holins

Ever since their discovery, holins have been considered useful tools in the field of biotechnology, in particular in areas such as food production and biological retention. Food production often involves biological transformation processes that depend on the spillage to the media of critical enzymes through a controlled or an uncontrolled lysis event. In this line, phage holin–endolysin pairs cloned under inducible promoters in plasmids of *Lactococcus lactic* have shown to satisfactorily modulate the efficiency of the lysis process with relevance in the dairy industry (de Ruyter et al. 1997; Labrie et al. 2004). This strategy can be also powerful for the effective elimination of food poisoning pathogens such as *Listeria monocytogenes, Campylobacter jejuni,* and *Salmonella* (Greer 2005). In fact, in August 2006, the Food and Drug Administration (FDA) approved the use of phages for the treatment of ready-to-eat meat, the process consisting of a mixture of six phages designed to be sprayed on the meat to eradicate strains of *Listeria monocytogenes* (Petty et al. 2007).

Similarly, the area of bioremediation centers its action on the exploitation of the metabolic power of different bacteria for environmental purposes (transforming a given contaminating compound, metabolizing a key substance, etc.), such that after executing their action, the bacteria must die. In this case, modifying such organisms with phages containing either a holin or the holin–endolysin pair under the control of an inducible promoter sensitive to the substance of interest provides a suitable controlled system.

6.3. Holins in Biomedicine

In biomedicine, holins can be exploited as universal lethal cargos. It must be mentioned that the ectopic expression of the S^λ holin in eukaryotic cells is cytotoxic, leading to cell death by a non-apoptotic mechanism (Agu et al. 2006, 2007). S^λ holin expression produces the disruption of the mitochondrial membrane potential and a massive vacuolization (Agu et al. 2006, 2007). This finding anticipates the feasibility of holins as antitumoral agents in gene therapy.

Regarding the potential of holins to combat infectious diseases, their use is still limited to phage therapy as a holin-modified bacteriophage.

In general, the use of phage treatments over antibiotic therapy offers the following advantages (Matsuzaki et al. 2005): (a) phages are efficient against multidrug-resistant pathogenic bacteria; (b) the phage specificity toward the targeted pathogenic bacteria ensures the lack of interference with the normal microbial flora; (c) the response to the appearance of phage-resistant bacterial mutants is fast, since the frequency of phage mutation is significantly higher than that of bacteria; (d) the developing costs of phage treatments are lower than those linked to antibiotics; and (e) there is a rarity of side effects.

The advantages of phage therapy are counteracted by several concerns, such as the following: (a) the rapid bacterial lysis may result in the release of a large amount of bacterial membrane-bound endotoxins; (b) some phages may encode toxins; (c) there is a lack of pharmacokinetic data; (d) phages may be neutralized by the immune system; and (e) conversion of lytic phages to lysogenic phages immunizes the bacteria to the attacks by lytic phages and may alter the bacterial virulence (Pálffy et al. 2009, and references therein). Some of the aforementioned concerns have been successfully addressed by different approaches; one of them relates particularly to the use of holin function. Incorporation of holins into therapeutical phages under the proper expression regulation provides the pathway for bacterial host inactivation in the absence of massive lysis. In other words, isolation of the holin expression from the cognate endolysin allows a fast cell death without major cell content spillage and membrane fragmentation. This strategy also reduces the probability of generation of toxin-coding phage progenies since cell death can be programmed to take place independently of phage cycle. Also, expanding the holin modification to a cocktail or combination of phages reduces the probability of immune neutralization ensuring the escape of active phages.

Thus, the selection and isolation of the holin function in terms of cytoplasmic membrane lesion without major lysis and switching off the phage life cycle dependency will provide a potent bactericidal action with significant improvements such as reduced immunogenicity, multiplicity of administration routes, feasibility in control quality, and thermal and chemical stability.

ACKNOWLEDGMENT

MG lab is supported by grants SAF2006-00418 and BFU2009-07971 from Ministerio de Ciencia e Innovación, and FOOD-CT2004-506579 from the European Union.

REFERENCES

Ambroggio E., L. A. Bagatolli, E. Goormaghtigh, J. Fominaya, and M. Gasset (2007) *Adv Planar Lipid Bilayers Liposomes* **5**, 1–23.

Agu C. A., R. Klein, J. Lengler, F. Schilcher, W. Gregor, T. Peterbauer, U. Bläsi, B. Salmons, W. H. Günzburg, and C. Hohenadl (2007) *Cell Microbiol* **9**, 1753–1765.

Agu C. A., R. Klein, S. Schwab, M. König-Schuster, P. Kodajova, M. Ausserlechner, B. Binishofer, U. Bläsi, B. Salmons, W. H. Günzburg, and C. Hohenadl (2006). *J Gene Med* **8**, 229–241.

Bläsi U., C.-Y. Chang, M. T. Zagotta, K. Nam, and R. Young (1990) *EMBO J* **9**, 981–989.

Bläsi U., P. Fraisl, C.-Y. Chang, N. Zang, and R. Young (1999) *J Bacteriol* **181**, 2922–2929.

Bläsi U. and R. Young (1996) *R. Mol Microbiol* **21**, 675–682.

De Ruyter P. G., O. P. Kuipers, W. C. Meijer, and W. M. de Vos (1997) *WM Nat Biotechnol* **15**, 976–979

Diaz E., M. Munthall, H. Lundsdorf, J. V. Holtje, and K. N. Timmis (1996) *Mol Microbiol* **19**, 667–81.

Garrett J. M. and R. Young (1982) *J Virol* **44**, 886–892.

Graschopf A. and U. Bläsi (1999a) *Mol Microbiol* **33**, 569–582.

Graschopf A. and U. Bläsi (1999b) *Arch Microbiol* **172**, 31–39.

Greer G. G. (2005) *J Food Prot* **68**, 1102–1111.

Gründling A., U. Bläsi, and R. Young (2000a) *J Biol Chem* **275**, 769–776.

Gründling A., U. Bläsi, and R. Young (2000b) *J Bacteriol* **182**, 6082–6090.

Haro A., M. Vélez, E. Goormaghtigh, S. Lago, J. Vázquez, D. Andreu, and M. Gasset (2003) *J Biol Chem* **278**, 3929–3936.

Johnson-Boaz R., C.-Y. Chang, and R. Young (1994) *Mol Microbiol* **13**, 495–504.

Krupovic M. and D. H. Bamford (2008) *Mol Microbiol* **69**, 781–783.

Labrie S., N. Vukov, M. J. Loessner, and S. Moineau (2004) *FEMS Microbiol Lett* **233**, 37–43.

Matsuzaki S., M. Rashel, J. Uchiyama, S. Sakurai, T. Ujihara, M. Kuroda, M. Ikeuchi, T. Tani, M. Fujieda, H. Wakiguchi, and S. Imai (2005) *J Infect Chemother* **11**, 211–219.

Pálffy R., R. Gardlink, M. Behuliak, L. Kadasi, J. Turna, and P. Celec (2009) *Mol Med* **15**, 51–59.

Park T., D. K. Struck, C. A. Dankenbring, and R. Young (2007) *J Bacteriol* **189**, 9135–9139.

Park T., D. K. Struck, J. F. Deaton, and R. Young. (2006) *Proc Natl Acad Sci U S A* **103**, 19713–19788.

Petty N. K., T. J. Evans, P. C. Fineran, and G. P. C. Salmond (2007) *Trends Biotechnol* **25**, 7–15.

Raab R., G. Neal, J. Garret, R. Grimaila, R. Fusselman, and R. Young (1986) *J Bacteriol* **167**, 1035–1042.

Raab R., G. Neal, C. Sohaskey, J. Smith, and R. Young (1988) *J Mol Biol* **199**, 95–105.

Ramanculov E. R. and R. Young (2001a) *Mol Microbiol* **41**, 575–583.

Ramanculov E. R. and R. Young (2001b) *Mol Genet Genomics* **265**, 345–353.

Ramanculov E. R. and R. Young (2001c) *Gene* **265**, 25–36.

Savva C. G., J. S. Dewey, J. Deaton, R. L. White, D. K. Struck, A. Holzenburg, and R. Young (2008) *Mol Microbiol* **69**, 784–793.

Smith D. L., D. K. Struck, J. M. Scholtzand, and R. Young (1998) *J Bacteriol* **180**, 2531–2540.

Srividhya K. V. and S. Krishnaswamy (2007) *J Biosci* **32**, 79–90.

Steiner M. and U. Bläsi (1993) *Mol Microbiol* **8**, 525–533.

Takác M., A. Witte, and U. Bläsi (2005) *Microbiology* **151**, 2331–2342.

Tran T. A. T., D. K. Struck, and R. Young (2005) *J Bacteriol* **187**, 6631–6640.

Wang I. N., D. L. Smith, and R. Young (2000) *Annu Rev Microbiol* **54**, 799–825.

Xu M., D. K. Struck, J. Deaton, I. N. Wang, and R. Young (2004) *Proc Natl Acad. Sci. USA* **101**, 6415–6420.

Xu M., A. Arulandu, D. K. Struck, S. Swanson, J. C. Sacchettini, and R. Young (2005) *Science* **307**, 113–117.

Young R. (1992) *Microbiol Rev* **56**, 430–481.

Young R. J. (2002) *Mol Microbiol Biotechnol* **4**, 21–36.

Young R. and U. Bläsi (1995) *FEMS Microbiol Rev* **17**, 191–205.

Young R., I. N. Wang, and W. D. Roof (2000) *Trends Microbiol* **8**, 120–128.

Zagotta M. T. and D. B. Wilson (1990) *J Bacteriol* **172**, 912–921.

CHAPTER 7

ANTI-STAPHYLOCOCCAL LYTIC ENZYMES

JAN BORYSOWSKI[1] and ANDRZEJ GÓRSKI[1,2]
[1]Department of Clinical Immunology, Transplantation Institute, Warsaw Medical University, Poland
[2]Laboratory of Bacteriophages, L. Hirszfeld Institute of Immunology and Experimental Therapy, Wrocław, Poland

1. INTRODUCTION

Staphylococcus aureus is currently one of the most challenging bacterial pathogens. It is the etiologic agent of a wide range of infections, which may be difficult to cure due to the ever-growing antibiotic resistance of staphylococci. Of particular concern are methicillin-resistant *S. aureus* (MRSA) strains, whose prevalence has alarmingly increased over the last two decades both in hospital and community settings (Boucher and Corey 2008). Furthermore, the recent emergence of *S. aureus* strains with reduced susceptibility to vancomycin, that is, vancomycin-intermediate *S. aureus* (VISA) and vancomycin-resistant *S. aureus* (VRSA), has dramatically limited available treatment options for some staphylococcal infections (Appelbaum 2007). These data clearly show an urgent need for the development of new anti-staphylococcal agents (Moreillon 2008).

One of the interesting classes of novel antibacterial agents that could be used in the prophylaxis and treatment of infections due to multi-drug-resistant staphylococci are lytic enzymes. These are enzymes that cleave covalent bonds in peptidoglycan, thus inducing rapid lysis of a bacterial cell (Parisien et al. 2008). Of great importance is that lytic enzymes can also kill MRSA (Climo et al. 1998; von Eiff et al. 2003)

Enzybiotics: Antibiotic Enzymes as Drugs and Therapeutics. Edited by Tomas G. Villa and Patricia Veiga-Crespo
Copyright © 2010 John Wiley & Sons, Inc.

and *S. aureus* strains with reduced susceptibility to vancomycin (Patron et al. 1999; Rashel et al. 2007).

The first application of lytic enzymes may be the prophylaxis of staphylococcal infections. For instance, these enzymes were shown to be highly effective in killing bacteria colonizing mucous membranes in an experimental model of colonization (Kokai-Kun et al. 2003). Thus, they could be used to clear staphylococci colonizing nasal mucosa, which, in some clinical settings, can be a starting point for serious infections (Wertheim et al. 2005). Another prophylactic use of lytic enzymes might be prevention of the colonization of catheters and implants by enzyme molecules coating their surface (Shah et al. 2004). Very encouraging results obtained in animal models of infections, including bacteremia, endocarditis, and ocular infections, suggest that lytic enzymes could also be used in the treatment of both topical and systemic staphylococcal infections in humans (Patron et al. 1999; Dajcs et al. 2001; Kokai-Kun et al. 2007). Another potential application of lysostaphin is the treatment of veterinary infections, for instance, bovine mastitis (Oldham and Daley 1991).

2. ANTI-STAPHYLOCOCCAL LYTIC ENZYMES

Anti-staphylococcal lytic enzymes specifically cleave covalent bonds in the peptidoglycan of both *S. aureus* and coagulase-negative staphylococci, thereby inducing their hypotonic lysis. The two major classes of anti-staphylococcal lytic enzymes are bacteriocins, especially lysostaphin, and staphylococcal phage lysins (a list of representative anti-staphylococcal lytic enzymes is shown in Table 7.1).

2.1. Bacteriocins

Bacteriocins are peptides or proteins produced by bacteria to inhibit the growth of other bacteria. While some authors suggest that bacteriolytic enzymes should in fact not be classified as bacteriocins, for the sake of simplicity, we will assume here that they belong to this class of proteins (Nes et al. 2007). The best studied anti-staphylococcal bacteriocin is lysostaphin, which was used as an antibacterial agent as early as the 1960s.

2.1.1. Lysostaphin

2.1.1.1. General Features   Lysostaphin is a plasmid-encoded extracellular bacteriolytic enzyme produced by *Staphylococcus simulans* biovar

TABLE 7.1. A List of Representative Anti-staphylococcal Lytic Enzymes

Enzyme Name	Class	Source	Activity against MRSA	Reference
Lysostaphin	Bacteriocin	*S. simulans*	+	Patron et al. (1999)
LasA	Bacteriocin	*P. aeruginosa*	+	Barequet et al. (2004)
MV-L	Lysin	Phage ØMR11	+	Rashel et al. (2007)
Ø11 phage lysin	Lysin	Phage Ø11	?[a]	Donovan et al. (2006a)
LysK	Lysin	Phage K	+	O'Flaherty et al. (2005)
PlyTW	Lysin	Phage Twort	?[a]	Loessner et al. (1998)
Ply187	Lysin	Phage 187	?[a]	Loessner et al. (1999)
LysH5	Lysin	Phage ΦH5	?[a]	Obeso et al. (2008)
Protein 17	Phage virion-associated enzyme	Phage P68	+	Takac and Blasi (2005)

[a]Activities of Ø11 phage lysin, PlyTW, Ply187, and LysH5 against MRSA have not been reported yet.

staphylolyticus. Its mature form consists of 246 amino acids and has a molecular weight of 25 kDa (Schindler and Schuhardt 1964; Thumm and Götz 1997). The lysostaphin molecule is composed of two main functional domains: an N-terminal enzymatic domain and a C-terminal cell wall-binding domain (Baba and Schneewind 1996; Kumar 2008). Lysostaphin is a zinc-containing endopeptidase that specifically cleaves the bonds between glycine residues in the inter-peptide cross-bridges of the staphylococcal peptidoglycan, resulting in hypotonic lysis of the bacterial cell (Fig. 7.1) (Francius et al. 2008). The antibacterial range of recombinant lysostaphin encompasses both *S. aureus* and coagulase-negative staphylococci (Climo et al. 1998; McCormick et al. 2006). However, its lytic activity against coagulase-negative staphylococci is generally weaker due to the different amino acid content of their cross-bridges. Specifically, replacement of the glycine residues, especially by serine residues, results in a decrease in lysostaphin susceptibility (Zygmunt et al. 1968; Kumar 2008). A significant feature of lysostaphin is its capacity for killing both multiplying and nondividing staphylococci (Kumar 2008). Main characteristics of lysostaphin are listed in Table 7.2.

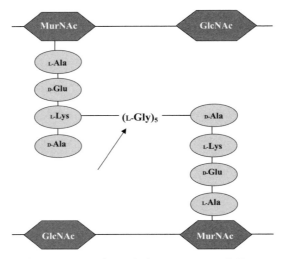

Figure 7.1. Schematic representation of the structure of *S. aureus* peptidoglycan. Peptidoglycan is a polymer composed of the glycan backbone and amino acids. The glycan backbone consists of alternating residues of *N*-acetylglucosamine (GlcNAc) and *N*-acetylmuramic acid (MurNAc). Attached to MurNAc are the tetrapeptide side chains cross-linked by the pentaglycine bridges. Lysostaphin specifically cleaves the pentaglycine cross bridges (marked with arrow), thereby inducing lysis of a bacterial cell.

TABLE 7.2. Main Characteristics of Lysostaphin

Source	*S. simulans*
Molecular mass	25 kDa
Enzymatic specificity	Endopeptidase
Antibacterial range	*S. aureus*, coagulase-negative staphylococci
Activity against MRSA	+
Activity against VISA	+
Capacity to disrupt staphylococcal biofilms	+

In the context of its potential applications as an anti-staphylococcal agent, it is very important that lysostaphin can rapidly lyse practically all *S. aureus* strains regardless of their antibiotic resistance. For instance, in a recent study, all of the 257 *S. aureus* clinical isolates tested, including 168 methicillin-sensitive *S. aureus* (MSSA) strains and 89 MRSA strains, were susceptible to lysostaphin, with minimum inhibitory concentrations (MICs) ranging from 0.03 to 2 μg/ml in the agar dilution assay (Yang et al. 2007). MSSA strains were lysed equally efficiently as MRSA strains. The minimum bactericidal concentrations (MBCs) for 17 selected strains ranged from 0.016 to 1 μg/ml. In another study, lysostaphin was found to kill all of the 429 clinical *S. aureus* isolates tested, including 210 isolates obtained from nasal swabs of *S. aureus* carriers and 219 isolates from blood of individuals with *S. aureus* bacteremia (von Eiff et al. 2003). No differences were found in the susceptibilities of the MSSA and MRSA isolates to lysostaphin, and the MBC values for 10 selected strains were 0.16 μg/ml.

Aside from planktonic cells, staphylococcal biofilms can also be eliminated by lysostaphin. This effect is specific to biofilms formed by lysostaphin-sensitive staphylococci (MSSA, MRSA, and *Staphylococcus epidermidis*). Lysostaphin's activity against staphylococcal biofilms was shown to be more potent than those of oxacillin, vancomycin, and clindamycin (Wu et al. 2003).

A very important problem is the development of an assay that could be used to determine the susceptibility of staphylococci to lysostaphin. This question was addressed in a study, in which four different methods for distinguishing lysostaphin-sensitive strains from lysostaphin-resistant ones were compared (MIC, MBC, turbidity, and disk diffusion assays) (Kusuma and Kokai-Kun 2005). The study was performed on 57 different *S. aureus* strains, including MSSA, MRSA, VISA, and mupirocin-resistant strains, and defined genetic mutants of *S. aureus*. As negative controls, three reference lysostaphin-resistant *S. aureus* strains were included. Although some inconsistencies were noted between individual assays when determining the degree of lysostaphin susceptibility, there was a high correlation between all four methods with respect to identifying lysostaphin-susceptible bacteria (all of the 57 strains examined were susceptible to lysostaphin by all four methods, whereas all reference strains were resistant). This study revealed that the simplest and most reproducible assay to determine lysostaphin susceptibility is the disk diffusion assay.

2.1.1.2. Applications   One of the main medical applications of lysostaphin may be the elimination of *S. aureus* nasal colonization. It was shown that staphylococci colonizing nasal mucous membrane can cause serious infections in some clinical settings (e.g., in patients undergoing major surgery) (Wertheim et al. 2005). Furthermore, clearance of *S. aureus* colonization was found to decrease the risk of developing infections and the horizontal spread of antibiotic-resistant staphylococci within some communities (Lee et al. 1999; Yano et al. 2000). Prophylactic use of antibiotics, especially mupirocin, as decolonizing agents is hindered by their relatively slow antibacterial effects and the development of bacterial resistance. Thus, lysostaphin, in view of its rapid anti-staphylococcal activity and a capacity for killing antibiotic-resistant strains, could be used as a means of prophylaxis of staphylococcal infections. Another advantage of lysostaphin over antibiotics is its narrow antibacterial range, which allows the specific elimination of staphylococci without adversely affecting other bacteria from normal microflora.

The most detailed study to evaluate the effectiveness of lysostaphin as a decolonizing agent was performed in a cotton rat model of *S. aureus* colonization (Kokai-Kun et al. 2003). In one of the experiments, rats were instilled intranasally with 10^9 colony forming units (cfu) of MRSA and, after several days, were topically administered lysostaphin formulated in a petrolatum-based cream. The highest eradication rate (100%) was obtained in animals that were administered three doses of 0.5% cream over 3 days. One dose of 0.5% lysostaphin cream eradicated colonization in 93% of the rats (in the remaining rats, the degree of colonization was very low). Further decreasing of the lysostaphin dose resulted in still lower eradication rates (55% and 20% for three doses and one dose of 0.125% cream, respectively). Importantly, one dose of 0.5% lysostaphin cream was more effective in clearing bacteria than one dose of 2% mupirocin ointment and 0.5% and 5% nisin cream. Furthermore, the antibacterial effect of lysostaphin was more rapid than that of mupirocin (bacteria were eradicated within 4 h after administration of a single dose of lysostaphin, whereas mupirocin eradicated colonization after three daily doses). The enzyme successfully eradicated nasal colonization by MSSA, MRSA, and mupirocin-resistant *S. aureus*. In another experiment, no recurrence of colonization was found over the 1-week observation period after eradication of staphylococci by lysostaphin. Interestingly, no lysostaphin-resistant staphylococci were isolated in any of the experiments, and an *in vitro*-isolated lysostaphin-resistant strain had a much lower colonizing capacity than its lysostaphin-sensitive counterpart.

It is also noteworthy that lysostaphin's capacity for clearing staphylococci colonizing nasal mucous membrane was evaluated in three other trials conducted in the late 1960s and early 1970s (Harris et al. 1967; Martin and White 1967; Quickel et al. 1971). However, the results of these studies were not as encouraging as those obtained in the above-discussed study performed on cotton rats, the major problem being relatively quick relapses of colonization. Two main reasons are currently ascribed to the inability of lysostaphin to eradicate the bacteria. First, the preparations of lysostaphin used at that time were not as reliable as those of the recombinant enzyme employed currently. Second, lysostaphin was administered to colonized individuals as a saline solution, which was shown to be ineffective in experiments performed on cotton rats. Therefore, the inadequate efficacy of lysostaphin found in earlier studies is not a solid argument against its use as a decolonizing agent.

One of the factors reducing the effectiveness of lysostaphin as a decolonizing agent could be the rapid clearance of enzyme molecules from the surface of the mucous membrane. Therefore, two different delivery systems were developed to extend the mucosal residence time of lysostaphin (Walsh et al. 2004). The first is based on a hydrophilic cream forming an emulsion with the secretions of the nasal mucosa. This formulation was shown to increase mucosal retention of the enzyme by 10-fold at 3 h post-administration and as much as 50-fold at 24 h post-administration compared with lysostaphin formulated in saline drops. It was also shown that administration of a single dose of the cream allows retention of the level of lysostaphin on the surface of the mucous membrane above the minimal bactericidal concentration for most staphylococcal strains for over 24 h. This implies that the enzyme could be effective when administered once daily. Furthermore, extended mucosal residence time of lysostaphin decreases the risk of developing resistance resulting from prolonged exposure of bacteria to a concentration of lysostaphin below its minimal therapeutic value. The other delivery system that was shown to extend the mucosal residence time of lysostaphin is based on the mucoadhesive polymers chitosan and sodium polystyrene sulfonate (SPSS). Both the hydrophilic cream and mucoadhesive formulations were more effective in clearing *S. aureus* nasal colonization than lysostaphin formulated in saline drops. It was proposed that the cream could be administered in the first instance to colonized individuals to clear staphylococci rapidly from the most anterior part of the nasal mucosa, where most bacteria are found. A spray containing a mucoadhesive formulation of the enzyme, which is cheaper and easier to use, could be applied next to prevent recolonization. Using a spray alone would likely be less effective because it is difficult to deliver to the most anterior part of the nose.

Another important prophylactic application of anti-staphylococcal lytic enzymes might be the prophylaxis of catheter-related bloodstream infections, which are most often caused by coagulase-negative staphylococci and *S. aureus*. The efficacy of lysostaphin in this regard was shown in an interesting study, which revealed that enzyme molecules can be readily adsorbed onto the catheter surface without losing their antibacterial activity (Shah et al. 2004). Lysostaphin adsorbed on the catheter surface could not only prevent *S. aureus* surface colonization but also kill bacteria in solution within the catheter lumen. While increasing the coating time resulted in a more potent antibacterial activity, a substantial effect was achieved even upon 5 min of coating of the catheter surface with enzyme. Importantly, lysostaphin adsorbed on the catheter surface maintained significant activity for at least 4 days. Furthermore, lysostaphin's antibacterial activity was not substantially decreased following incubation of catheters with human serum for 24 h, which shows that serum proteins do not adversely affect the activity of enzyme adsorbed on the catheter surface. In view of the clinical importance of catheter-related infections, the results of this study have very important implications. First, they indicate that lysostaphin may provide a novel means of preventing catheter colonization by both *S. aureus* and, possibly, coagulase-negative staphylococci. Second, they show that a significant anti-staphylococcal effect could be maintained relatively easily by health-care workers by periodic flushing of the catheter with a solution of enzyme.

The other general application of lysostaphin in clinical medicine could be the treatment of staphylococcal infections, both topical and systemic. Thus far, the therapeutic effectiveness of lysostaphin has been evaluated in experimental models of bacteremia (Kokai-Kun et al. 2007), endocarditis (Patron et al. 1999), neonatal infections (Oluola et al. 2007), and some ocular infections, especially endophthalmitis (Dajcs et al. 2001) and keratitis (Dajcs et al. 2000). The enzyme was also proposed as a potential means of eliminating staphylococci from infected wounds (Gunn and Hengesh 1969; Huan et al. 1994).

In a rabbit model of aortic valve endocarditis induced by a clinical MRSA isolate, the therapeutic efficacy of lysostaphin was compared with that of vancomycin (Climo et al. 1998). It was found that lysostaphin administered three times daily at a dose of 5 mg/kg sterilized staphylococcal vegetations in 10 of 11 rabbits, with a mean reduction in bacterial vegetation counts of $8.5 \log_{10}$ cfu/g of tissue compared with untreated controls. Vancomycin given twice daily at a dose of 30 mg/ kg did not sterilize any vegetation and decreased staphylococcal counts by only $4.8 \log_{10}$ cfu/g of tissue. Although lysostaphin administered once daily was less efficacious than when given three times daily, the combination of lysostaphin and vancomycin was more effective than either treatment alone. Similar results were obtained in a rabbit model of aortic valve endocarditis caused by a clinical isolate of *S. aureus* with reduced susceptibility to glycopeptides (GISA), in which lysostaphin was significantly more effective than vancomycin in terms of the sterilization of aortic valve vegetation (Patron et al. 1999).

It is noteworthy that the results obtained in the treatment of *S. aureus* systemic infection in mice suggest that repeated administration of lower doses of lysostaphin during a short period of time may be more effective than the administration of one large dose of the enzyme (bolus) (Kokai-Kun et al. 2007).

The antibacterial effects of lysostaphin were also evaluated in a rat model of neonatal infections (Oluola et al. 2007). To that end, suckling rats were inoculated subcutaneously with 2×10^6 cfu of MSSA. Lysostaphin was administered intraperitoneally (i.p.) to separate groups of pups at 30 min or 3 h after staphylococcal challenge. Both early and late lysostaphin treatment significantly improved the survival of the pups compared with the control group. The effectiveness of lysostaphin in this model was comparable with those of vancomycin and oxacillin. It was also found that lysostaphin treatment did not adversely affect weight gain by the pups.

Lysostaphin was also investigated as a potential means of treating some ocular infections, especially endophthalmitis and keratitis. In a rabbit model of MRSA endophthalmitis, lysostaphin was injected intravitreally at 8 or 24 h postinfection (Dajcs et al. 2001). Both the early and late lysostaphin treatment resulted in a significant decrease in MRSA counts in the vitreous humor. Furthermore, administration of lysostaphin at 8 h postinfection led to sterilization of 88% of the eyes, while injection of the enzyme at 24 h after bacterial challenge resulted in sterilization of 50% of the eyes compared with the 0% sterilization rate in the untreated rabbits. It was also shown that only the eyes of the rabbits that received lysostaphin 8 h postinfection had less pronounced inflammatory changes, while lysostaphin administered 24 h postinfection failed to prevent the development of severe inflammation in ocular tissues.

Similar results were obtained in a study in which endophthalmitis was induced by intravitreal injection of different coagulase-negative staphylococcal species (McCormick et al. 2006). At 8 h postinfection, lysostaphin was administered at a dose of 250 µg by the same route, and the numbers of viable staphylococci were determined in the vitreous humor. It was found that lysostaphin significantly reduced the colony forming unit counts compared with untreated eyes for most of the strains tested, including two methicillin-sensitive *S. epidermidis* strains, three *Staphylococcus warneri* strains, four *Staphylococcus simulans* strains, two *Staphylococcus cohnii* strains, and two *Staphylococcus capitis* strains. For the remaining strains, that is, two methicillin-resistant *S. epidermidis* strains, one *S. cohnii* strain, and two *Staphylococcus haemolyticus* strains, the reduction in the colony forming unit counts in the vitreous humor was very small and insignificant. An interesting finding of this study was a discrepancy between the MIC values for some strains and their susceptibility to lysostaphin *in vivo*. For example, one *S. capitis* strain was fairly susceptible to the enzyme *in vivo* in spite of a very high MIC value. On the other hand, one methicillin-resistant *S. epidermidis* strain was not susceptible to lysostaphin treatment *in vivo* despite a relatively low MIC value.

In a rabbit model of MRSA keratitis, the antibacterial effects of lysostaphin were compared with those of vancomycin (Dajcs et al. 2000). Both drugs were administered topically. When applied early (4–9 h postinfection), lysostaphin sterilized 100% of the corneas, while vancomycin did not sterilize any. Although when administered later (10–15 h postinfection) the enzyme failed to sterilize any cornea, its effectiveness was significantly higher (approximately 100,000-fold) than that of vancomycin in terms of reducing colony forming unit per cornea values. Importantly, lysostaphin did not exert any deleterious effects on ocular tissues, as graded by the slit-lamp examination. In a separate experiment performed using an *S. aureus* mutant lacking the capability to induce corneal epithelial erosion, lysostaphin was shown to penetrate the intact cornea efficiently and kill the bacteria. As in previous experiments, the decrease in colony forming unit per cornea was significantly higher following lysostaphin administration than upon vancomycin treatment.

An interesting study showed that lysostaphin can also cure ocular infections in immunized animals (Dajcs et al. 2002). In this study, rabbits were immunized against lysostaphin by the subcutaneous, intranasal, or intraocular route. In separate experiments, the efficacy of lysostaphin was subsequently evaluated in MRSA keratitis and endophthalmitis. In both models, a single topical dose of the enzyme significantly decreased bacterial counts in ocular tissues, and in rabbits with endophthalmitis, it was sufficient to completely clear the bacteria, as indicated by sterile vitreous humor in the eyes of the treated animals. In spite of the generation of neutralizing antibodies, the results obtained in the immunized rabbits were essentially the same as those found in nonimmunized animals. Furthermore, no adverse reactions were noted following the administration of lysostaphin to immunized rabbits, and slit-lamp examination (SLE) scores were comparable in both groups of animals. These data suggest that lysostaphin could be administered repeatedly in the treatment of ocular infections while maintaining substantial biological activity.

Another field of lysostaphin application could be the prophylaxis and treatment of veterinary infections, for example, bovine mastitis. Three major ways of preventing and/or treating bovine mastitis have been proposed: (a) topical administration of recombinant lysostaphin, (b) the introduction of the lysostaphin gene into cells of the mammary glands by means of a gene therapy vector, and (c) the generation of infection-resistant transgenic animals expressing the lysostaphin gene in their mammary glands. The first approach relies on the intramammary administration of recombinant lysostaphin to infected animals. In one such study, 30 dairy cattle were infected with *S. aureus* and subsequently administered, through the teat canal, recombinant lysostaphin dissolved in phosphate-buffered saline (PBS), sodium cephapirin in PBS, or a commercial preparation of cephapirin (Oldham and Daley 1991). It was found that the minimum effective therapeutic dose of lysostaphin was 100 mg. Administration of one 100-mg dose of the enzyme resulted in maintaining the lysostaphin MBC in milk for up to 36–48 h. That lysostaphin is fairly stable in milk was also shown in another experiment in which incubation of the enzyme in milk at 37 °C for 72 h did not lead to a significant decrease in lysostaphin's antibacterial activity. The cure rate in animals that received lysostaphin at the dose of 100 mg was 20% compared with 29% and 57% for sodium cephapirin in PBS and its commercial preparation, respectively. In fact, although after the administration of lysostaphin no staphylococci were detected in the majority of the animals, the overall efficacy of the treatment was decreased by relatively quick relapses, likely caused by bacteria internalized by cells of the mammary glands. Furthermore, in view of the large difference in therapeutic efficacy between cephapirin in PBS and its commercial preparation (containing the antibiotic formulated in a peanut oil base), the authors suggested that lysostaphin could also be much more efficacious in an improved formulation.

The second approach to preventing or treating mastitis relies on the introduction of the lysostaphin gene to the mammary gland cells of individual animals. The effectiveness of this approach was evaluated by Fan et al., who employed a replication-deficient human adenoviral vector containing a modified lysostaphin gene. In the first experiment, the adenoviral vector was used to transduce bovine mammary gland epithelial cells *in vitro*, which resulted in the production and release of functional lysostaphin into cell culture medium. While the lysostaphin concentration in the culture medium reached 0.8 µg/ml, its biological activity was at a level of 20% of that typical of recombinant lysostaphin produced by bacteria. In the second major experiment, the same vector, administered by intramammary infusion, was used to transduce mammary gland cells of nonlactating goats *in vivo*. Lysostaphin expression in the mammary glands did not appear to adversely affect the health of the animals. The enzyme's concentration in the secretions of the mammary glands ranged from 0.2 to 0.6 µg/ml on day 1 after infusion and from 0.9 to 1.1 µg/ml on day 2 postinfusion. These concentrations were fairly low, resulting in inadequate antibacterial activity of lysostaphin, as measured in spot-on-lawn assays (Fan et al. 2002).

In a separate study, it was shown that the primary factor accounting for the inadequate persistence of lysostaphin expression in the mammary cells of goats is a potent immune response to both the enzyme and the adenoviral vector (Fan et al. 2004). After intramammary infusion of the vector, the lysostaphin concentration in the secretions of the mammary glands rose gradually to reach its maximum level approximately 1 week postinfusion. Then the lysostaphin concentration dropped sharply, returning to basal levels between days 14 and 21 postinfusion. A specific humoral response developed fairly rapidly, as manifested by high titers of IgG reactive to lysostaphin and adenovirus in both blood and the secretions of the mammary glands. Antibodies reactive to the enzyme were detectable in blood of the goats on day 11 postinfusion and remained at a high level for 42 days, while those reactive to adenovirus developed within 1 week after infusion. That lysostaphin is neutralized largely by specific antibodies was clearly evident in one goat in which the lowest titer of anti-lysostaphin antibodies was accompanied by the highest concentration of the enzyme in mammary gland secretions. The authors of the two studies concluded that gene therapy does not appear to be a feasible method of preventing bovine mastitis due to inadequate persistence of lysostaphin expression resulting from the strong humoral immune response to both the enzyme and the vector.

The most sophisticated method of preventing mastitis is the generation of staphylococcal infection-resistant transgenic animals expressing the lysostaphin gene in their mammary glands. The effectiveness of this strategy was evaluated in two studies, one performed on mice and the other on cows.

In the first study, three lines of transgenic mice secreting lysostaphin into their milk were generated, a low-expressing line, a medium-expressing line, and a high-expressing line, in which the milk contained lysostaphin in concentrations of ~0.06, ~0.13, and ~0.8 mg/ml, respectively (Kerr et al. 2001). Lysostaphin expression was shown to be restricted essentially to cells of the mammary gland. The enzyme did not appear to adversely affect the physiology of the transgenic mice, the structure of their mammary glands, or the quality of their milk. Specifically, both the total protein content and the general pattern of milk proteins were comparable to those found in non-transgenic mice, which indicates that lysostaphin does not cleave milk proteins. Furthermore, transgenic mice were fertile and lysostaphin did not impair their lactation. While the antibacterial activity of lysostaphin produced by eukaryotic cells was 5–10 times lower than that of the enzyme obtained from bacteria, milk from transgenic mice did contain considerable staphylolytic activity. To evaluate the efficacy of the studied approach, the mammary glands of lactating transgenic mice were inoculated with 10^4 cfu of *S. aureus*. Twenty-four hours after inoculation, staphylococci were enumerated in mammary gland homogenates. It was found that 100% of the glands of the high-expressing line and ~40% of those of the other two lines were completely resistant to infection. In contrast, ~77% of the glands of the non-transgenic mice were severely infected and contained over 10^8 cfu of staphylococci per gland, while a further 18% contained a moderate bacterial load (10^5–10^8 cfu/gland). It was also shown that the degree of resistance to staphylococcal challenge correlated with the lysostaphin content in the mammary glands. Importantly, even the glands of transgenic mice containing viable staphylococci retained their normal structure, while the majority of glands of non-transgenic mice had apparent signs of inflammation in histopathologic examination. It is worth pointing out that the acute infection induced in this study is not likely to develop spontaneously. Therefore, it is possible that even the transgenic line with the lowest level of lysostaphin expression would be resistant to infection in natural settings. It needs to be stressed that the transgenic mice generated in this study will secrete lysostaphin only during the lactating period. However, resistance to infection during the non-lactating period can be achieved by fusing the lysostaphin gene with a different regulatory region. The risk of developing resistance to lysostaphin may be reduced by simultaneous use of more than one antibacterial protein.

In the other study, transgenic cows were generated in which the expression of the lysostaphin gene under the control of an ovine β-lactoglobulin promoter was restricted to mammary gland secretory epithelium (Wall et al. 2005). The lysostaphin concentration in the milk of the transgenic animals ranged from 0.9 to 14 mg/ml, maintaining a constant level over most of the lactation period. Remarkably, the lysostaphin concentration in the milk of the highest expressing transgenic cow was at a level of 1% of the concentration found in the highest expressing transgenic mice in the previous study. While in the spot-on-lawn assay the lysostaphin from the milk of the transgenic cows had only 15% of the biological activity of that produced by bacteria, its concentration in the milk of the highest expressing cow was sufficient to completely block the growth of staphylococci. No differences were found in the volume of milk produced by the transgenic and non-transgenic cows, and lysostaphin did not appear to exert any effect on milk proteins. The lysostaphin expression in the mammary glands of the transgenic cows resulted in substantial resistance to infection by *S. aureus*. Following intramammary infusion of staphylococci, only 14% of the mammary glands of the transgenic cows become infected compared with 71% of the glands of the non-transgenic animals. Interestingly, the degree of resistance to bacterial challenge appeared to correlate with the level of lysostaphin expression: the cow with the highest lysostaphin expression was completely resistant to infection, while that with the lowest lysostaphin concentration in milk was the most susceptible. Furthermore, unlike the non-transgenic controls, the transgenic cows had no signs of inflammation, as measured by the percentage of somatic cells in the milk, the level of acute-phase blood proteins, and body temperature.

2.1.1.3. Synergy with Other Antibacterial Agents   In several studies, synergy was noted between lysostaphin and other antibacterial agents, including lytic enzymes (lysozyme and LysK), cationic antimicrobial peptides, and some antibiotics. While lysozyme itself practically did not act on staphylococci, it substantially reduced the lysostaphin MIC values for all of the 84 *S. aureus* strains tested and 151 strains belonging to nine different coagulase-negative staphylococcal species. A particularly large (200-fold) decrease in the lysostaphin MIC value was noted for *S. cohnii*, while the MIC values for the other species were reduced 16- to 45-fold (Cisani et al. 1982). Synergy was also observed against MRSA for a combination of lysostaphin and LysK lysin (Becker et al. 2008).

In another study, synergy in the effects against *S. aureus* was shown for combinations of lysostaphin and three different cationic antimicrobial peptides: ranalexin, magainin 2, and dermaseptin s3 (Graham and Coote 2007). The synergistic antibacterial effect was restricted to lysostaphin-susceptible staphylococci and was not found with *Escherichia coli* or *Enterococcus faecalis*. Synergy most likely resulted from lysostaphin-mediated peptidoglycan degradation facilitating the access of membrane-active peptides to the surface of the plasma membrane. Synergistic inhibition of bacterial growth was also achieved upon exposure of both MSSA and MRSA to filter paper disks impregnated with lysostaphin and ranalexin. These results have important clinical implications, as they show that dressings impregnated with both agents could be used to clear *S. aureus* from infected wounds.

Some studies revealed that lysostaphin can also act synergistically with different antibiotics both *in vitro* and *in vivo*. For example, synergy was shown for a combination of lysostaphin and different β-lactam antibiotics against oxacillin-resistant *S. aureus* (ORSA) and oxacillin-resistant *S. epidermidis in vitro* and in a rabbit model of endocarditis (Climo et al. 2001; Kiri et al. 2002). Moreover, lysostaphin exhibited synergistic bactericidal activity with β-lactam antibiotics, bacitracin, and polymyxin B against both MSSA and MRSA *in vitro* (Polak et al. 1993). In another study, lysostaphin was found to act synergistically with gentamicin against MRSA *in vitro* (LaPlante 2007).

2.1.1.4. Immunogenicity   One of the major problems associated with the therapeutic use of proteins is their immunogenicity, resulting largely from the induction of a specific humoral response (De Groot and Scott 2007). Specific antibodies could decrease the therapeutic effectiveness of lytic enzymes particularly following their repeated injection. However, some studies have unexpectedly shown that such antibodies in fact do not neutralize, but at most moderately reduce the antibacterial activity of lysostaphin *in vivo* (Climo et al. 1998; Dajcs et al. 2002).

It is noteworthy that similar results were obtained in experiments involving other lytic enzymes that exerted significant antibacterial activity both *in vitro* and *in vivo* in spite of the presence of neutralizing antibodies (for details, see Chapter 1).

It is also very important that protein immunogenicity can be reduced by means of conjugation of polyethylene glycol (PEG) molecules (PEGylation), resulting in an improvement in pharmacokinetic features. The beneficial effects of PEGylation of proteins include, among others, a decrease in their ultrafiltration and uptake by dendritic cells as well as blocking access of proteolytic enzymes and specific antibodies. PEGylation was also evaluated as a potential means of improving lysostaphin pharmacokinetics (Walsh et al. 2003). Although PEGylated lysostaphin could still lyse staphylococci, its lytic capacity was decreased compared with unmodified enzyme. However, lysostaphin with a lower degree of PEGylation was capable of lysing *S. aureus* cells equally effectively as, though slightly more slowly than, non-PEGylated enzyme. Predictably, PEG conjugation considerably improved the pharmacokinetic features of lysostaphin, especially the serum half-life and the total serum drug concentration. Furthermore, PEGylated lysostaphin had an even more than 10-fold reduced binding affinity to specific antibodies compared with unconjugated enzyme. However, to what extent the improved pharmacokinetics of lysostaphin will translate into its more efficient *in vivo* antibacterial activity remains to be verified.

2.1.1.5. Resistance to Lysostaphin   Thus far, four mechanisms have been identified that may account for the development of staphylococcal resistance to lysostaphin. The first results from mutations in the *femA* gene. Functional inactivation of FemA leads to the formation of mono-glycine cross-bridges in the staphylococcal peptidoglycan, which make bacteria insensitive to lysostaphin. However, such mutants become sensitive to methicillin and hypersusceptible to other classes of antibi-otics (Kusuma et al. 2007). In a rabbit model of ORSA endocarditis, coadministration of nafcillin suppressed the development of lysostaphin resistance (Climo et al. 2001). Furthermore, the emergence of resis-tance dependent on this mechanism results in a reduction in bacterial fitness and virulence both *in vitro* and *in vivo* (Kusuma et al. 2007).

A decrease in staphylococcal susceptibility to lysostaphin may also be mediated by mutations in the *lyrA* gene (Gründling et al. 2006). While the function of LyrA protein has not been identified as yet, it was shown that the disruption of the *lyrA* gene results in a fourfold increase in the lysostaphin MIC value.

The third mechanism associated with lysostaphin resistance is based on the incorporation of amino acids other than glycine into the cross-bridges (Kusuma et al. 2007). This modification eliminates the glycine–glycine bonds, the natural target of lysostaphin, from the peptidoglycan. The incorporation of other amino acids, especially serine, into the cross-bridges can be mediated by three different plasmid genes (*lif*, *epr*, and *epr*-like genes found in *S. simulans* biovar *staphylolyticus*, *S. capitis* EPK1, and *Staphylococcus sciuri*, respectively).

Lysostaphin resistance may also be mediated by mutations in the *sspC* gene encoding an inhibitor of the secreted protease staphopain B (Shaw et al. 2005). However, the exact mechanism accounting for the development of lysostaphin resistance following functional inactivation of SspC has not been elucidated.

2.1.2. Other Bacteriocins Aside from *S. simulans*, other bacterial species can also produce staphylolytic enzymes. These include, among others, *Pseudomonas aeruginosa*, *Aeromonas hydrophilia*, *Streptomyces griseus*, *S. capitis*, and *S. epidermidis* (Sugai et al. 1997). It appears that at least some of these could be used as anti-staphylococcal agents. An example of such an enzyme is LasA protease, also termed staphyloly-sin. This enzyme is produced by *P. aeruginosa* and its efficacy was shown in a rabbit model of both MSSA and MRSA keratitis (Barequet et al. 2004).

2.2. Bacteriophage Lysins

Endolysins or lysins make up another class of anti-staphylococcal lytic enzymes (Borysowski et al. 2006; Fischetti 2008). These are dsDNA bacteriophage-encoded peptidoglycan hydrolases that are produced in phage-infected bacterial cells to degrade peptidoglycan, thereby ensuring the release of progeny virions (lysins are described in more detail in Chapter 1). Below are characterized lysins derived from staphylococcal phages, including MV-L, LysK, PlyTW, Ply187, and *S. aureus* Ø11 phage lysin.

MV-L is a lysin encoded by *S. aureus* phage ØMR11 (Rashel et al. 2007). This enzyme contains two enzymatic domains, the first of which is homologous to the cysteine, histidine-dependent aminohydrolases/peptidases (CHAP) domain (endopeptidase) and the other to the Ami-2 domain (amidase). *In vitro*, MV-L rapidly lysed all 13 *S. aureus* strains tested, including nine MRSA strains and a lysostaphin-resistant MSSA strain (MRSA and MSSA strains were lysed with comparable efficacy). While MV-L was also capable of lysing two VRSA strains and seven VISA strains under growing conditions, the other 11 VISA strains could only be lysed under nongrowing conditions. Importantly, a synergistic effect against a VISA strain was noted for the combination of MV-L and vancomycin. Interestingly, unlike lysostaphin, MV-L did not display activity against *S. epidermidis*. The inability of MV-L to kill *S. epidermidis* may prove advantageous because this species can inhibit the growth of *S. aureus*. In a murine model of nasal colonization, one topical dose of lysin administered to mice 60 h after intranasal MRSA inoculation completely cleared bacteria in one of nine mice and substantially reduced the colonization in the remaining eight animals. In a murine model of MRSA bacteremia, injection of a single dose of MV-L 60 min after a lethal dose of MRSA reduced the mortality rate to 40%, whereas administration of the enzyme either simultaneously with or 30 min after MRSA rescued 100% of the mice.

Another anti-staphylococcal lytic enzyme is lysin encoded by *S. aureus* Ø11 bacteriophage (Navarre et al. 1999; Donovan et al. 2006a). This enzyme is composed of an N-terminal CHAP domain, an amidase domain, and a C-terminal cell wall-binding SH3b domain, all of which are required for maximal lytic activity. Deletion of either enzymatic domain decreased the efficiency of bacterial lysis, whereas deletion of the cell wall-binding domain practically abolished the antibacterial activity of the enzyme. Ø11 lysin was also found to be capable of eliminating the biofilm of *S. aureus*, but not that of *S. epidermidis*, on artificial surfaces (Sass and Bierbaum 2007). The effects of Ø11 against both staphylococcal cells and *S. aureus* biofilm were essentially comparable to those of lysostaphin. In the turbidity reduction assay, the antibacterial range of the lysin was shown to encompass several major staphylococcal mastitis pathogens, including both *S. aureus* and coagulase-negative staphylococci (*Staphylococcus chromogenes*, *S. epidermidis*, *Staphylococcus hyicus*, *S. simulans*, *S. warneri*, and *Staphylococcus xylosus*). Since the enzyme is considered a potential means of treating bovine mastitis, it is very important that it retains its antibacterial activity at the physiological pH and Ca^{2+} concentration found in bovine milk.

LysK is a lysin derived from the polyvalent staphylococcal phage K (O'Flaherty et al. 2005). The antibacterial range of this enzyme was restricted to *S. aureus* and coagulase-negative staphylococci. Importantly, the enzyme was capable of killing, though less efficiently, MRSA, heteroVRSA (hVRSA), and VRSA, as well as teicoplanin-resistant staphylococcal strains.

PlyTW is an amidase encoded by the *S. aureus* bacteriophage Twort (Loessner et al. 1998). It is an interesting enzyme in that its full-length form has a substantially lower lytic activity than that of the truncated N-terminal fragment, even though it is the C-terminal domain that is responsible for cell wall binding. In this regard, it is noteworthy that some lysins encoded by phages specific to other bacterial species are also more active in their N-truncated forms. These findings imply that the activity of at least some anti-staphylococcal lytic enzymes could be increased relatively easily by removing their C-terminal domains.

Another anti-staphylococcal lysin is Ply187 encoded by *S. aureus* bacteriophage 187 (Loessner et al. 1999). The N-terminus of this lysin contains a putative amidase domain, whereas the exact function of the C-terminal domain is unknown. Despite displaying homology to the putative glucosaminidase domains of the Atl staphylococcal autolysins, it was shown to have no enzymatic activity. It is unlikely to be an essential cell wall-targeting domain either, because the C-truncated form of the enzyme exhibits higher lytic activity than its intact form. It is possible that Ply187, like the Atl autolysins, is posttranslationally cloven into two separate polypeptides.

LysH5 is a lysin derived from *S. aureus* bacteriophage ΦH5 (Obeso et al. 2008). This enzyme contains three putative domains: an N-terminal CHAP domain, an amidase domain situated in the central part of the protein, and a C-terminal cell wall-binding SH3b domain. LysH5 exhibited maximal lytic activity in a pH range between 6 and 7 and in a temperature range between 30 and 45 °C. In turbidity reduction assays, all of the 52 *S. aureus* strains tested of both bovine and human origin as well as 25 clinical *S. epidermidis* strains were found susceptible to LysH5 lytic activity. Interestingly, strains of bovine origin isolated from animals with mastitis were lysed more efficiently than clinical isolates of human origin. The enzyme was reported as a potential biopreservative specifically preventing *S. aureus* growth in milk and dairy products. Therefore, its antibacterial activity was tested in milk. At both higher and lower *S. aureus* contamination levels of milk (10^6 and 10^3 cfu/ml, respectively), LysH5 reduced staphylococcal counts to undetectable levels within 4 h. In the context of potential application of LysH5 as a biopreservative, it is very important that its lytic activity is specific to staphylococci and the enzyme is not capable of lysing any other bacteria, including lactic acid bacteria.

All of the above-discussed lysins are encoded by *S. aureus* phages. However, phages of coagulase-negative staphylococci can also be a source of new lytic enzymes. For instance, endolysin genes were identified in two recently sequenced *S. epidermidis* bacteriophages (Daniel et al. 2007). It is very likely that specific lysins could be obtained for all major coagulase-negative staphylococcal species.

Aside from lysins, bacteriophages encode peptidoglycan hydrolases that are integral components of the bacteriophage virion. The major task of these enzymes is to locally degrade peptidoglycan to enable the phage genome to be injected into the bacterial cell. An example of such an enzyme is the minor structural protein 17 of the *S. aureus* phage P68 (Takac and Blasi 2005). Of the 35 clinical *S. aureus* isolates examined (16 of which were oxacillin-resistant), 26 were susceptible to protein 17's lytic effect. Of the 16 oxacillin-resistant strains, 15 were sensitive. These results suggest that these enzymes could also be used as antistaphylococcal agents.

2.3. Chimeric Lytic Enzymes

The vast majority of lytic enzymes are composed of separate enzymatic domains and a cell wall-binding domain. This enables individual domains of different enzymes to be exchanged to construct chimeric enzymes with new features (e.g., a broader antibacterial range). For instance, Manoharadas et al. (2009) coupled the N-terminal enzymatic CHAP domain derived from *S. aureus* phage P68 endolysin (Lys16) to the C-terminal cell wall-binding domain of the P17 minor coat protein (a component of the P68 phage virion) (Manoharadas et al. 2009). This chimeric enzyme was developed because Lys16 in its intact form could not be purified due to poor solubility. Unlike Lys16, the chimeric P16-17 was soluble and successfully lysed *S. aureus in vitro*. Moreover, the enzyme augmented the antimicrobial activity of gentamicin *in vitro*, most likely by increasing the penetration of this antibiotic through the staphylococcal cell wall, which enabled the effective dose of gentamicin to be reduced.

Two other chimeric lytic enzymes were recently developed by Donovan et al., who fused the mature form of lysostaphin to either the native B30 lysin (an anti-streptococcal enzyme) or its N-terminal enzymatic CHAP domain. Both fusion proteins were found to maintain antibacterial ranges typical of the parental enzymes, as they could lyse *S. aureus* and several streptococcal species, including *Streptococcus agalactiae*, *Streptococcus dysgalactiae*, and *Streptococcus uberis* (the four major etiologic agents of bovine mastitis). To evaluate the capability of the fusion constructs to kill bacteria in milk, the enzymes were preincubated for 30 min at 37 °C with whey proteins and their antibacterial activity was subsequently measured in the turbidity reduction assay (whole milk could not be used in this experiment because it interfered with the measurement of optical density in the assays). It was found that preincubation with whey proteins resulted in a substantial decrease in lytic activity of both constructs against *S. aureus* and *S. agalactiae*. Interestingly, only a slight reduction of antibacterial activity was noted for the native form of lysostaphin. While both constructs also lysed some lactic acid bacteria used in the production of cheese and yogurt, they were completely inactivated by pasteurization (63 °C for 30 min), as was the native form of B30 lysin. Therefore, the capability of the B30 constructs to kill some cheese-making bacteria in fact does not preclude their use as a means of preventing or treating mastitis in dairy cattle (Donovan et al. 2006b).

3. CONCLUDING REMARKS

The growing prevalence of antibiotic resistance in staphylococci has generated considerable interest in the use of lytic enzymes as novel anti-staphylococcal agents. The most representative examples of anti-staphylococcal lytic enzymes are lysostaphin and staphylococcal phage lysins. In view of their unique capacity to rapidly cleave staphylococcal peptidoglycan, these enzymes constitute very efficient anti-staphylococcal agents capable of killing both *S. aureus* and coagulase-negative staphylococci. Of paramount importance is that lytic enzymes can also lyse MRSA and *S. aureus* strains with reduced susceptibility to vancomycin. Interestingly, in at least some cases, the development of lysostaphin resistance in MRSA results in an increase in susceptibility to different antibiotics, including methicillin, and a decrease in staphylococcal fitness and virulence. Coadministration of a lytic enzyme with an antibiotic can reduce the risk of developing lysostaphin resistance and increase the efficacy of treatment. An important clinical application of lytic enzymes may be different forms of prophylaxis of staphylococcal infections, especially the elimination of bacteria colonizing nasal mucous membrane and the prevention of catheter colonization. A number of experimental studies have also shown high efficacy of lytic enzymes in the treatment of both topical and systemic staphylococcal infections even in immunized animals.

ACKNOWLEDGMENTS

This work was supported by the Ministry of Science and Higher Education Grant No. 2 P05B 012 30 and Warsaw Medical University Intramural Grants 1MG/W1/08 and 1MG/N/2008.

REFERENCES

Appelbaum P. C. (2007) *International Journal of Antimicrobial Agents* **30**, 398–408.

Baba T. and O. Schneewind (1996) *The EMBO Journal* **15**, 4789–4797.

Barequet I. S., G. J. Ben Simon, M. Safrin, D. E. Ohman, and E. Kessler (2004) *Antimicrobial Agents and Chemotherapy* **48**, 1681–1687.

Becker S. C., J. Foster-Frey, and D. M. Donovan (2008) *FEMS Microbiology Letters* **287**, 185–191.

Borysowski J., B. Weber-Dabrowska, and A. Gorski (2006) *Experimental Biology and Medicine* **231**, 366–377.

Boucher H. W. and G. R. Corey (2008) *Clinical Infectious Diseases* **46 (Suppl. 5)**, S344–S349.

Cisani G., P. E. Varaldo, G. Grazi, and O. Soro (1982) *Antimicrobial Agents and Chemotherapy* **1982**, 531–535.

Climo M. W., K. Ehlert, and G. L. Archer (2001) *Antimicrobial Agents and Chemotherapy* **45**, 1431–1437.

Climo M. W., L. R. Patron, B. P. Goldstein, and G. L. Archer (1998) *Antimicrobial Agents and Chemotherapy* **42**, 1355–1360.

Dajcs J. J., E. B. H. Hume, J. M. Moreau, A. R. Caballero, B. M. Cannon, and R. J. O'Callaghan (2000) *Investigative Ophthalmology & Visual Science* **41**, 1432–1436.

Dajcs J. J., B. A. Thibodeaux, D. O. Girgis, M. D. Shaffer, S. M. Delvisco, and R. J. O'Callaghan (2002) *Investigative Ophthalmology & Visual Science* **43**, 3712–3716.

Dajcs J. J., B. A. Thibodeaux, E. B. H. Hume, X. Zheng, G. D. Sloop, and R. J. O'Callaghan (2001) *Current Eye Research* **22**, 451–457.

Daniel A., P. E. Bonnen, and V. A. Fischetti (2007) *Journal of Bacteriology* **189**, 2086–2100.

De Groot A. S. and D. W. Scott (2007) *Trends in Immunology* **28**, 482–490.

Donovan D. M., M. Lardeo, and J. Foster-Frey (2006a) *FEMS Microbiology Letters* **265**, 133–139.

Donovan D. M., S. Dong, W. Garrett, G. M. Rousseau, S. Moineau, and D. G. Pritchard (2006b) *Applied and Environmental Microbiology* **72**, 2988–2996.

Fan W., K. Plaut, A. J. Bramley, J. W. Barlow, and D. E. Kerr (2002) *Journal of Dairy Science* **85**, 1709–1716.

Fan W., K. Plaut, A. J. Bramley, J. W. Barlow, S. A. Mischler, and D. E. Kerr (2004) *Journal of Dairy Science* **85**, 602–608.

Fischetti V. A. (2008) *Current Opinion in Microbiology* **11**, 393–400.

Francius G., O. Domenecho, M. P. Mingeot-Leclercq, and Y. F. Dufrene (2008) *Journal of Bacteriology* **190**, 7904–7909.

Graham S. and P. J. Coote (2007) *Journal of Antimicrobial Chemotherapy* **59**, 759–762.

Gründling A., D. M. Missiakas, and O. Schneewind (2006) *Journal of Bacteriology* **188**, 6286–6297.

Gunn L. C. and J. Hengesh (1969) *Review of Surgery* **26**, 214.

Harris R. L., A. W. Nunnery, and H. D. Riley (1967) *Antimicrobial Agents and Chemotherapy* **11**, 110–112.

Huan J. N., Y. L. Chen, and S. D. Ge. (1994) *Zhonghua Wai Ke Za Zhi* **32**, 244–245 (article in Chinese).

Kerr D. E., K. Plaut, A. J. Bramley, C. M. Williamson, A. J. Lax, K. Moore, K. D. Wells, and R. J. Wall (2001) *Nature Biotechnology* **19**, 66–70.

Kiri N., G. Archer, and M. W. Climo (2002) *Antimicrobial Agents and Chemotherapy* **46**, 2017–2020.

Kokai-Kun J. F., T. Chanturiya, and J. J. Mond (2007) *Journal of Antimicrobial Chemotherapy* **60**, 1051–1059.

Kokai-Kun J. F., S. M. Walsh, T. Chanturiya, and J. J. Mond (2003) *Antimicrobial Agents and Chemotherapy* **47**, 1589–1597.

Kumar J. K. (2008) *Applied Microbiology Biotechnology* **80**, 555–561.

Kusuma C., A. Jadanova, T. Chanturiya, and J. F. Kokai-Kun (2007) *Antimicrobial Agents and Chemotherapy* **51**, 475–482.

Kusuma C. M. and J. F. Kokai-Kun (2005) *Antimicrobial Agents and Chemotherapy* **49**, 3256–3263.

LaPlante K. L. (2007) *Diagnostic Microbiology and Infectious Disease* **57**, 413–418.

Lee Y.-L., T. Cesario, A. Pax, C. Tran, A. Ghouri, and L. Thrupp (1999) *Age and Ageing* **28**, 229–232.

Loessner M. J., S. Gaeng, and S. Scherer (1999) *Journal of Bacteriology* **181**, 4452–4460.

Loessner M. J., S. Gaeng, G. Wendlinger, S. K. Maier, and S. Scherer (1998) *FEMS Microbiology Letters* **162**, 265–274.

Manoharadas S., A. Witte, and U. Blasi (2009) *Journal of Biotechnology* **139**, 118–123.

Martin R. R. and A. White (1967) *The Journal of Laboratory and Clinical Medicine* **70**, 1–8.

McCormick C. C., J. J. Dajcs, J. M. Reed, and M. E. Marquart (2006) *Current Eye Research* **31**, 225–230.

Moreillon P. (2008) *Clinical Microbiology and Infection* **14 (Suppl. 3)**, 32–41.

Navarre W. W., H. Ton-That, K. F. Faull, and O. Schneewind (1999) *The Journal of Biological Chemistry* **274**, 15847–15856.

Nes I. F., D. B. Diep, and H. Holo (2007) *Journal of Bacteriology* **189**, 1189–1198.

Obeso J. M., B. Martinez, A. Rodriguez, and P. Garcia (2008) *International Journal of Food Microbiology* **128**, 212–218.

O'Flaherty S., A. Coffey, W. Meaney, G. F. Fitzgerald, and R. P. Ross (2005) *Journal of Bacteriology* **187**, 7161–7164.

Oldham E. R. and M. J. Daley (1991) *Journal of Dairy Science* **74**, 4175–4182.

Oluola O., L. Kong, M. Fein, and L. E. Weigman (2007) *Antimicrobial Agents and Chemotherapy* **51**, 2198–2200.

Parisien A., B. Allain, J. Zhang, R. Mandeville, and C. Q. Lan (2008) *Journal of Applied Microbiology* **104**, 1–13.

Patron R. L., M. W. Climo, B. P. Goldstein, and G. L. Archer (1999) *Antimicrobial Agents and Chemotherapy* **43**, 1754–1755.

Polak J., L. P. Della, and P. Blachburn (1993) *Diagnostic Microbiology and Infectious Disease* **17**, 265–270.

Quickel K. E. Jr., R. Selden, J. R. Caldwell, N. F. Nora, and W. Schaffner (1971) *Applied Microbiology* **22**, 446–450.

Rashel M., J. Uchiyama, T. Ujihara, Y. Uehara, S. Kuramoto, S. Sugihara, K. Yagyu, A. Muraoka, M. Sugai, K. Hiramatsu, K. Honke, and S. Matsuzaki (2007) *Journal of Infectious Diseases* **196**, 1237–1247.

Sass P. and G. Bierbaum (2007) *Applied and Environmental Microbiology* **73**, 347–352.

Schindler C. A. and V. T. Schuhardt (1964) *Proceedings of the National Academy of Sciences of the USA* **51**, 414–420.

Shah A., J. Mond, and S. Walsh (2004) *Antimicrobial Agents and Chemotherapy* **48**, 2704–2707.

Shaw L. N., E. Golonka, G. Szmyd, S. J. Foster, J. Travis, and J. Potempa (2005) *Journal of Bacteriology* **187**, 1751–1762.

Sugai M., T. Fujiwara, T. Akiyama, M. Ohara, H. Komatsuzawa, S. Inoue, and H. Suginaka (1997) *Journal of Bacteriology* **179**, 1193–1202.

Takac M. and U. Blasi (2005) *Antimicrobial Agents and Chemotherapy* **49**, 2934–2940.

Thumm G. and F. Götz (1997) *Molecular Microbiology* **23**, 1251–1265.

von Eiff C., J. F. Kokai-Kun, K. Becker, and G. Peters (2003) *Antimicrobial Agents and Chemotherapy* **47**, 3613–3615.

Wall R. J., A. M. Powell, M. J. Paape, D. E. Kerr, D. D. Bannerman, V. G. Pursel, K. D. Wells, N. Talbot, and H. W. Hawk (2005) *Nature Biotechnology* **23**, 445–451.

Walsh S., J. Kokai-Kun, A. Shah, and J. Mond (2004) *Pharmaceutical Research* **21**, 1770–1775.

Walsh S., A. Shah, and J. Mond (2003) *Antimicrobial Agents and Chemotherapy* **47**, 554–558.

Wertheim H. F., D. C. Melles, M. C. Vos, W. van Leeuwen, A. van Belkum, H. A. Verbrugh, and J. L. Nouwen (2005) *The Lancet Infectious Diseases* **5**, 751–762.

Wu J. A., C. Kusuma, J. J. Mond, and J. F. Kokai-Kun (2003) *Antimicrobial Agents and Chemotherapy* **47**, 3407–3414.

Yang X.-Y., C.-R. Li, R.-H. Lou, Y-M. Wang, W.-X. Zhang, H.-Z. Chen, Q.-S. Huang, Y.-X. Han, J.-D. Jiang, and X.-F. You (2007) *Journal of Medical Microbiology* **56**, 71–76.

Yano M. Y., Y. Doki, M. Inoue, T. Tsujinaka, H. Shiozaki, and M. Moden (2000) *Surgery Today* **30**, 16–21.

Zygmunt W. A., H. P. Browder, and P. A. Tavormina (1968) *Applied Microbiology* **16**, 1168–1173.

CHAPTER 8

MEMBRANE-TARGETED ENZYBIOTICS

MARÍA GASSET

Instituto de Química-Física "Rocasolano," Consejo Superior de Investigaciones Científicas, Madrid, Spain

1. INTRODUCTION

Coevolution of host and pathogen has dictated a variety of complex survival mechanisms in the host that are either germline encoded (innate immunity) or acquired (adaptative immunity). Innate immunity embodies the battery of host weapons designed to react fast, almost instantly, on invasion. The existence of antimicrobial activities in fluids and tissues of animals was recognized as early as around 1850, but it was not until the period between 1920 and 1950 that some of the substances were isolated and assayed against Gram-negative and Gram-positive bacteria (Skarnes and Watson 1957). The following research steps led to the establishment of cutoff properties and the biological sources of the antimicrobial substances, which together with isolation and purification efforts yielded the first group of antimicrobial peptides (AMPs). In this line, Bowman isolated and purified the insect cecropin (Steiner et al. 1981); Zasloff (1987) succeeded with the amphibian manganin, and Lehrer achieved the purification of human defensin (Ganz et al. 1990). Now, more than 800 AMPs have been isolated, purified, and characterized, both structurally and functionally, and constitute the battery of host defense peptides (Boman 1995, 2003; Hancock 1997; Andreu and Rivas 1998; Vizioli and Salzet 2002; Zasloff 2002; Brogden et al. 2003; Lehrer 2004; Brogden 2005; Peschel and Sahl 2006).

Host defense peptides are endogenous molecules that are mobilized shortly after microbial infection and act rapidly to neutralize a broad range of pathogens (Fig. 8.1). These peptides, commonly referred to as

Enzybiotics: Antibiotic Enzymes as Drugs and Therapeutics. Edited by Tomas G. Villa and Patricia Veiga-Crespo
Copyright © 2010 John Wiley & Sons, Inc.

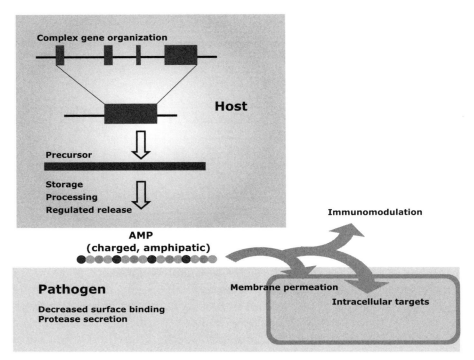

Figure 8.1. Mechanism of action of AMPs. AMPs constitute a complex and ancient battery of weapons tailored for best efficiency against pathogens. AMPs are gene encoded and synthesized as inactive pre-pro-forms. The presence of the pre- and pro-signals dictates the targeting and the infection-induced proteolytic maturation for optimal active unit production. Coevolution with pathogen survival strategies (production of proteases, charge neutralization, etc.) has probably dictated their flexibility to interact and to interfere with a wide range of molecules expanding the number and nature of targets and subsequently their ultimate functionality.

AMPs, exhibit bactericidal, fungicidal, virucidal, and tumoricidal properties (Table 8.1). Importantly, as opposed to other fields of biological science, AMPs have been in the reign of protein chemistry (isolation, purification, sequencing, and solid-phase synthesis), and the study of their gene organization, expression, and processing has been delayed in time (Boman 2003). AMPs are a class of antimicrobial agents with promising applications across the spectrum of human and animal disease. They do not simply target a single species of microbial pathogen but work on two fronts. First, AMPs can kill a broad range of microbial pathogens directly by perturbation of their membranes or by permeating them and attacking an internal target. Second, AMPs can selectively modulate innate host defenses. Such an integrated mode of

TABLE 8.1. AMP Classification

Classes	Name	Sequence	Activity	Entry
Anionic	Maximin H5	ILGPVLGLVSDTLDDVLGIL	B	AP00497
	Anionic peptide SAAP	DDDDDD	B	AP00528
	Gm anionic pept 1	EADEPLWLYKGDNIERAPTTADHPILPSIIDDV KLDPNRRYA	B,F	AP00749
	Gm anionic pept2	ETESTPDYLKNIQQQLEEYTKNFNTQVQNAFD SDKIKSEVNNFIESLGKILNTEKKEAPK	B,F	AP00754
Cationic Linear α-helical	Cecropin A	RWKVFKKIEKVGRNIRDGVIKAAPAIEVLGQA KAL	B,F	AP00127
	Magainin 1	GIGKFLHSAGKFGKAFVGEIMKS	B	AP00771
	Buforin II	TRSSRAGLQFPVGRVHRLLRK	B,F	AP00308
	Brevinin-I	FLPVLAGIAAKVVPALFCKITKKC	B,V,C	AP00074
	Esculentin-IA	GIFSKLGRKKIKNLLISGLKNVGK EVGMDVVRTGIDIAGCKIKGEC	B,V,C	AP00080
	Dermaseptin B2	GLWSKIKEVGKEAAKAAAKAAGKAALGAVS EAV	B,F	AP00001
	Pleurocidin	GWGSFFKKAAHVGKHVGKAALTHYL	B	AP00166
	Seminalplasmin	SDEKASPDKHHRFSLSRYAKLANRLANPKLLE TFLSKWIGDRGNRSV	B;F;C	AP00234
	SMAP-29	RGLRRLGRKIAHGVKKYGPTVLRIIRIAG	B,F	AP00155
	CAP18	GLRKRLRKFRNKIKEKLKKIGQKIQGFVPKLA PRTDY	B	AP00418
	LL37	LLGDFFRKSKEKIGKEFKRIVQRIKDFLRNLV PRTES	B,V,C,L	AP00310

TABLE 8.1 *Continued*

Classes	Name	Sequence	Activity	Entry
Enriched in specific amino acids	Abaecin (Pro-rich)	YVPLPNVPQPGRRPFPTFPGQGPFNPKIKW PQGY	B	AP00002
	Drosocin (Pro- and Arg-rich)	GKPRPYSPRPTSHPRPIRV	B	AP00172
	Prophenin (Pro- and Phe-rich)	FPPPNVPGPRFPPPNFPGPRFPPPNFPGPRFPPPN FPGPRFPPPNFPGPPFPPPIFPGPWFPPPPPFRP PPFGPPRFP	B	AP00689
	Hymenoptaecin (Gly-rich)	EFRGSIVIQGTKEGKSRPSLDIDYKQRVYDKN GMTGDAYGGLNIRPGQPSRQHAGFEFGKE YKNGFIKGQSEVQRGPGGRLSPYFGINGGF RF	B	AP01213
	Holotricin (Gly- and Pro-rich)	YGPGDGHGGGHGGGHGGGHGNGQGGGHG HGPGGGFGGGHGGGHGGGGRGGGGSGG GGSPGHGAGGGYPGGHGGGHHGGYQTH GY	F	AP00404
	Indolicidin (Trp-rich)	ILPWKWPWWPWRR	B,F,V	AP00150
	Histatin (His-rich)	RKFHEKHHSHREFPFYGDYGSNYLYDN	B,F	AP00799
Containing disulfide bonds	Brevinin (1-disulfide bond)	FLPVLAGIAAKVVPALFCKITKKC	B,V, C	AP00074
	Protegrin (2-disulfide bonds)	RGGRLCYCRRRFCVCVGR	B,V,F	AP00195
	β-defensin HNP-1 (3-disulfide bonds)	ACYCRIPACIAGERRYGTCIYQGRLWAFCC	B,F,V,C	AP00176

AMPs have been classified attending to structural features as charge, structure, dominance of a specific amino acid, and presence of a number of disulfide bonds. This classification allows the dual location of some AMPs (clavanin A can be placed both in the linear cationic α-helical groups and also in the His- and Phe-rich peptides). A new and growing class of AMPs has been added to include those peptide sequences sharing the function and produced in the processing of unrelated proteins. Examples for each class are included and referred to as the entry code of the AMP database (http://aps.unmc.edu/AP/main.php). The activities recorded for the selected examples are marked as B, antibacterial; C, cytotoxic; F, antifungal; V, antiviral; L, lectin-like (LPS binding).

action is without question an attractive alternative to the standard use of antimicrobials. The potential application of AMPs as novel therapeutic antibiotics fueled an enormous body of research. Hundreds of artificial AMPs have been synthesized, characterized, and investigated by biophysical and biochemical means. Various AMPs have shown potential as therapeutics for sepsis and bacterial infections, although the practical use of these peptides *in vivo* still presents problems that must be overcome.

2. CLASSES OF AMPS: CATEGORIES FROM STRUCTURAL FEATURES

AMPs are gene encoded and are expressed in mammals in a variety of cells as mast cells, neutrophiles, monocyte/macrophages, epithelial cells, and keratinocytes. They are usually synthesized as a larger precursor that requires one or more proteolytic activation steps for the release of the biologically active and mature peptide. AMPs comprise a diverse group of short peptides (6–59 amino acids), with an overall charge (mostly positive but in some others negative) and a substantial proportion of hydrophobic residues (often $\geq 30\%$). There is an enormous diversity of sequences and similarity is found only within defined groups of AMPs from closely related species. The immense diversity of AMPs within a given species, which has been argued to be a result of their function as well as the difference in pathogenic challenges, anticipates that these peptides may have overlapping roles. Thus, it is unsurprising that animal models engineered for lacking a single one of these peptides may not display a strong phenotype of alteration (Huang et al. 2007; Wang et al. 2008).

AMPs are categorized by their function, but they are classified according to their structural features. The parameters used for classification are the net and type of charge, the predominance of a specific amino acid in their sequences, the presence or absence of disulfide bonds, and their number and the secondary structure, which are depicted in Table 8.1 together with some examples.

2.1. Anionic AMPs

The first group of AMPs is composed of those peptides exhibiting negative charge. These peptides have very short sequences (Table 8.1). Among them as an example are the short peptides similar to the charge-neutralizing pro-peptides of zymogens found in the bronchoalveolar lavage of different mammalians (Brogden 2005).

2.2. Cationic Linear α-Helical AMPs

Linear sequences displaying a net positive charge and the capacity to fold into α-helical structures constitute the second class of AMPs, which contains the largest number of members. Members of this group exhibit a net positive charge for the recognition and interaction with polyanions. Their sequence also displays an accentuated amphiphilicity shown by the distribution of a basic residue every other four (i, i + 3 distribution). This sequence organization permits conformational transitions from disorder states in their free form to a helical fold upon interactions allowing the manifestation of the amphiphilicity (salt, counter ion partners, etc.).

2.3. AMPs with Sequences Featured by the Relative Abundance of Specific Residues

The second group is formed by those AMPs that have the dominance of a specific amino acid in their sequence. Depending on the dominant amino acid, these AMPs are classified into several subgroups. Peptides from this group are linear and lack a preferential folding tendency. For example, histatin is highly rich in His residues; tripticin and indolicin are rich in Trp; diptericins and attacins are rich in Gly; drosocin, apidaecin, formaecin, PR-39, and prophenin are rich in Pro (Brogden 2005; Li et al. 2006).

2.4. Disulfide Bond Containing AMPs

The presence of both disulfide bonds and a folding with stable β-strands are the molecular descriptors for the fourth group of AMPs. This group accounts for cyclic and open-ended cyclic AMPs that contain from two to eight Cys residues engaged in disulfide bridges. Depending on the pairing of the Cys residues, these AMPs may adopt a β-sheet conformation with three strands (most vertebrate defensins), β-hairpin-like (thanatin, androctonin, gomesin, tachyplesin, brevinins, and esculetins), or mixed α-helical/β-sheet structures (plant, invertebrate, and some mammalian defensins) (Bulet et al. 2004; Lehrer 2004).

2.5. AMPs Resulting from the Proteolytic Processing of Unrelated Proteins

Given the structural versatility of AMPs, the presence of similar sequences embedded in larger and unrelated proteins that could be released under proteolysis was considered as an additional source of these compounds. These considerations led to the addition of a new class gathering those peptides that are fragments of larger proteins with unrelated functions (carriers, storage, etc.). Members of this AMP group are lactoferricin derived from lactoferrin, casocidin 1 produced from casein, and fragments from ovoalbumin and α-lactoalbumin, among others (Zucht et al. 1995; Pellegrini 2003; Haug et al. 2007).

2.6. Precursor Design Groups Different AMP Classes: The Cathelicidin Family

Analysis of the precursors from which AMPs are released led to the discovery of common patterns of organization. Cathelicidins are a family of vertebrate-specific AMP precursors exhibiting unique bipartite features, which are identified by an evolutionarily conserved N-terminal cathelin-like domain (CLD) of 99–114 residues followed by a heterogeneous C-terminal antimicrobial domain (AMD or active AMP) of 12–100 residues (Zanetti 2005). The heterogeneity of the AMD is reflected by its structural diversity, since it includes all three major folding types of AMPs: that is, linear cationic α-helical, Cys-containing peptides with β-sheet folds, and peptide-rich Pro, Arg, and Trp residues.

The cathelicidin conserved construction mirrors the gene organization and the subsequent polypeptide biogenesis. The cathelicidin genes are translated as precursors in the cytoplasmic granules of neutrophils with signal peptides that are removed co-translationally to yield the proforms. These proforms are inactive tandem fusions of the CLD and the AMD (Zhu 2008). Upon microbial infection, the proforms are further proteolytically processed to release the functionally active AMD (AMP) and the CLD with a yet unclear function.

3. AMPS AND THEIR BIOLOGICAL SOURCES: DIVERSITY AT THE DEFENSE FRONT

Given their action in defense, AMPs are present in all living organisms at the defense borders. Interestingly, despite their antimicrobial action, AMPs produced by bacteria were the first members to be isolated and characterized (Hirsch and Mattick 1949). These AMPs, also called bacteriocins, do not protect against infection in the classical sense but contribute to the survival of individual bacterial cells by killing other cells competing for nutrients (Riley 1998). Among bacteriocins, the most studied members are nisin, produced by *Lactococcus lactis*, and mersacidin, produced by *Bacillus* spp., which are commonly used in the food production industry (Chatterjee et al. 1992; Jenssen et al. 2006). Plants are also a source of AMPs, since they constitute a fundamental force in the defense against infection by bacteria and fungi. Plant AMPs are formed entirely by β-sheet-forming peptides, divided into thionins and defensins (García-Olmedo et al. 1998; Broekaert et al. 1995). In invertebrates, which lack the vertebrate adaptive immune system, AMPs are key elements to counteract invading pathogens. AMPs are found in hemolymph, in phagocytic cells, and in certain epithelial cells, and they are either expressed constitutively or induced in response to a pathogen (Lemaitre et al. 1996; Iwanaga and Kawabata 1998; Imler and Bulet 2005). AMPs have been also isolated from a wide variety of vertebrate species, including fish, amphibians, and mammals. In this case, AMPs are found at sites that routinely encounter pathogens, such as the skin and mucosal surfaces, and within the immune cell granules, among other sites (Yang et al. 2004).

3.1. AMPs from Invertebrates

The first α-helical AMP isolated from an insect was a cecropin (Boman and Hultmark 1987). This molecule was isolated from the hemolymph of bacteria-challenged diapausing pupae of the giant silk moth. Since the first report, more than 60 cecropins and cecropin-like molecules have been isolated (Otvos 2000). Briefly, cecropins and cecropin-like molecules from *Diptera* and *Lepidoptera* consist of 29–42 amino acid residues, the shortest being ceratotoxin, a peptide isolated from the female accessory reproductive glands of the dipteran medfly *Ceratitis capitata*, and the longest one being stomoxyn, a 42-residue peptide isolated from the midgut of the cyclical vector of trypanosomes, the stable fly *Stomoxys calcitrans* (Boman 1995; Bulet et al. 2004). With a few exceptions, cecropins from insects have two major characteristics: the presence of a Trp residue in position 1 or 2 and an amidated C-terminal residue, which correlates with the selectivity of their action and the cationicity of the peptide.

Venom glands of invertebrates are also a source of AMPs (Bulet et al. 2004). In insects, the α-helical melittin and cabrolin are the major components of bee venoms. In addition, ponericins were isolated from venom glands of a predatory ant. Two α-helical AMPs named hadrurin and opistoporin were isolated from the venom of scorpions. Spider venoms contribute oxyopinins and cupiennins, the latter with high antimicrobial, hemolytic, and insecticidal activity.

Prochordates contribute to the AMP collection with styelins, clavanins, and halocidin. Styelins (Phe-rich 31- to 32-residue peptides) and clavanins (His-rich 23-residue peptides) isolated from hemocytes and pharyngeal tissues are C-terminally amidated peptides that resemble magainins and cecropins, respectively (Lee et al. 1997; Taylor et al. 2000; Jang et al. 2002). Among the styelins, styelin D displays extensive posttranslational modifications including dihydroxy-Arg, dihydroxy-Lys, 3,4-dihydroxy-Phe and a Br-Trp (Taylor et al. 2000). Interestingly, unlike most of the α-helical AMPs, styelins and clavanins maintain their activity in the presence of a high salt concentration, but they are also rather cytotoxic to eukaryotic cells (Lee et al. 1997). Halocidin is a heterodimer composed of two different subunits containing 18 and 15 amino acid residues linked covalently by a single disulfide bond found in the tunicate *Halocynthia aurantium* (Jang et al. 2002).

The simple disulfide bond stabilized β-hairpin structural organization is not widely distributed among invertebrates, thanatin being the only member documented (Fehlbaum et al. 1996; Bulet et al. 2004). Thanatin, produced by the fat body of experimentally infected bugs, has no sequence similarity with other insect defense peptides, but it is similar to AMPs isolated from frog skin secretions as brevinin. Thanatin includes an N-terminal domain with a large structural variability linked to a well-conformed C-terminal cationic loop or insect box, which is defined by the two Cys residues and the hydrophilic residues localized at the two opposite sites. Interestingly, while a synthetic all-D-enantiomer is inactive against Gram-negative and some Gram-positive bacteria, the all-D-thanatin preserves the antifungal activity of the natural thanatin (all-L), which indicates a certain stereoselective pressure in the antimicrobial activity.

3.2. AMPs from Vertebrates

Different species of fishes have provided a large collection of AMPs (Bulet et al. 2004). The mucosal surface of the skin in specific mucus glands of soles yielded pardaxins, which feature two helical segments linked by a Pro hinge and are effective against Gram-positive and Gram-negative bacteria at μM concentrations with limited hemolytic activity (Lazarovici et al. 1986; Oren and Shai 1996). The skin mucus layer of carp yielded two hydrophobic AMPs of 27 and 31 kDa with ion channel properties (Lemaitre et al. 1996). Parasin is a 19-residue AMP that is derived from the N-terminal processing by a specific protease upon injury of the catfish histone H2A (Park et al. 1998b). Parasin is almost identical to buforin I and to the N-terminal domain of several histone H2A and tends to adopt preferentially extended structures (Park et al. 1988). Its antimicrobial activity is about 12–100 times more potent than that of magainin 2 against a wide range of microorganisms and lacks evident hemolytic activity. The skin secretions of the rainbow trout permitted the isolation of oncorhyncin II and III (Smith et al. 2000; Fernandes et al. 2004). Oncorhyncin II is a 69-residue peptide corresponding to the C-terminal fragment of histone H1, whereas oncorhyncin III is a 6.7-kDa AMP that resembles the nonhistone chromosomal protein H6.

Amphibian skin glands and the stomach mucosa have provided more than 500 members, comprising a true arsenal of AMPs (Rinaldi 2002). Magainins, isolated from the skin secretion of *Xenopus laevi*, are indubitably the prototype of the α-helical AMPs (Zasloff 1987). Other amphibian α-helical AMPs are fragments of proxenopsin and prolevitide hormones, bombinins, dermaseptins, adenoregulin, phylloxin, caerins, frenatins, maculatins, citropins, aureins, kassinateurins, and temporins, among others (Bulet et al. 2004). All these molecules show remarkable differences in sequences, and the diversity of the structures has been used as a taxonomic tool. These AMPs are about 10–30 amino acids long. Approximately 50% of them are C-terminally amidated, while a few of them present a rather nonconventional posttranslational modification, D-amino acids as D-Leu or D-alloIle. Most of these linear AMPs exhibit little secondary structure in water, but in contact with the negatively charged membranes, they easily fold into amphipathic helical structures (Shai 1999). Compared with the linear AMPs from arthropods that adopt a double α-helical secondary structure, most of the amphibian linear peptides adopt a single α-helix. The amphibian AMPs are almost universally synthesized and are stored in the neuroendocrine granular glands in the skin, from where they are released in a holocrine manner following stress, adrenergic stimulation, or injury. Interestingly, some of them are also produced in the gastric mucosa and accumulate in the mucus coating the stomach surface (Bulet et al. 2004). The activity spectrum of the amphibian AMPs is broad. They can kill a large variety of aerobic and anaerobic bacteria, clinical isolates of multiresistant human pathogens, yeast, and filamentous fungi, and they are also efficient against some viruses, tumor cells, and protozoa. However, some of them have a rather limited activity. The efficacy of amphibian AMPs is markedly increased by the considerable synergy existing in their action. Considering their toxicity against erythrocytes, most of them are highly hemolytic. However, this activity is rather low or even nonexistent for some. A detailed study of the relationship between peptide sequence and antimicrobial properties was performed on magainins and other linear small-sized AMPs from amphibians (Bulet et al. 2004). Different from the previous linear amphibian AMPs, a β-hairpin fold was observed for a large number of amphibian AMPs, including brevinins esculentins, gaegurins, rugosins, ranatuerins, and tigerinins (Bulet et al. 2004). Some exhibit a common structural motif named as Rana box, which consists of a C-terminal loop stabilized by a disulfide bond, which, if opened upon reduction of the single disulfide bridge, results in loss of activity.

In mammals, the most abundant AMPs are indubitably the α- and β-defensins. However, several linear AMPs forming α-helices belonging to the cathelicidin family (Zanetti 2005) are also found in neutrophil granules. In these cases, the inactive pro-peptides are released exclusively on neutrophil activation for their processing by an extracellular elastase, allowing the liberation of the C-terminal active domain (Zanetti 2005). Human cathelicidin (hCAT-18/LL-37) is found in secretory granules of neutrophils, the mouth, tongue, esophagus, epithelia, submucosal glands of the airway, and the genitourinary tract (vagina, epididymis, and seminal plasma) and also is expressed in skin keratinocytes during inflammatory disorders such as psoriasis.

The only AMP found in mammals with a cyclic β-sheet scaffold is the dodecapeptide (RLCRIVVIRVCR) bactenecin isolated from bovine neutrophils (Romeo et al. 1988). This highly cationic peptide has a nonapeptide ring between the two Cys residues resembling the Rana box of tigerinins. However, in this case, the reduction of disulfides does not abolish the activity but produces a drastic change in the activity spectrum. The reduced bactenecin shows marked selectivity to Gram-positive bacteria with almost a total loss of activity against Gram-negative strains, whereas the native bactenecin is more active on Gram-negative strains.

The only AMPs isolated from a vertebrate with two disulfide bridges are the cathelicidin protegrins (Zanetti 2005). Five isoforms of protegrins (PG-1–PG-5) were identified under latent state in porcine neutrophil granules. Protegrins share 17% and 25% homology with gomesin and androctonin, respectively. In aqueous solution, PG-1 forms a two-stranded antiparallel β-sheet, with the two strands connected by a β-turn and both N- and C-terminal ends essentially disordered. The presence of the β-sheet and the amphiphilic character are key structural elements for both the stability of the protegrin structure and the biological efficiency. Linear analogs and variants with polar amino acids in the hydrophobic face have a reduced activity, especially in the presence of physiological ionic strength concentrations. An extensive structure–function relationship has been conducted by IntraBiotics

Pharmaceuticals, Inc. (Palo Alto, CA, USA) on several hundred of protegrin analogs to determine the precise relationship between the primary/secondary structures and the antimicrobial properties. From these studies, the analog IB-367 was selected initially for clinical development as a topical agent to prevent the oral mucositis (see below).

Hepcidins, which adopt a β-hairpin-like structure stabilized by four internal disulfide bridges, have been isolated from various species of fish and mammals (Shi and Camus 2006). The first representative, the human hepcidin, also referred to as liver-expressed AMP-1 (LEAP-1), was independently isolated from urine, plasma ultrafiltrate, and liver. All hepcidins have eight Cys residues, engaged in four internal disulfide bridges. Human hepcidin adopts a distorted β-hairpin-like structure with a two-stranded β-sheet stabilized by a Cys scaffold, showing a vicinal disulfide bridge localized at the turn of the β-sheet (Cys1–Cys8, Cys2–Cys7, Cys3–Cys6, and Cys4–Cys5). Interestingly, several experiments in mice, including knockout mice, suggest that hepcidin can play a pivotal role in maintaining iron homeostasis.

Leukocytes and epithelial cells of birds and mammals produce more than 100 vertebrate defensins (Wong et al. 2007). These defensins can be classified in α- and β-subfamilies according to the precursor and gene structures as well as to the connectivity of the six Cys residues of their sequence. α-Defensins with chains of 29–35 residues displayed Cys1–Cys6, Cys2–Cys4, and Cys3–Cys5 disulfide pairing. They are abundant in the azurophil granules of the neutrophils, in certain populations of macrophages, in cytoplasmic granules of small intestinal Paneth cells, and in vaginas, fallopian tubes, and ectocervixes. On the contrary, β-defensins contain 36–42 residues with a Cys1–Cys5, Cys2–Cys4, and Cys3–Cys6 disulfide pairing. The β-defensins are present in epithelial tissues, including genitourinary tract organs, skin, respiratory passages, and alveolar macrophages, and in granulocytes. Both α- and β-defensins have a similar structure based on three β-strands (βββ), including a β-hairpin, which align to form an antiparallel β-sheet. Interestingly, the scaffold αβββ has been reported for human β-defensin-2 and β-defensin-3 (Bulet et al. 2004).

θ-Defensins are purely macrocyclic 18-residue peptides that contain three internal disulfide bridges. RTD-1 was the first θ-defensin to be isolated from neutrophils and monocytes of the rhesus monkey (Tang et al. 1999). Two more θ-defensins, RTD-2 and RTD-3, were later found at a lower abundance than RTD-1 (Tang et al. 2002). RTD-1 to RTD-3 are formed by the head-to-tail ligation of two nonapeptides derived from similar 76-residue, α-defensin-related precursors named demidefensin-1 to demidefensin-3. RTD-1 is formed by the ligation of demidefensin-1 and demidefensin-2, while RTD-2 and RTD-3 are the homodimers of demidefensin-1 and demidefensin-3, respectively.

4. MODES OF ACTION OF AMPS

Although the exact mechanism of action of AMPs, as a family or as singular members, remains a matter of controversy, there is a consensus that they have an eclectic range of functions involving a complex set of indirect protective actions as well as a direct antimicrobial killing activity. The unique nature of AMPs derived from their structural features (amino acid composition, charge, amphipaticity, size, etc.) allows them to easily attack membrane bilayers with the vast majority causing their damage (Powers and Hancock 2003; Zhang and Falla 2006). With very few exceptions, the AMP activities are almost universally dependent on the interaction with the pathogen cell membrane.

4.1. Indirect Roles: Immunomodulatory Activities of AMPs

AMPs have been shown to have a broad range of immunomodulatory properties, including the modulation of pro-inflammatory signals, the expression of hundreds of genes in different cells, the direct chemoattraction of immune cells, and the induction of chemokines (Hancock and Sahl 2006; Mookherjee and Hancock 2007; Eliasson and Egesten 2008; Kolls et al. 2008).

Innate immunity is triggered when conserved bacterial signatures, as for instance the lipopolysaccharide (LPS), interact with host pattern recognition receptors including toll-like receptors (TLRs). After TLR binding, effector mechanisms that assist in the prevention or resolution of modest infections are stimulated. However, the overstimulation of innate immunity leads to the harmful septic syndrome. In this complex landscape of signals, AMPs attenuate the initial triggers of inflammatory reactions through TLR by binding to molecules as LPS, lipopoteichoic acid (LTA), and CpGs (Mookherjee and Hancock 2007 and references therein). In addition to their capacity to modulate inflammatory triggers, AMPs promote chemotaxis of immune effector cells and through their chemokine activity recruit leukocytes to the site of infection (Dürr and Peschel 2002). Also, some AMPs orchestrate a variety of signals involved in enhancing the ability of dendritic cells to capture and present antigens (Davidson et al. 2004), the suppression of apoptosis in neutrophiles prolonging their life span (Nagaoka et al. 2006), and the activation of the expression of genes related with defense reactions (Scott et al. 2002), among others.

In addition to these indirect functions in defense, some AMPs have acquired the capacity to play alternative roles, which expands the functional repertoire and may have important implications in their adaptation for pharmacological use. One example of these alternative roles is the angiogenesis modulation performed by PR39, an AMP that gets across cell membranes (Li et al. 2000).

4.2. Direct Roles: The Killing Action of AMPs

The direct antimicrobial action of AMPs starts with the recognition of the pathogens. For this process, AMPs generally make use of their positive charge and interact with the anionic polymers of the pathogen membrane. For a long time, this interaction was only seen as the obligatory initial step for the irreversible and selective disruption of the cell membranes (Westerhoff et al. 1989; Shai 1995, 1999; Oren and Shai 1998; Huang 2000, 2004). However, in some cases, such membrane interaction takes place, but instead of leading to membrane damage, it results in the passage of the AMP to the intracellular compartment. In these cases, the major action of the AMP is indeed executed intracellularly through the interference with other targets.

4.2.1. Action at the Membrane: Permeability Barrier Disruption
Independent of the existence of additional targets, AMPs start their action at the lipid bilayer of the cell (Brogden 2005; Peschel and Sahl 2006). Virtually any peptide sequence displaying the charge and amphiphilicity of AMPs interacts with membranes and perturbs their isolation properties. Biophysical studies using model membranes have shown that AMPs bind liposomes into physically distinct states and that the parameters featuring this process as lipid requirements (dictating specificity and selectivity of the event), concentration range (determining relative efficiency), kinetic features, and the extent of the effects are highly dependent on the peptide entity (Oren and Shai 1998; Epand and Vogel 1999; Matsuzaki 1999; Shai 1999; Huang 2000; Yang et al. 2001).

Once the peptides are at the membrane phase, they can execute their perturbing action. Conceptually, the permeation of a membrane requires the break of the bilayer continuity and the formation of a transmembrane pore, a hollow cavity across the bilayer with the walls recovered by hydrophilic groups for allowing the passage of hydrophilic compounds (Fig. 8.2). For doing so, peptides must first bind to the surface of the membrane, then associate, and insert into the hydrophobic core of the lipid bilayer to build a hydrophilic cavity. As a function of the magnitude of lipid bilayer damage, the membrane

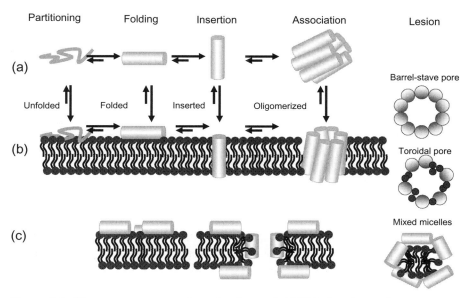

Figure 8.2. Elementary steps and mechanisms of AMP-induced membrane permeation. Peptide amphiphilicity determines the existence of two pathways: (a) one in the aqueous phase and (b,c) another at the membrane phase. Partitioning to the membrane is linked to a folding event that results in a structured peptide lying at the interface of the bilayer. When the surface density of the peptide reaches the critical value, an orientational transition from the I-state to the S-state takes place. Adoption of a transmembrane configuration with varying tilt angles with respect to the bilayer normal endows the definition of a hydrophilic cavity through the hydrophobic core of the bilayer. The nature of the groups involved in the wall lining, only amino acid side chains or amino acid side chains and lipid head groups determine the pore as barrel stave-like or toroidal-like, respectively (b). In other cases, the transmembrane orientation is a simple intermediate in a detergent-like action, which involves the formation of mixed micelles and the complete destruction of the bilayer entity (c).

permeation is described by one of the following mechanisms: carpet-like, barrel-stave, electroporation, or transient sinking-rafting (Oren and Shai 1998; Epand and Vogel 1999; Matsuzaki 1999; Shai 1999; Huang 2000, Yang et al. 2001; Brogden 2005). Despite the mechanism used, peptide binding at the surface produces a membrane tension through the expansion of the membrane area and the thinning of the acyl chain region (Huang 2000). In fact, the bilayer thickness decreases with the peptide surface density until reaching a critical value above which the permeability barrier breaks. This critical density correlates with a change in orientation of the peptide with respect to the normal membrane (from perpendicular to parallel), which is known as the I-state to the S-state transition (Huang 2000; Huang et al. 2004).

Peptides acting according to the carpet-like mechanism bind massively to the membrane surface. The massive binding causes the disruption of the interfacial properties that eventually triggers drastic vesicle damage. This mechanism, also referred to as detergent-like, mimics the micellarization of bilayers exhibited by detergents and occurs with the quantitative efflux of entrapped compounds without size discrimination. AMPs following the barrel-stave mechanism bind to the membranes and then insert into the bilayer. Upon insertion, the different molecules self-assemble into proper transbilayer oligomers with a fixed stoichiometry. The stoichiometry defines the pore diameter and therefore the existence of an upper limit for the size of the compound to which the bilayer becomes permeable. If the lining of the pore is made by a lipid monolayer interacting with the peptide, then the pore is known as a thoroidal pore (Oren and Shai 1998; Huang 2000; Yang et al. 2001). Some peptides are able to create an electrostatic potential across the bilayer sufficient for the generation of a pore by electroporation. This mode of action requires the achievement of a high charge density and therefore is restricted to highly positively charged sequences (Miteva et al. 1999). Alternatively, the binding and sinking of amphipatic peptides can imbalance the lipid structure and cause the formation of transient pores lethal for the pathogens (Pokorny and Almeida 2004; Chang et al. 2006).

4.2.2. Intracellular Targets
Apart from the membrane action that eventually leads to deleterious lytic events, a growing number of AMPs enter the cells without major membrane damage and interfere with the cellular machinery (Fig. 8.1, Table 8.2). In these cases, this intracellular action is the mechanism of killing. For instance, some Bac7 fragments display a significant bactericidal action in the absence of membrane-permeating activity (Gennaro and Zanetti 2000). Apiadecin is translocated by a permease/transporter-mediated mechanism and, once in the cytoplasm, carries a variety of inhibitory actions (Otvos 2000). Histatins can display a cell-penetrating activity and then target mitochondrial ATP synthesis and reactive oxygen metabolism (De Smet and Contreras 2005; Luque-Ortega et al. 2008). Buforin II translocates across the bacterial membrane and binds to DNA and RNA (Brogden 2005). Pyrrhocidin enters the cells and binds to DnaK chaperone, inhibiting its ATPase activity and leading to the accumulation of misfolded proteins (Kragol et al. 2001). Others as some anionic peptides get across the membranes and provoke the flocculation of intracellular content (Brogden et al. 1996).

TABLE 8.2. Membrane and Intracellular Modes of Killing of Antimicrobial Peptides

Site of Action	Activity Mechanism	Examples	References
Membrane	Toroidal pore	LL37	Henzler-Wildman et al. (2003)
	Carpet-like destabilization	Cecropin	Gazit et al. (1995)
	Barrel-stave distortion	Alamethicin	Yang et al. (2001)
Intracellular targets	Protein synthesis inhibition	HNP-1	Lehrer et al. (1989)
	Cell wall synthesis inhibition	Mersacidin	Brotz et al. (1998)
	Nucleic acid synthesis inhibition	Pleurocidin	Patrzykat et al. (2002)
	Cytoplasmic membrane septum formation disortion	Indolicidin	Subbalakshmi and Sitaram (1998)
	Enzymatic activity inhibition	Histatin	Luque-Ortega et al. (2008)
	Media flocculation	Anionic	Brogden et al. (1996)
	Nucleic acid binding	Buforin II	Park et al. (1998a)

Killing of pathogens by AMPs involves multiple processes. The first step is executed at the membrane and can yield either its permeation or the passage through. For the peptides getting across the membrane, independent of their action at the membrane, intracellular targets become available and their recognitions convert into the principal mechanism of pathogen death.

5. ACTIVITY AND SPECIFICITY OF AMPS

Despite the large volume of data available on the site of production, range of activity, and structure–function relationships, little is known about the basis of peptide activity and specificity, which in turn relate to the basis of their efficiency after million of years of host defense and pathogen attack coevolution. Parameters such as size, sequence (charge, hydrophilic–hydrophobic balance, and distribution), and structure appear to be tailored for best action (Gennaro and Zanetti 2000; Tossi et al. 2000; Yount and Yeaman 2004; Klüver et al. 2006). However, the manipulation of any of those properties always allows the improvement of a single activity (membrane binding, membrane perturbation, etc.), suggesting that AMPs may be the result of a balanced optimization for allowing the best function multiplicity (Rodríguez-Hernández et al. 2006; Nishida et al. 2007).

Nowadays, the belief is that AMPs have evolved as a consensus molecule for interactions. Their charge, amphiphilicity, and structural flexibility confers these peptides the perfect flexible design for the recognition of and interaction with multiple biological systems. For instance, AMPs that have lost the membrane-perturbing activity (either by a pathogen reaction or by their own sequence) have gained the ability to get across the cell membrane and to exploit their design features for interference with cellular components. In this sense, cationic amphipatic peptides designed for binding to and condensing DNA for their transport in transfection experiments are also membrane-perturbing peptides (Fominaya et al. 2000). In fact, these peptides bind efficiently to both DNA and acid lipids, each polyanion acting as a competitive inhibitor for the other (Fominaya et al. 2000; Ambroggio et al. 2007). Thus, it seems plausible to postulate that the activity and specificity of the large number of AMPs is the result of a perfect balance of multiple interactions that result from the coordination of parameters such as concentration and environmental conditions.

Another important issue in the functional antimicrobial outcome of AMPs is the consideration of their combined action. In this line, the studies of Niu et al. (2008) using a recombinant heat- and acid-stable AMP containing the sequences of metalnikowin-2A, SMAP-29, protegrin-1, and scorpion defensin have shown an activity that compared favorably with ampicillin and that was synergistic with polymixin B and streptomycin.

6. AMPS AS THERAPEUTICAL AGENTS

The use of AMPs to combat undesired infections is still under debate after some trial failures. Table 8.3 summarizes some of the records reported by Gordon et al. in 2005 in a follow-up study of some synthetic compounds derived from AMPs. From the data in this table, it can be concluded that despite the many attractive qualities displayed by AMPs, the challenges for successful development are still considerable and are intimately related to the limits of our knowledge.

6.1. Factors Affecting the Therapeutic Use of AMPs

As described previously, the action of AMPs involves a complex balance of interactions that are highly dependent on the local concentration at the desired site, which relates to the intrinsic stability, the dosing parameter, and the targeting strategy.

By virtue of their structural features, an obvious disadvantage of natural peptides as AMPs is their sensitivity to proteases. In general, AMPs are the optimal substrates for trypsin-like and chymotrypsin-like proteases, and thereof they are labile agents in biological media. Proteolytic degradation of AMPs constitutes a twofold disadvantage.

TABLE 8.3. Follow-up of the Clinical Development of Some AMPs

AMP	Synthetic Compound	Clinical Trial Outcome
Magainin 2 (frog)	Pexiganan (MSI-78)	Failed to gain FDA approval for the absence of advantage over conventional antibiotics in a Phase III trial for the topical treatment of diabetic foot ulcers
Protegrin (pig)	Iseganan (IB-367)	Failed two Phase III trials as mouth rinse for stomatitis in high-risk patients; failed Phase III trial as aerosolized for the treatment of ventilator-associated pneumonia
Indolicidin (bovine)	Omiganan (MBI-226)	Failed Phase III trial as topical formulation for the prevention or reduction of catheter-related bloodstream infections
Indolicidin (bovine)	MBI 594AN	Phase IIb trial showed efficacy as topical formulation for acne.
Histatins (human)	P113 P113D	Completed Phase II trial as mouth rinse for oral candidiasis in HIV patients
BPI (human)	XMP.629	Failed Phase III trial as topical use for acne
BPI (human)	Neuprex (rBPI21)	Failed Phase III trial as adjunctive parenteral formulation to reduce mortality in pediatric meningococcemia

Data in this table have been adapted from Gordon et al. (2005) and depict a summarized state in the clinical development and its follow-up of AMP-based compounds. BPI = bactericidal-permeability-increasing protein.

On one hand, their degradation reduces the effective concentration, which in turn decreases the efficiency of their action. On the other hand, proteolysis can yield truncated forms with unexpected side effects. This issue has been experimentally addressed, resulting in the development of several strategies (McPhee et al. 2005). For instance, AMPs can gain intrinsic protease resistance when the natural L-amino acids are replaced by the corresponding D-amino acids. Alternatively, the introduction of non-peptidic bonds reduces if not abrogates their sensitivity to proteases. A third strategy to decrease AMP proteolytic degradation is the use of physical shields, such as those provided by liposomes and chemical modifications.

A second limiting factor for AMP exploitation is the dosing issue. From their mode of action, AMPs' activity is in most cases concentration dependent (requires an oligomerization event); their therapeutic development was initiated for practical cases, allowing the direct delivery to the infected site. Such accessibility provides the best scenario for an aggressive dosing aimed at achieving high local concentrations.

However, differences in context between the natural function of a given AMP as part of innate immunity compared to its repeated application in higher concentrations have raised a number of important issues regarding toxicity and appearance of unexpected secondary activities. Regarding their potential toxicity, as part of innate immunity, AMP function is controlled by several processes including expression levels, storage in granules, and synthesis as inactive pro-peptides requiring enzymatic activation for biological activity (as in the case of cathelicidin) (Jenssen et al. 2006). These features underline the importance of preservation of a minimum basal concentration as well as the potential activation of undesired routes at high concentrations.

AMP *in vivo* activity is optimized not only for concentration but also for the site of action. In this sense, combination with other effector molecules with overlapping antimicrobial functions can yield additive and synergistic effects or even the manifestation of secondary activities (Li et al. 2000; Boman 2003). An example of this strategy is the targeted killing designed by Eckert et al. (2006). In this work, the AMP novispirin G10 was synthetically fused to a linker sequence followed by the targeting peptide KH (KKHRKHRKHRKH) to display increased bacterial activity and specificity against *Pseudomonas mendocina*.

6.2. Experience in the Clinical Application of AMPs

As the problem of emergence of bacterial resistance to current antibiotic drugs continues to grow, there has been considerable interest in the development of alternative anti-infectious therapies. Given the origin of the AMPs and their mode of action, development of resistance to AMPs was believed to be unlikely. Nevertheless, various pathogens have developed resistance through different mechanisms, such as the modification of the cell wall or expression of proteases (Barak et al. 2005). Even with this potential limitation, to date, several AMP-based compounds have been developed and entered into clinical trials. The results from these trials are somehow discouraging due to the limited success, and none of the AMP-based compounds has succeeded in getting Food and Drug Administration (FDA) approval for clinical use (Table 8.3).

6.3. New Strategies and Potential for Pharmacological Exploitation of AMPs

To date, most advances in the field of AMP optimization have been limited by the expense of production. However, the fast development of high-throughput approaches together with the design of solid cell systems and clear-cut assays for different molecular targets may offer unprecedented alternatives. Using cellulose sheets as support and a peptide spot-synthesizing robot has allowed the preparation of a peptide array. This array in combination with a sensitive enzymatic assay has allowed the discovery of an optimized octapeptide with an excellent broad spectrum activity in a record time with important reduced costs (Hilpert et al. 2005). Other investigations include the search of AMPs containing nonstandard amino acids (the so-called peptaibols) as well as the development of polymers based on non-peptidic scaffolds (or peptidomimetic compounds). Reduction of cost has been also attempted by using biotechnological approaches, but apart from exceptional cases, they have failed either in the levels of production or in the purification stages.

An exceptional therapeutic potential in infectious diseases of AMPs that looks very promising is that provided by the exploitation of their immunostimulatory properties (Yang et al. 2002; Finlay and Hancock 2004; Scott et al. 2007). In this strategic action, the optimization of that secondary activity will allow the enhancement of the natural innate immunity without the toxicity of their function. In this line and still under experimentation is the case of BMAP-18 (Haines et al. 2009), which was derived from the bovine myeloid AMP 27 (BMAP-27) in an effort to reduce the cytotoxicity to mammalian cells. BMAP-18 inhibits the LPS-induced tumor necrosis factor alpha (TNF-α) secretion and also kills trypanosomatid parasites and consequently constitutes an adequate candidate for therapy.

Despite their capacity to kill pathogens, AMPs display a plethora of side functions that can be exploited for biomedical purposes. Among the different activities, the capacity to interfere with the viability of tumoral cells is commanding much attention (Mader et al. 2006). This antitumoral activity is most probably the result of the altered membrane composition of the transform cells, for example, the higher content of phosphatidylserine. Magainins lyse many types of cells, with the toxic concentration being an order of magnitude higher for non-malignant cells (Jacob and Zasloff 1994; Maloy and Kari 1995; Imura et al. 2008). This activity has also been described for defensins and cecropins (Suttmann et al. 2008; Xu et al. 2008). Moreover, the alteration in the expression pattern of human β-defensins 1 and 2 has underlined the role of defensins in the pathogenesis of some carcinomas (Gambichler et al. 2006).

In summary, although AMPs are recognized as essential components of natural host innate immunity against microbial challenge, their use as a new class of drugs and their value as external therapeutic agents still require more work for their successful application.

ACKNOWLEDGMENT

MG lab is supported by grants SAF2006-00418 and BFU2009-07971 from Ministerio de Ciencia e Innovación, and FOOD-CT-2004-506579.

REFERENCES

Ambroggio E., L. A. Bagatolli, E. Goormaghtigh, E. J. Fominaya, and M. Gasset (2007) *Advances in Planar Lipid Bilayers and Liposomes* **5**, 1–23.

Andreu D. and L. Rivas (1998) *Biopolymers* **47**, 415–433.

Barak O., J. R. Treat, and W. D. James (2005) *Advances in Dermatology* **21**, 357–374.

Boman H. G. (1995) *Annual Reviews in Immunology* **13**, 61–92.

Boman H. G. (2003) *Journal of Internal Medicine* **254**, 197–215.

Boman H. G. and D. Hultmark (1987) *Annual Reviews in Microbiology* **41**, 103–126.

Broekaert W. F., F. R. Terras, B. P. Cammue, and R. W. Osborn (1995) *Plant Physiol* **108**, 1353–1358.

Brogden K. A. (2005) *Nature Reviews. Microbiology* **3**, 238–250.

Brogden K. A., M. Ackermann, P. B. McCray, and B. F. Tack (2003) *International Journal of Antimicrobial Agents* **22**, 465–478.

Brogden K. A., A. J. De Lucca, J. Bland, and S. Elliott (1996) *Proceedings of the National Academy of Sciences of the United States of America* **93**, 412–416.

Brotz H., G. Bierbaum, K. Leopold, P. E. Reynolds, and H. G. Sahl (1998) *Antimicrobial Agents and Chemotherapy* **42**, 154–160.

Bulet P., R. Stöcklin, and L. Menin (2004) *Immunology Reviews* **198**, 169–184.

Chang D. I., E. J. Prenner, and H. J. Vogel (2006) *Biochimica et Biophysica Acta* **1758**, 1184–1202.

Chatterjee S., D. K. Chatterjee, R. H. Jani, J. Blumbach, B. N. Ganguli, N. Klesel, M. Limbert, and G. Seibert (1992) *The Journal of Antibiotics* **45**, 839–845.

Davidson D. J. A., A. J. Currie, G. S. Reid, D. M. Bowdish, K. L. McDonald, R. C. Mae, R. E. W. Hancock, and D. P. Seepert (2004) *The Journal of Immunology* **172**, 1146–1156.

De Smet K. and R. Contrearas (2005) *Biotechnol Lett* **27**, 1337–1347.

Dürr M. and A. Peschel (2002) *Infection and Immunity* **70**, 6515–6517.

Eckert R. F., D. K. Qi, H. J. Yarbrough, M. H Anderson, and W. Shi (2006) *Antimicrobial Agents and Chemotherapy* **50**, 1480–1488.

Eliasson M. and A. Egesten (2008) *Contrib Microbiol* **15**, 101–117.

Epand R. M. and H. J. Vogel (1999) *Biochimica et Biophysica Acta* **1462**, 11–28.

Fehlbaum P., P. Bulet, S. Chernys, J. P. Briand, J. P. Roussel, L. Letellier, C. Hetru, and J. A. Hoffmann (1996) *Proceedings of the National Academy of Sciences of the United States of America* **93**, 1221–1225.

Fernandes J. M., G. Molle, G. D. Kemp, and V. J. Smith (2004) *Developmental and Comparative Immunology* **28**, 127–138.

Finlay B. B. and R. E. Hancock (2004) *Nature Reviews. Microbiology* **2**, 497–504.

Fominaya F., M. Gasset, R. Garcia, F. Roncal, J. P. Albar, and A. Bernad (2000) *The Journal of Gene Medicine* **2**, 455–464.

Gambichler T., M. Skrygan, J. Huyn, F. G. Bechara, M. Sand, P. Altmeyer, and A. Kreuter (2006) *BMC Cancer* **6**, 163.

Ganz T., M. E. Selsted, and R. I. Lehrer (1990) *European Journal of Haematology* **44**, 1–8.

García-Olmedo F., A. Molina, J. M. Alamillo, and P. Rodríguez-Palenzuela (1998) *Biopolymers* **47**, 479–491.

Gazit E., A. Boman, H. G. Boman, and Y. Shai (1995) *Biochemistry* **34**, 11479–11488.

Gennaro R. and M. Zanetti (2000) *Biopolymers* **55**, 31–49.

Gordon Y. J., E. G. Romanowski, and A. M. McDermott (2005) *Current Eye Research* **30**, 505–515.

Hancock R. E. W. (1997) *Lancet* **349**, 418–422.

Hancock R. E. W. and H. G. Sahl (2006) *Nature Biotechnology* **24**, 1551–1557.

Haines L. R., J. M. Thomas, A. M. Jackson, B. A. Eyford, M. Razavi, C. N. Watson, B. Gowen, R. E. W. Hancock, and T. W. Pearson (2009) *PLoS Neglected Tropical Diseases* **3**, e373.

Haug B. E., M. B. Strøm, and J. S. Svendsen (2007) *Current Medicinal Chemistry* **14**, 1–18.

Henzler-Wildman H. K. A., D. K. Lee, and A. Ramamoorthy (2003) *Biochemistry* **42**, 6545–6558.

Hilpert K., R. Volkmer-Engert, T. Walter, and R. E. W. Hancock (2005) *Nature Biotechnology* **23**, 1008–1012.

Hirsch A. and A. T. Mattick (1949) *Lancet* **2**, 190–193.

Huang H. W. (2000) *Biochemistry* **39**, 8347–8352.

Huang H. W., F. Y. Chen, and M. T. Lee (2004) *Physical Review Letters* **92**, 198304.

Huang L. C., R. Y. Reins, R. L. Gallo, and A. M. McDermott (2007) *Investigative Ophthalmology & Visual Science* **48**, 4498–4508.

Imler J. L. and P. Bulet (2005) *Chemical Immunology and Allergy* **86**, 1–21.

Imura Y., N. Choda, and K. Matsuzaki (2008) *Biophys J* **95**, 5757–5765.

Iwanaga S. and S. Kawabata (1998) *Frontiers in Bioscience* **3**, 973–984.

Jacob L. and M. Zasloff (1994) *Ciba Foundation Symposium* **186**, 197–216.

Jang W. S., K. N. Kim, Y. S. Lee, M. H. Nam, and I. H. Lee (2002) *FEBS Letters* **521**, 81–86.

Jenssen H., P. Hamill, and R. E. W. Hancock (2006) *Clinical Microbiology Reviews* **19**, 491–511.

Klüver E., K. Adermann, and A. Schulz (2006) *Journal of Peptide Science* **12**, 243–257.

Kolls J. K., P. B. McCray, and Y. R. Chan (2008) *Nature Reviews. Immunology* **8**, 829–835.

Kragol G., S. Lovas, G. Varadi, B. A. Condie, R. Hoffman, and L. Otvos (2001) *Biochemistry* **40**, 3016–3026.

Lazarovici P., N. Primor, and L. M. Loew (1986) *Journal of Biological Chemistry* **261**, 16704–16713.

Lee I. H., Y. Cho, and R. I. Lehrer (1997) *Infection and Immunity* **65**, 2898–2903.

Lehrer R. I. (2004) *Nature Reviews. Microbiology* **2**, 727–738.

Lehrer R. I., A. Barton, K. A. Daher, S. S. Harwig, T. Ganz, and M. E. Selsted. (1989) *Journal of Clinical Investigation* **84**, 553–561.

Lemaitre B., E. Nicolas, L. Michaut, J. M. Reichart, and J. A. Hoffmann (1996) *Cell* **86**, 973–983.

Lemaitre C., N. Orange, P. Saglio, N. Saint, J. Gagnon, and G. Molle (1996) *European Journal of Biochemistry* **240**, 143–149.

Li P., M. Post, R. Volk, Y. Gao, M. Li, C. Metais, K. Sato, J. Tsai, W. Aird, R. D. Rosenberg, T. G. Hampton, F. Selke, P. Carmeliet, and M. Simons (2000) *Nature Medicine* **6**, 46–55.

Li W. F., G. X. Ma, and X. X. Zhou (2006) *Peptides* **27**, 2350–2359.

Luque-Ortega J. R., W. van't Hof, E. C. Veerman, J. M. Saugar, and L. Rivas (2008) *FASEB Journal* **22**, 1817–1828.

Mader J. S., D. Smyth, J. Marshall, and D. W. Hoskin (2006) *American Journal of Pathology* **169**, 1753–1766.

Maloy W. L. and U. P. Kari (1995) *Biopolymers* **37**, 105–122.

Matsuzaki K. (1999) *Biochemistry et Biophysics Acta* **1462**, 1–10.

McPhee J. B., M. G. Scott, and R. E. W. Hancock (2005) *Combinatorial Chemistry & High Throughput Screening* **8**, 257–272.

Miteva M., M. Andersson, A. Karshikoff, and G. Otting (1999) *FEBS Letters* **462**, 155–158.

Mookherjee N. and R. E. W. Hancock (2007) *Cellular and Molecular Life Sciences* **64**, 922–933.

Nagaoka I., H. Tamura, and M. Hirata (2006) *Journal of Immunology* **176**, 3044–3052.

Nishida M., Y. Imura, M. Yamamoto, S. Kobayashi, Y. Yano, and K. Matsuzaki (2007) *Biochemistry* **46**, 14284–14290.

Niu M., X. Li, J. Wei, R. Cao, B. Zhou, and P. Chen (2008) *Protein Expression and Purification* **57**, 95–100.

Oren Z. and Y. Shai (1996) *European Journal of Biochemistry* **237**, 303–310.

Oren Z. and Y. Shai (1998) *Biopolymers* **47**, 451–463.

Otvos L., Jr. (2000) *J Pept Sci* **6**, 497–511.

Park C. B., H. S. Kim, and S. C. Kim (1998a) *Biochemical and Biophysical Research Communications* **244**, 253–257.

Park I. Y., C. B. Park, M. S. Kim, and S. C. Kim (1998b) *FEBS Letters* **437**, 258–262.

Patrzykat A., C. L. Friedrich, L. Zhang, V. Mendoza, and R. E. Hancock (2002) *Antimicrobial Agents and Chemotherapy* **46**, 605–614.

Pellegrini A. (2003) *Current Pharmaceutical Design* **9**, 1225–1238.

Peschel A. and S. G. Sahl (2006) *Nature Reviews. Microbiology* **4**, 529–536.

Pokorny A. and P. F. Almeida (2004) *Biochemistry (Mosc.)* **43**, 8846–8857.

Powers J. P. and R. E. Hancock (2003) *Peptides* **24**, 1681–1691.

Riley M. A. (1998) *Annual Reviews in Genetics* **32**, 255–278.

Rinaldi A. C. (2002) *Current Opinion in Chemical Biology* **6**, 799–804.

Rodríguez-Hernández M. J., J. Saugar, F. Docobo-Pérez, B. G. de la Torre, M. E. Pachón-Ibáñez, A. García-Curiel, F. Fernández-Cuenca, D. Andreu, L. Rivas, and J. Pachón (2006) *Journal of Antimicrobial Chemotherapy* **58**, 95–100.

Romeo D., B. Skerlavaj, M. Bolognesi, and R. Gennaro (1988) *Journal of Biological Chemistry* **263**, 9573–9575.

Scott M. G., D. J. Davidson, M. R. Gold, D. Bowdish, and R. E. W. Hancock (2002) *Journal of Immunology* **169**, 3883–3891.

Scott M. G., E. Dullaghan, N. Mookherjee, N. Glavas, M. Waldbrook, A. Thompson, A. Wang, K. Lee, S. Doria, P. Hamill, J. J. Yu, Y. Li, O. Donini, M. M. Guarna, B. B. Finlay, J. R. North Jr. and R. E. Hancock (2007) *Nature Biotechnology* **25**, 465–472.

Shai Y. (1995) *Trends in Biochemical Sciences* **20**, 460–464.

Shai Y. (1999) *Biochimica et Biophysica Acta* **1462**, 55–70.

Shi J. and A. C. Camus (2006) *Developmental and Comparative Immunology* **30**, 746–755.

Skarnes R. C. and D. W. Watson (1957) *Bacteriology Reviews* **21**, 273–294.

Smith V. J., J. M. Fernandes, S. J. Jones, G. D. Kemp, and M. F. Tatner (2000) *Fish & Shellfish Immunology* **10**, 243–260.

Steiner H., D. Hultmark, A. Engstrom, H. Bennich, and H. G. Boman (1981) *Nature* **292**, 246–248.

Subbalakshmi C. and N. Sitaram (1998) *FEMS Microbiology Letters* **160**, 91–96.

Suttmann H., M. Retz, F. Paulsen, J. Harder, U. Zwergel, J. Kamradt, B. Wullich, G. Unteregger, M. Stöckle, and J. Lehmann (2008) *BMC Urology* **8**, 5.

Tang Y. Q., M. R. Yeaman, and M. E. Selsted (2002) *Infection and Immunity* **70**, 6524–6533.

Tang Y. Q., J. Yuan, G. Osapay, K. Osapay, D. Tran, C. J. Miller, A. J. Ouellette, and M. E. Selsted (1999) *Science* **286**, 498–502.

Taylor S. W., A. G. Craig, W. H. Fischer, M. Park, and R. I. Lehrer (2000) *Journal of Biological Chemistry* **275**, 38417–38426.

Tossi A., L. Sandri, and A. Giangaspero (2000) *Biopolymers* **55**, 4–30.

Tran D., P. A. Tran, Y. Q. Tang, J. Yuan, T. Cole, and M. E. Selsted (2002) *Journal of Biological Chemistry* **277**, 3079–3084.

Vizioli J. and M. Salzet (2002) *Trends in Pharmacological Sciences* **23**, 494–496.

Wang G., E. F. Stange, and J. Wehkamp (2008) *Expert Review of Anti-Infective Therapy* **5**, 1049–1057.

Westerhoff H. V., D. Juretić, R. W. Hendler, and M. Zasloff (1989) *Proceedings of the National Academy of Sciences of the United States of America* **86**, 6597–6601.

Wong J. H., L. Xia, and T. B. Ng (2007) *Current Protein & Peptide Science* **8**, 446–459.

Xu N., Y. S. Wang, W. B. Pan, B. Xiao, Y. J. Wen, X. C. Chen, L. J. Chen, H. X. Deng, J. You, B. Kan, A. F. Fu, D. Li, X. Zhao, and Y. Q. Wei (2008) *Molecular Cancer Therapeutics* **7**, 1588–1597.

Yang D., A. Biragyn, L. W. Kwak, and J. J. Oppenheim (2002) *Trends in Immunology* **23**, 291–296.

Yang D., A. Biragyn, D. M. Hoover, J. Lubkowski, and J. J. Oppenheim (2004) *Annu Rev Immunol* **22**, 181–215.

Yang L., T. A. Harroun, T. M. Weiss, L. Ding, and H. W. Huang (2001) *Biophysics Journal* **81**, 1475–1485.

Yount N. Y. and M. R. Yeaman (2004) *Proceedings of the National Academy of Sciences of the United States of America* **101**, 7363–7368.

Zanetti M. (2005) *Current Issues in Molecular Biology* **7**, 179–196.

Zasloff M. (1987) *Proceedings of the National Academy of Sciences of the United States of America* **84**, 5449–5453.

Zasloff M. (2002) *Nature* **415**, 389–395.

Zhang L. and T. J. Falla (2006) *Expert Opinion on Pharmacotherapy* **6**, 653–663.

Zhu S. (2008) *Trends Microbiol* **16**, 353–360.

Zucht H. D., M. Raida, K. Adermann, H. J. Mägert, and W. G. Forssmann (1995) *FEBS Letters* **372**, 185–188.

CHAPTER 9

DESIGN OF PHAGE COCKTAILS FOR THERAPY FROM A HOST RANGE POINT OF VIEW

LAWRENCE D. GOODRIDGE

Department of Animal Sciences, Colorado State University,
Fort Collins, CO

1. INTRODUCTION

1.1. Initial Steps in Phage Infection

A discussion surrounding the design of phage cocktails for effective implementation of phage therapeutic strategies must begin with a discussion of phage-bacterial host range. The susceptibility of a bacterium to phage infection is primarily dependent on the ability of the phage to attach to receptors on the cell. A successful attachment may then facilitate entry of the phage nucleic acid into the cell, which may then lead to replication and release of new phage progeny (Lindberg 1973). While there are several mechanisms employed by bacteria to resist phage infection (Dinsmore and Klaenhammer 1995; Allison and Klaenhammer 1998), a primary method is to stop phage attachment through alteration of phage receptors. As such, phage cocktails should be rationally designed taking into account the possibility of the development of bacterial resistance through receptor mutation. In addition, phage cocktails should be designed with the understanding that the biological diversity that is observed within bacterial species means that there may be strains within a given species that are quite adept at resisting phage infection. Therefore, phage cocktails should include phages that have the ability to alter their host specificity *in situ*.

Enzybiotics: Antibiotic Enzymes as Drugs and Therapeutics. Edited by Tomas G. Villa and Patricia Veiga-Crespo
Copyright © 2010 John Wiley & Sons, Inc.

Several approaches may be used to design phage cocktails with the ability to infect all bacterial strains within a given species. These approaches can be broadly grouped into three sections including the combination of large numbers of natural phages that use different receptors to infect bacteria within the same species, the use of dual specificity phages that can infect bacteria within a given species using more than one bacterial receptor, and the combination of phages that have expanded host ranges through genetic manipulation, or phages that can naturally expand their host ranges *in situ*.

2. RATIONAL DESIGN OF PHAGE COCKTAILS

A major concern regarding the use of phage therapy in the treatment of infections is the development of phage-resistant bacteria. In earlier phage resistance work, Luria and Delbrück (1943) made the general observation that a decrease in turbidity occurred within several hours upon attack of a pure bacterial culture by a phage. However, continued incubation of the culture resulted in the growth of a bacterial variant that was resistant to the attacking phage. The susceptibility of a bacterium to phage infection is primarily dependent on whether the phage can attach to receptors on the cell (Lindberg 1973). Many constituents of the bacterial cell surface act as receptors for phage. These include pili, flagella, and capsule (Edwards and Meynell 1968; Stirm 1968; Lindberg 1973). For example, bacterial capsule has been shown to be a receptor for the Vi or K phages, which infect *Salmonella* and *Escherichia coli* (Stirm 1968). Phages also use flagella of both Gram-positive and Gram-negative cells as receptors (Edwards and Meynell 1968). Phage adsorption to teichoic acid of Gram-positive bacteria has also been observed (Coyette and Ghuysen 1968). In Gram-negative bacteria, the components of the O antigen and outer core oligosaccharide (OS) of the lipopolysaccharide (LPS) can serve as phage receptors (Montag et al. 1990). Phages also employ various outer membrane proteins (OMPs) (OmpA, OmpC, OmpF, LamB, Tsx, OmpP, FadL) as receptors (Wilson et al. 1970; Montag et al. 1990; Heller 1992; Hashemolhosseini et al. 1994a, b; Henning and Hashemolhosseini 1994). Several groups have developed strategies for developing phage cocktails that are designed to address the issue of phage resistance. Tanji et al. (2004) described a method used to develop a phage cocktail to control the foodborne pathogen *E. coli* O157:H7. In this work, the researchers isolated 26 phages that were able to infect *E. coli* O157:H7. The phages were isolated from bovine, swine, poultry samples, and sewage. Each phage was plaqued on an appropriate *E. coli* O157:H7 isolate, and nine phages causing visible lysis during liquid growth were used to produce phage-resistant *E. coli* O157:H7 cells. The remaining 17 phages (those that did not cause complete lysis) were screened for their ability to lyse the phage-resistant cells, and phages that effected lysis were selected. Based on the lysis profiles, several cocktails of phages were developed, and these cocktails were shown to delay the development of phage-resistant cells, when compared with the use of a single phage. For example, the researchers showed that deletion of OmpC from the *E. coli* O157:H7 cells facilitated the emergence of resistant bacterial cells after 8 h to one phage, designated as SP21. Alternatively, alteration of the LPS profile of *E. coli* O157:H7 facilitated cell resistance to phage SP22, which was observed following 6 h of incubation. However, when a mixture of both phages was used to infect the *E. coli* O157:H7 cells, the emergence of phage-resistant cells was not observed for 30 h. This work indicated that the combination of two phages, one that used an OMP (OmpC) as a receptor and one that employed an LPS component as its receptor, was able to significantly delay the emergence of phage resistance cells, as compared with if either phage used alone (Tanji et al. 2004).

Following up on this work, Tanji et al. (2005) developed a three-phage cocktail and investigated its efficacy in controlling *E. coli* O157:H7 cells using *in vitro* and *in vivo* models. Three phages, SP15, SP21, and SP22, were selected from the 26-phage stock developed during the previous study (Tanji et al. 2004). Addition of one or two phages to *E. coli* O157:H7 growing in batch culture reduced the turbidity of the culture. However, an increase in the turbidity occurred due to the appearance of phage-resistant cells. In contrast, the addition of a three-phage mixture (SP15-21-22) did not lead to an increase in culture turbidity under aerobic growth conditions. Under anaerobic conditions, a slight increase in culture turbidity was observed after SP15-21-22 addition.

A chemostat continuous culture system was developed under anaerobic conditions to optimize the titer of phage cocktail and frequency of the addition of the cocktail for controlling *E. coli* O157:H7 cells. The authors observed a 5-log decrease in the *E. coli* O157:H7 concentration after the addition of the phage cocktail (10^9 plaque forming units [PFU]/ml). While an increase in cell concentration was observed after 1 day of incubation, repeated addition of the phage cocktail was sufficient to reduce the cell concentration. Based on the *in vitro* experiments, the three-phage cocktail was orally administered to mice, which had been previously inoculated with *E. coli* O157:H7. The *E. coli* O157:H7 and phage concentration in the feces was monitored for 9 days following phage addition. High titers of phage were detected in the feces when the phage cocktail was continuously administrated. The *E. coli* O157:H7 concentration in the feces was reduced during the 9-day test period. The authors concluded that repeated oral administration of the SP15-21-22 cocktail effectively reduced the *E. coli* O157:H7 concentration in the feces and gastrointestinal tract of mice (Tanji et al. 2005).

Chase et al. (2005) used a receptor modeling procedure to produce a phage cocktail to reduce *E. coli* O157:H7 in cattle. This approach allowed the researchers to produce a phage cocktail that contained phages that used multiple different receptors on the cell surface. The rationale behind this approach was based on the results of the Tanji et al. (2004) study, which showed that two phages that used different receptors delayed the formation of phage resistance cells. In the Chase et al. (2005) study, it was hypothesized that the combination of more than two phages that each used different bacterial receptors would further delay the formation of phage-resistant cells. A total of 56 different phages, with varying degrees of specificity for *E. coli* O157:H7, were screened on different *E. coli* strains to determine the receptors used by each phage. A series of *E. coli* K-12 OMP mutants and a series of *E. coli* LPS outer core OS mutants were used for the phage receptor studies. Four pairs of *E. coli* K-12 isogenic mutants were used, and these strains were isogenic with respect to OmpF, OmpA, FadL, and Tsx. Additionally, two LamB strains that were not isogenic but had mutations that did not affect the outer membrane were evaluated for phage receptor specificity. Two pairs of OmpC isogenic mutants were included as negative controls. In addition to the OMP strains, 12 *E. coli* LPS outer core OS mutant strains were used to assess the usefulness of *E. coli* LPS as a receptor for the phages. Four of these strains represented four of the five different core types (R1 to R4) (Heinrichs et al. 1998) of *E. coli*. These strains are deficient in O antigen and were used to assess the usefulness of the outer core OS region of the LPS as a receptor for the phages. *E. coli* K-12 was also included in the host range study. This strain represents the fifth core type (K-12) of *E. coli* (Heinrichs et al. 1998). The remaining eight strains were sequentially deficient in the genes (the *waa* operon) responsible for producing the enzymes that add carbohydrates to the growing outer core OS region of the outer membrane LPS. These strains were used to determine which carbohydrate residues within the LPS could serve as a receptor for the phages. Each phage was individually tested against each bacterial isolate. The specificity data was used to generate a database of the receptors that each phage used. Some phages used more than one receptor. The database was used to construct a phage cocktail consisting of 37 different phages. The 37-phage cocktail was screened against 58 *E. coli* O157:H7 isolates, by placing a drop of phage lysate (10^{11} PFU/ml) on a lawn of each individual bacterial isolate. After overnight incubation, the zone of clearing was checked for the presence of bacterial colonies, which would indicate bacterial mutants that were resistant to the cocktail. The cocktail produced complete clearing on all of the isolates, and no resistant colonies were observed. In addition, the phage cocktail lysate was dropped on an agar plate containing a lawn of a mixture of all 58 of the *E. coli* O157:H7 isolates. After overnight incubation, no colonies were observed within the zone of clearing, indicating the absence of phage-resistant cells.

The cocktail was examined for its ability to reduce the presence of *E. coli* O157:H7 in bovine fecal slurries and in calves. In an anaerobic *in vitro* model, the phage cocktail completely eliminated a strain mixture of 10^4 cfu/ml of *E. coli* O157:H7 from bovine fecal slurries, within 4 h. To evaluate the phage cocktail in live animals, a total of 14 Black Angus calves ranging from 4 to 6 months of age were orally inoculated with 10^8 cfu *E. coli* O157:H7. The *in vivo* experiments consisted of two trials. The first trial evaluated ileal samples and the second trial evaluated fecal samples for the presence of *E. coli* O157:H7 and phages. In the first trial, a significant decline in the concentration of *E. coli* O157:H7 in the ileal samples was observed at 8 h ($P = 0.05$). However, the concentration of *E. coli* O157:H7 increased back to the concentration of the control samples at 16 h. In the second trial, shedding of *E. coli* O157:H7 decreased significantly in the treated group ($P = 0.05$) at 24 h. Similar to the ileal samples, an increase in the concentration of *E. coli* O157:H7 was observed at 36 h in the fecal samples. The increases in cell concentration were associated with a decrease in phage concentration. None of the *E. coli* O157:H7 cultured from ileal or fecal samples showed resistance to the phage cocktail.

These results highlight the ability of the 37-phage cocktail to eliminate *E. coli* O157:H7 in fecal slurries, and to reduce the concentration of *E. coli* O157:H7 in calves, without the formation of phage-resistant mutants. The receptor modeling approach appears to be effective in the design of phage cocktails to reduce the presence of *E. coli* O157:H7 in animals. As with the work of Tanji et al. (2005), continuous dosing of the phage cocktail to the calves may have eliminated the increase in concentration of the *E. coli* O157:H7 cells following the initial decrease.

Other research groups have evaluated the use of phage cocktails to decrease various bacterial pathogens in live animals. Callaway et al. (2006) anaerobically isolated phage that targeted *E. coli* O157:H7 from fecal samples collected from commercial feedlot cattle in the central United States. The spectrum of activity of the phages was determined, and the phages were combined to form a cocktail of phage for *in vivo* studies. When a 21-phage cocktail was inoculated into sheep artificially contaminated with *E. coli* O157:H7, intestinal populations of *E. coli* O157:H7 were decreased ($P < 0.05$) in the cecum and rectum. When sheep in a follow-up study were dosed with 10^5, 10^6, or 10^7 PFU of the phage cocktail, the addition of 10^5 PFU (multiplicity of infection [MOI] of 10) was the most effective at reducing *E. coli* O157:H7 populations throughout the gastrointestinal tract. Collectively, the results indicate that properly selected phages can be used to reduce *E. coli* O157:H7 in food animals. The authors concluded that phage could be an important part of an integrated foodborne pathogen reduction program.

Kudva et al. (1999) isolated *E. coli* O157 antigen-specific phages and tested them to determine their ability to lyse laboratory cultures of *E. coli* O157:H7. A total of 53 bovine or ovine fecal samples were enriched for phage, and 5 of these samples were found to contain lytic phages that grew on *E. coli* O157:H7. Three phages, designated KH1, KH4, and KH5, were further evaluated. At 37 or 4 °C, a mixture of these O157-specific phages lysed all of the *E. coli* O157 cultures tested and none of the non-O157 *E. coli* or non-*E. coli* cultures tested. These results required culture aeration and a high MOI. Without aeration, complete lysis of the bacterial cells occurred only after 5 days of incubation and only at 4 °C. Phage infection and plaque formation were influenced by the nature of the host cell O157 LPS. For example, strains that did not express the O157 antigen or expressed a truncated LPS were not susceptible to plaque formation or lysis by phage. In addition, strains that expressed abundant mid-range molecular-weight LPS did not support plaque formation but were lysed in liquid culture.

While the above studies suggest that phage resistance can be overcome through the use of well-designed cocktails, it should be noted that phage-resistant bacterial mutants can be produced even when phage cocktails are employed to reduce the presence of a given bacterium. For example, O'Flynn et al. (2004) evaluated three lytic phages (e11/2, e4/1c, PP01), and a cocktail of all three phages for their ability to lyse *E. coli* O157:H7 *in vivo* and *in vitro*. Phage e11/2, pp01, and the cocktail resulted in a 5-log reduction of *E. coli* O157:H7 within 1 h at 37 °C. However, phage-resistant mutants emerged following the challenge. The mutant cells had a growth rate that approximated that of the parental O157 strain, although they exhibited a smaller, more coccoid cellular morphology. The frequency (10^{-6} cfu) of phage-resistant mutant formation was observed to be similar for e11/2, pp01, and the phage cocktail, while bacterial mutants insensitive to e4/1c occurred at a higher frequency (10^{-4} cfu). The authors observed that the phage-resistant mutants commonly reverted to phage sensitivity within 50 generations. In an initial meat trial experiment, the phage cocktail completely eliminated *E. coli* O157:H7 from the beef meat surface in seven of nine cases. The authors concluded that given that the frequency of phage-resistant mutant development was low for two of the phages, and the fact that the mutants reverted to phage sensitivity, phage-resistant mutant formation should not hinder the use of these phages as biocontrol agents, particularly since low levels of the pathogen are typically encountered in the environment (O'Flynn et al. 2004). Nevertheless, as described above, the use of rationale design methods to develop phage cocktails should reduce the possibility of the development of phage-resistant bacteria.

Andreatti Filho et al. (2007) used cocktails of four different phages obtained from commercial broiler houses (CB4Ø) and 45 phages from a municipal wastewater treatment plant (WT45Ø) to effect reduction of *Salmonella enterica* serovar *enteritidis in vitro* and in experimentally infected chicks. In one experiment, an *in vitro* crop assay was conducted with selected phage concentrations (10^5–10^9 PFU/ml) to determine the ability to reduce *S. enteritidis* in the simulated crop environment. After 2 h at 37 °C, CB4Ø or WT45Ø reduced the *S. enteritidis* concentration by 1.5 or 5 log, respectively, as compared with a control. However, the CB4Ø cocktail did not affect total *S. enteritidis* recovery after 6 h (the concentration returned to pre-phage cocktail treatment levels), whereas WT45Ø resulted in up to a 6-log reduction of *S. enteritidis*. In another experiment, day-of-hatch chicks were challenged orally with 9×10^3 cfu/chick of *S. enteritidis* and treated via oral gavage with 1×10^8 CB4Ø PFU/chick, 1.2×10^8 WT45Ø PFU/chick, or a combination of both, 1 h after challenge with *S. enteritidis*. All treatments significantly reduced *S. enteritidis* recovered from cecal tonsils at 24 h as compared with untreated controls, but no significant differences were observed at 48 h following treatment. The authors concluded that the data suggested that some phages can be efficacious in reducing *S. enteriditis* colonization in poultry during a short period, but with the phages and methods presently tested, persistent reductions were not observed (Andreatti Filho et al. 2007). As described with the Chase et al. (2005) study, continuous dosing of the phages may have led to sustained reduction of the bacterial concentration. Toro et al. (2005) used a combination of three *Salmonella*-specific bacteriophages (BPs) and competitive exclusion (CE) bacteria to reduce *Salmonella* colonization in experimentally infected chickens. A cocktail of the phages was administered orally to the chickens several days prior and after *Salmonella* challenge but not simultaneously. The phages that comprised the cocktail were readily isolated from the feces of the phage-treated chickens approximately 48 h after administration. A CE product consisting of a defined culture of seven different microbial species was used either alone or in combination with phage cocktail treatment. CE was administered orally at hatch. *Salmonella* counts in the intestine, ceca, and a pool of liver/spleen were evaluated in *Salmonella*-challenged chickens treated with the phage cocktail or with the cocktail and CE. In both trials 1 and 2, a beneficial effect of the phage treatment on weight gain performance was evident. A reduction in *Salmonella* counts was detected in the cecum and ileum of phage-, CE-, and phage/CE-treated chickens as compared with nontreated birds. In trial 1, phage treatment reduced *Salmonella* counts to marginal levels in the ileum and reduced counts sixfold in the ceca. The CE and phage treatments showed differences in the reduction of *Salmonella* counts after challenge between specimens obtained at days 4 and 14 post-challenge in the ceca, liver/spleen, and ileum during trial 2. The results of this work indicate that a combination of a phage cocktail and a CE product was capable of reducing *Salmonella* colonization of experimentally infected chickens (Toro et al. 2005).

Other researchers have developed phage cocktails to treat non-foodborne infections. For example, McVay et al. (2007) developed an *in vivo* model based on mice compromised by a burn wound injury followed by fatal infection with *Pseudomonas aeruginosa*. The mice were administered a single dose of a *P. aeruginosa* phage cocktail consisting of three different phages by three different routes including intramuscular (i.m.), subcutaneous (s.c.), or intraperitoneal (i.p.) administration. The results of these studies indicated that a single dose of the *P. aeruginosa* phage cocktail could significantly decrease the mortality of thermally injured, *P. aeruginosa*-infected mice (from 6% survival without treatment to 22% to 87% survival with treatment) and that the route of administration was particularly important to the efficacy of the treatment, with the i.p. route providing the most significant (87%) protection. The pharmacokinetics of phage delivery to the blood, spleen, and liver suggested that the phages administered by the i.p. route were delivered at a higher dose, were delivered earlier, and were delivered for a more sustained period of time than the phages administered by the i.m. or s.c. routes, which may explain the differences in the efficacies of these three different routes of administration (McVay et al. 2007).

Researchers at the Eliava Institute in Tbilisi, Republic of Georgia, have spent decades developing a comprehensive phage collection, and the institute currently produces a range of phage preparations in a variety of pharmaceutical forms that can be administered topically, orally, rectally, by inhalation, or by injection (Hanlon 2007). For example, "Pyophage" is the commercial name given to a cocktail of phages active against staphylococci, streptococci, *P. aeruginosa*, *Proteus* spp., and *E. coli* that can be used for the treatment and prophylaxis of purulent wound infections (Markoishvili et al. 2002). It is also used in surgery (both pre- and postoperatively), burn wounds, osteomyelitis, skin infections, and eye and ear infections. This phage mixture is also used in a commercial wound dressing product called "PhagoBioDerm" (Markoishvili et al. 2002), which is a novel biodegradable polymer based on polyester amides impregnated both with the phage cocktail and the antibiotic ciprofloxacin. The dressing has been used successfully to treat infected wounds including those containing multidrug-resistant *Staphylococcus aureus* (Markoishvili et al. 2002; Jikia et al. 2005). The Eliava Institute has a long history of treating gastrointestinal infections and has developed an 11-phage cocktail active against six different species of *Salmonella* as well as a 17-phage cocktail effective against a broad range of gastrointestinal pathogens (Hanlon 2007).

3. PHAGE COCKTAILS AND DIAGNOSTICS

Phage cocktails have also been used for diagnostic purposes. In one diagnostic scenario, the phages are used to reduce the presence of background flora that interferes with the recovery of the target bacteria. Strategic Diagnostics Incorporated (Newark, DE; www.sdix.com) has developed an assay called RapidChek® SELECT™ *Salmonella*, in which the assay enrichment media is supplemented with phage as a selective agent, which reduces levels of background flora, allowing *Salmonella* to grow freely. Kumar et al. (2006) employed a phage cocktail to reduce background flora during processing of sputum samples for isolation of *Mycobacterium* spp. Such a treatment is needed because the mechanical pressure exerted during centrifugation and the chemical pressure experienced when sputum specimens are processed, leave the tubercle bacilli in the sputum unsuitable for rapid detection especially in phage-based assays. As such, overnight incubation of *Mycobacterium tuberculosis* in broth is mandatory to allow the bacteria to resuscitate. During this time, the surviving colonizing flora grows faster and overgrows the tubercle bacilli interfering with tuberculosis diagnosis. In the work of Kumar et al. (2006), phages capable of killing 14 different species representing the normal flora contained within sputum samples were isolated from soil and sewage samples and characterized. Sputum samples were treated with a cocktail of three phages capable of killing most of the 14 representative organisms while not infecting mycobacteria. The addition of the cocktail controlled all of the background species in 54 tested sputum samples, significantly reducing the colony forming units of the background flora, when the samples were plated on solid media, although a few discrete colonies were observed on the plates. The isolation of phages capable of controlling the surviving organisms and inclusion of these phages in the phage cocktail mixture should lead to the effective control of background microflora in sputum samples during isolation of *Mycobacterium* spp. (Kumar et al. 2006).

The research described above has highlighted the possibilities and successes associated with the use of phage cocktails to control bacterial infections, and to detect bacterial pathogens in clinical and environmental samples. Still, the development of phage cocktails for effective use in phage therapy interventions remains a time- and labor-consuming endeavor. It is evident that acceptable results can only be achieved through thorough selection of phages with the required host ranges to control the bacterial species in question. Continuous monitoring of the phage cocktail, and alteration of the cocktail over time will likely be required to ensure success (Krylov 2001). A series of Polish clinical trials during the 1980s (Slopek et al. 1981a, b, 1984; Weber-Dabrowska et al. 1987) demonstrated the principles of effective phage therapy control of infections. The collective results of these studies identified several trends. For example, the researchers observed that better results were obtained when phage cocktails were tailored to each patient being treated for a given disease, as opposed to when a single phage cocktail was utilized. Overall, positive treatment outcomes were observed in 90% of the cases. The fact that commercial preparations containing pre-chosen phage mixtures were less efficient than the mixtures prepared for each patient is a potential disadvantage of phage therapy approaches that use large numbers of naturally occurring phages. The composition of a phage cocktail usually depends on the bacterial strains of a species predominating in a certain geographic region, and the proportion of the pathogenic strains affected by the most robust commercial phage cocktails never exceeds 80% under clinical situations, meaning that at least 20% of strains are not affected by the phage therapy (Krylov 2001). Also, great care should be taken when screening phage cocktails against bacterial strains isolated from a patient. Typically, the entire cocktail is screened against each strain, which does not allow for a determination of how many phages within the cocktail are effective at causing bacterial lysis. It is possible that only one of the phages within the cocktail is active against the target bacteria, which could lead to rapid development of phage-resistant bacterial mutants. Therefore, all phages in the cocktail should be individually screened against the target bacterial strains (Krylov 2001). Even when appropriate characterization of the phages within a cocktail is accomplished, the mixture can still fail to eliminate pathogenic bacteria, especially in the case of chronic infections requiring long-term treatment, during which phage-resistant bacteria can appear (Krylov 2001). Therefore, phage cocktails should contain phages capable of responding *in vivo* to the appearance of phage-resistant bacteria. The inclusion of dual specificity phages is one approach that provides phage cocktails with a built-in ability to respond to the presence of phage-resistant mutants.

4. DUAL SPECIFICITY PHAGES

It is clear that broad host range phages should be selected for phage therapy applications, and these phages should be combined into cocktails to delay the chance of the development of phage-resistant bacterial mutants. If broad host range phages cannot be isolated, an alternative approach is to take advantage of phages that can use more than one receptor on the bacterial cell surface. There are many examples of dual specificity phages described in the literature. Perhaps, the most characterized phages that display dual specificity belong to the T-even family. Approximately 170 phages with morphologies similar to T4 have been identified (Ackermann and Krisch 1997; Ackermann 1998; Tétart et al. 2001). These T-even phages have been isolated on a wide range of bacterial hosts that grow in diverse environments (Ackermann and Krisch 1997; Ackermann 1998; Tétart et al. 2001). Studies have been conducted to characterize the nature of the dual component receptors in several T-even phages. For example, each of the T-even-type phages (including the type phage, T4) binds to two different receptors on the bacterial cell surface in a sequential manner. The first reversible step is mediated by the long tail fibers, which bind to LPS or to OMPs. It is this step that determines the host-receptor specificity of the T-even phages (Beumer et al. 1984). Yu et al. (1998) characterized the T-even phage AR1 and showed that it can use OmpC as a receptor. Goodridge et al. (2003) used various *E. coli* OMP and LPS mutants to characterize the receptors of phage AR1, and showed that in addition to OmpC, phage AR1 can also utilize *E. coli* outer core OS that has a terminal Glc residue or Gal residue including

LPS outer core OS types R1 (Gal), R3 (Glc), and R4 (Gal), but not core OS types that have a terminal GlcNAc residue (core types K-12 [GlcNAc]), and R2 (GlcNAc). Other T-even phages have been shown to possess dual receptors. Beher and Pugsley (1981) showed that the T-even phage SSI could use either LamB or OmpC as receptors. Similarly, Morona et al. (1985) showed that the T-even phage Ox2 can give rise to host range mutant phages that use OmpA and OmpC as receptors unlike their parental phage. Earlier, Moreno and Wandersman (1980) observed that both OmpC and LamB could serve as receptors for host range mutants of another T-even phage, Tula, which employs OmpF as its receptor. The researchers described an elegant method in which to isolate the mutants, which initially involved plating the wild-type phage Tula on an *E. coli* strain that produced very low amounts of the OmpF and LamB proteins. The resulting phage, designated as TP2 was capable of using either OmpF or OmpC as its receptor. When Tula was plated on an OmpF mutant, phage TP5 was created, which has tri-specificity and utilized OmpF, OmpC, or LamB as a receptor. Plating Tula on other mutant strains produced other phages capable of using OmpF or LamB to adsorb to the host cell (phage TP1), or the OmpC protein as a receptor (TP6). The authors concluded that the ability of the phages to adapt to using different receptors could have important selective value. For example, this aspect of T-even biology could be used to create phages with expanded host ranges using a directed receptor approach, and a cocktail of such phages should vastly reduce the possibility of resistant phage isolates, while at the same time addressing the problem of bacteria within the same species that express varied cell surface proteins, which determines phage susceptibility.

A novel dual specificity phage was described by Scholl et al. (2001). This phage, designated as phage K1-5, is unique, because it carries two capsule-specific enzymatic tail proteins, an endosialidase and a lyase, allowing it to attack and replicate in both K1 and K5 strains of *E. coli*. One tail protein found on phage K1-5 (the lyase protein) is similar to that of phage K5 (specific for the K5 polysaccharide capsule), and a second tail protein found on this phage (the endosialidase) is similar to a tail protein found in phage K1E (specific for the K1 polysaccharide capsule). In addition, the genomic region encoding these proteins is almost identical to the genomic construct found in the *Salmonella* phage SP6, which codes for a protein that binds to the *Salmonella* O antigen (Scholl et al. 2002). The lytic phage SP6 encodes a tail protein with a high degree of sequence similarity to the tail protein of the biologically unrelated lysogenic *Salmonella* phage P22. It has also been reported that the SP6 tail gene is flanked by an upstream region that contains a promoter and a downstream region that contains a putative Rho-independent transcription terminator, giving it a cassette or modular structure almost identical to the structure of the tail genes of coliphages K1E, K5, and K1-5 (Scholl et al. 2002). It was concluded that phages SP6, K1-5, K5, and K1E are very closely related but have different tail fiber proteins, which allows each phage to have different host specificities (Scholl et al. 2002). The observation of a similar tail genome motif in both the *Salmonella* phage SP6 and the coliphages K1E, K5, and K1-5 indicates that this genomic construct might serve in the development of a modular phage platform that could operate over a wide bacterial host range. The presence of such host range platforms is described more fully below.

Other mechanisms have been found that permit the expansion of the bacterial host range of phage based on the use of more than one bacterial receptor. One such mechanism is site-specific recombination systems that permit phage to switch between alternative tail fiber proteins. Recombination sites for DNA invertases and recombination site-like sequences have been observed to flank gene segments conferring the specificity of a given phage for its host receptors. When combined with the properties of DNA inversion, it is possible that the site-specific recombination enzymes could be responsible for the exchange of host range determinants. Sandmeier (1994) extensively reviewed the scientific literature in this area, providing insight into the putative role of DNA inversion enzymes in the recombination of tail fiber gene segments between genomes.

Another mechanism by which phages modulate their host range *in situ* entails the use of a reverse transcriptase to generate variation in tail fiber proteins. Liu et al. (2002) identified a group of temperate phages that generate diversity in a gene, designated major tropism determinant (mtd), which specifies tropism for receptor molecules on host *Bordetella* species. The infectious cycles of *Bordetella* subspecies, which cause respiratory infections in humans and other mammals, is controlled by the BvgAS signal transduction system (Uhl and Miller 1996). The Bvg1 phase, which is necessary for bacterial colonization, is characterized by a high level of BvgAS activity. In the Bvg2 phase, BvgAS is inactive; the Bvg2 phase has been shown to be adapted to *ex vivo* growth and survival. Liu et al. (2002) identified several temperate phages present in clinical isolates of *Bordetella bronchiseptica* that displayed a marked tropism for Bvg1 as opposed to Bvg2 phase bacteria. The efficiency of plaque formation of a representative phage, designated BPP-1, was 10^6-fold higher on a Bvg1 phase wild-type *B.*

bronchiseptica isolate than on an isogenic Bvg2 phase-locked strain (DbcgS). Adsorption assays indicated that the BPP-1 receptor is specifically expressed in the Bvg1 phase. Mutagenesis of loci encoding Bvg1 phase surface factors showed that deletion of prn, a gene that encodes the adhesion pertactin, eliminated BPP-1 adsorption and decreased phage plaquing to a level similar to that observed on Bvg2 phase cells. Complementation of prn was sufficient to confer full infectivity by BPP-1, indicating that pertactin is the primary receptor for BPP-1. Interestingly, while the BPP-1 efficiency of plaquing decreased by a factor of 10^6 on Bvg2 phase cells, plaques that did form had a normal morphology. Since the formation of a plaque requires multiple rounds of phage infection and multiplication, this observation suggested that a tropism switch had occurred. Two types of tropic variants were subsequently identified. The first variant switched tropism to favor Bvg2 phase *Bordetella*. The second variant formed plaques with nearly equal efficiency on Bvg1 or Bvg2 phase strains. The fact that tropism correlated with specific adsorption to the respective bacteria indicated that these phages have evolved a mechanism for adapting to cell surface alterations that occur during the infectious cycles of their hosts (Liu et al. 2002). Several studies have indicated that during serial passages of phages and their bacterial hosts, the phages are not constrained in their ability to coevolve with bacteria. For example, long-term antagonistic coevolution of *Pseudomonas fluorescens* and a phage were observed over multiple cycles of defense (i.e., development of phage-resistant bacterial mutants) and counter-defense (phages isolated from two subsequent transfers showed consistently greater infectivity to bacteria) (Buckling and Rainey 2002). Similar data have also been obtained using a T-even phage, PP01, and its bacterial host, *E. coli* O157:H7 (Mizoguchi et al. 2003). A series of bacterial mutants were observed when the phage and bacteria were grown in continuous culture over a period of 200 h. The mutants differed in colony morphology, the nature of the PP01 receptors OmpC and LPS, and phage susceptibility. Phage PP01 responded to the presence of the bacterial mutants by broadening its host range. The system eventually reached a coexistence of PP01 and *E. coli* O157:H7, both at high concentrations, and the system continued to evolve. Collectively, the dynamics of both interacting systems were largely determined by the trade-offs between resistance to phage, which is usually costly from a metabolic standpoint (Brüssow 2005), and competitiveness with the parental strain for limiting resources. The inclusion of such phages that can sense the metabolic state of their host bacteria and switch their receptors accordingly within a host cocktail should, in theory, decrease the chances of phage resistance developing, since such resistance would likely decrease the virulence and survival of the bacteria in the host.

Therefore, it is possible that in a live animal, the development of phage-resistant isolates would coincide with a loss in fitness and virulence, rendering the contaminating bacteria unable to cause illness. For example, in the case of K1- or LPS-specific phages, the most likely resistant mutants will be deficient in capsule or LPS. Several studies have shown that, in both cases, the mutant bacterial isolates are less virulent. Smith and Huggins (1982) inoculated calves with a K1 strain of *E. coli*, followed by the addition of a K1-specific phage. A low number of phage resistant *E. coli* strains were isolated from the calf intestine, but due to the loss of the K1 antigen, the strain had become avirulent in mice. Therefore, great care must be taken when interpreting the results of phage therapy *in vitro* studies, since *in vitro* predator–prey studies do not accurately reflect *in vivo* conditions. Such studies cannot account for the complexity of the phage–host interaction in the natural environment. In pure culture experiments, the only competitor of the phage-resistant cell is its phage susceptible ancestor cell, which is counter-selected in the presence of a phage. However, in the natural environment, the phage-resistant clone must also compete with many other strains that are not subject to the pressure of the infecting phage (Brüssow 2005). This principle was demonstrated by Harcombe and Bull (2005), who conducted an *in vitro* study of *E. coli*, *Salmonella enterica* serovar *typhimurium*, and coliphages T5 and T7. The researchers showed that, while resistant mutants could be obtained from pure *E. coli* cultures, when the *E. coli* cells were mixed with the *S. typhimurium* cells, the resistant *E. coli* cells were outcompeted by the *Salmonella*.

5. GENETIC MANIPULATION OF HOST RANGE

In addition to the natural ability of certain phages to expand their host ranges, lytic phages can be engineered to produce recombinant phages with desired host ranges. In this scenario, the host range of a respective phage would be modified by replacing its tail fiber loci with that of another phage that has the required host range.

Several phages are capable of altering their host range via a site-specific recombination system that inverts the sequence that determines the host range (Plasterk et al. 1983). For example, phage Mu is a broad host range phage capable of forming plaques on several bacterial species (Harshey 1988). The Mu genome contains tail fiber gene S, which encodes a site recognized by a Mu-encoded invertase, with a second site beyond the gene in inverted orientation. Depending on the orientation of the genomic segment between the two sites, the tail fiber has alternative carboxyl termini that encode different specificities (Plasterk et al. 1983). T-even phages such as AR1 use an alternative strategy to obtain adhesion diversity. The T-even phages exchange their adhesion domains with related ones from other phages, possibly via a specialized recombination system (Plasterk et al. 1983). Homologous regions within tail fiber genes have been identified in phage Mu; phages P1 and P2; the T-even phages Tula, Tulb, and T4; and phage lambda (Haggard-Ljungquist et al. 1992). The similarities in the tail fiber genes of phages from different families provide evidence that recombination events are occurring between unrelated phages, an observation that is compatible with the modular theory of phage evolution (Botstein 1980). Several researchers have demonstrated the utility of such a method to produce chimeric phages with the ability to infect bacteria that were previously resistant to the phage. These studies have been accomplished using T-even phages. The adsorption specificity of phages in the T4-like virus genus is determined by the protein sequence near the tip of the long tail fibers (the distal tail fiber locus) (Tétart et al. 1998). The tail fiber adhesin domains are located within different genes in closely related phages of the T-even type. For example, in phage T4, the adhesin sequence is encoded by the C-terminal domain of the large tail fiber gene (gene 37), while phage T2 encodes the adhesin as a separate gene product (gp) (gene 38) that binds to the tip of the T2 tail fibers (34). Three groups of T-even phages that possess the tail fiber organization of phage T2 have been identified, including the T2-like, T6-like, and Ac3-like sequences (Tétart et al. 1998). The sequences of genes 37 and 38 vary to differing degrees among these phages. Tétart et al. (1996, 1998) were the first to show that T4-like phages have conserved regions flanking their host range genes, and that these conserved sequences could be used to amplify the host range genes (genes 37 and 38) of many T4-like phages. For the T4-like phages, there are two completely different organizations of the distal tail fiber locus—the T2 and T4 forms. The sequences of gene 37 and gene 38 vary to differing degrees among the T4-like phages, while those of gene 36 and gene t, which flank the locus, are conserved (Yu et al. 2000). Using the conserved regions of gene 36 and gene t, Tétart et al. (1996) amplified a DNA segment of two *Yersinia* T4-like phages that contained genes 37 and 38, cloned each fragment into a plasmid, and used a phage T4 mutant that had amber mutations in genes 37 and 38 (am37, am38) to infect cells carrying each of the plasmids. Chimeric T4 phages, which replaced their distal tail fiber locus with the plasmid-borne analog, were isolated and shown to form plaques on *Yersinia* (Tétart et al. 1996). Since then, several other chimeric host range mutants of phage T4 have been constructed, by utilizing the same strategy (Tétart et al. 1998). Following on Tétart's work, Yoichi et al. (2005) changed the specificity of phage infection, by exchanging gps 37 and 38, expressed at the tip of the long tail fiber of T2, with those of PP01 phage, an *E. coli* O157:H7 specific phage. Homologous recombination between the T2 phage genome and a plasmid encoding the region around genes 37–38 of PP01 occurred in transformant *E. coli* K12 cells. The recombinant T2 phage, named T2ppD1, carried PP01 gp37 and 38 and infected the *E. coli* O157:H7. However, T2ppD1 could not infect *E. coli* K12, the original host of T2, or its derivatives. The host range of T2ppD1 was the same as that of PP01. Infection of T2ppD1 produced turbid plaques on a lawn of *E. coli* O157:H7 cells. The binding affinity of T2ppD1 to *E. coli* O157:H7 was weaker than that of PP01. Yoichi et al. (2005) concluded that, in addition to the tip of the long tail fiber, exchange of gps expressed in the short tail fiber may be necessary for tight binding of recombinant phage. The transfer of tail fiber loci among T-even phages represents a host range exchange platform system. Such an expanded host range platform phage could provide for versatility and save time and effort compared with that required for the isolation and characterization of completely new phages for each bacterial strain.

6. CONCLUSION

The use of phages to control bacterial infections has progressed to the point where effective methods have been developed that address the issues of broad bacterial infectivity and phage resistance. These methods include the combination of broad host range phages and phages that can alter their host range *in situ* within cocktails that can be used to treat bacterial disease. Alternatively, the host ranges of phages can be genetically manipulated to create phages with desired host ranges. The use of well-developed phage cocktails represents a practical solution to the rapid emergence of new infectious diseases. Phage cocktails can be developed more rapidly than new antibiotics, and with less cost. A potential challenge to the advance of phage therapy is regulation of the treatment. All of the phages in a given cocktail will need to be appropriately characterized before they can be used in clinical treatment, and the manner in which the cocktails are developed will also affect regulation. For example, the use of recombinant phages in a cocktail might complicate regulation because of the use of genetically modified organisms. Regardless, the use of rational and logical methods to produce phage cocktails that are capable of treating bacterial diseases, and the publication of studies indicating the effective use of phage cocktails as antimicrobials will ensure the continuing interest in the use of phage as novel biocontrol agents.

REFERENCES

Ackermann H. W. (1998) *Advances in Virus Research* **51**, 135–201.

Ackermann H. W. and H. M. Krisch (1997) *Archives of Virology* **142**, 2329–2345.

Allison G. E. and T. R. Klaenhammer (1998) *International Dairy Journal* **8**, 207–226.

Andreatti Filho R. L., J. P. Higgins, S. E. Higgins, G. Gaona, A. D. Wolfenden, G. Tellez, and B. M. Hargis (2007) *Poultry Science* **86**, 1904–1909.

Beher M. G. and A. P. Pugsley (1981) *Journal of Virology* **38**, 372–375.

Beumer J., E. Hannecart-Pokorni, and C. Godard (1984) *Bulletin De L'Institut Pasteur (Paris)* **82**, 173–253.

Botstein D. (1980) *Annals of the New York Academy of Science* **354**, 484–490.

Brüssow H. (2005) *Microbiology* **151**, 2133–2140.

Buckling A. and P. B. Rainey (2002) *Proceedings of the Royal Society of London. Series B* **269**, 931–936.

Callaway T. R., T. S. Edrington, A. D. Brabban, E. S. Kutter, R. C. Anderson, and D. J. Nisbet (2006) *Proceedings of International Conference on Perspectives of Bacteriophage Preparation*, 72–83.

Chase J., N. Kalchayanand, and L. D. Goodridge (2005) Use of bacteriophage therapy to reduce Escherichia coli O157:H7 concentrations in an anaerobic digestor that simulates the bovine gastrointestinal tract. Institute of Food Technologists Annual Meeting and Food Expo, New Orleans, LA, abstract 108-6.

Coyette J. and J. M. Ghuysen (1968) *Biochemistry* **7**, 2385–2389.

Dinsmore P. K. and T. R. Klaenhammer (1995) *Molecular Biotechnology* **4**, 297–314.

Edwards S. and G. G. Meynell (1968) *The Journal of General Virology* **2**, 443–444.

Goodridge L. D., A. Gallaccio, and M. W. Griffiths (2003) *Applied and Environmental Microbiology* **69**, 5364–5371.

Haggard-Ljungquist E., C. Halling, and R. Calendar (1992) *Journal of Bacteriology* **174**, 1462–1477.

Hanlon G. W. (2007) *International Journal of Antimicrobial Agents* **30**, 118–128.

Harcombe W. R. and J. J. Bull (2005) *Applied and Environmental Microbiology* **71**, 5254–5259.

Harshey R. M. (1988) Phage Mu. In *The Bacteriophages*, vol. **1**, ed. R. Calendar, 193–294. New York: Plenum Press.

Hashemolhosseini S., Z. Holmes, B. Mutschler, and U. Henning (1994a) *Journal of Molecular Biology* **240**, 105–110.

Hashemolhosseini S., D. Montag, L. Kramer, and U. Henning (1994b) *Journal of Molecular Biology* **241**, 524–533.

Heinrichs D. E., M. A. Monteiro, M. B. Perry, and C. Whitfield (1998) *The Journal of Biological Chemistry* **273**, 8849–8859.

Heller K. J. (1992) *Archives in Microbiology* **158**, 235–248.

Henning U. and S. Hashemolhosseini (1994) Receptor Recognition by T-Even-Type Coliphages. In *Molecular Biology of Bacteriophage T4*, ed. J. D. Karam, 291–298. Washington, DC: American Society for Microbiology.

Jikia D., N. Chkhaidze, E. Imedashvili, I. Mgaloblishvili, G. Tsitlanadze, R. Katsarava, J. Glenn Morris Jr., and A. Sulakvelidze (2005) *Clinical and Experimental Dermatology* **30**, 23–26.

Krylov V. N. (2001) *Russian Journal of Genetics* **37**, 715–730.

Kudva I. T., S. Jelacic, P. I. Tarr, P. Youderian, and C. J. Hovde (1999) *Applied and Environmental Microbiology* **65**, 3767–3773.

Kumar V., S. Balaji, N. S. Gomathi, P. Venkatesan, G. Sekar, K. Jayasankar, and P. R. Narayanan (2006) *Journal of Microbiological Methods* **68**, 536–542.

Lindberg A. A. (1973) *Annual Reviews of Microbiology* **27**, 205–241.

Liu M., R. Deora, S. R. Doulatov, M. Gingery, F. A. Eiserling, A. Preston, D. J. Maskell, R. W. Simons, P. A. Cotter, J. Parkhill, and J. F. Miller (2002) *Science* **295**, 2091–2094.

Luria S. E. and M. Delbrück (1943) *Genetics* **28**, 491–511.

Markoishvili K., G. Tsitlanadze, R. Katsarava, J. G. Morris Jr., and A. Sulakvelidze (2002) *International Journal of Dermatology* **41**, 453–458.

McVay C. S., M. Velásquez, and J. A. Fralick (2007) *Antimicrobial Agents and Chemotherapy* **51**, 1934–1938.

Mizoguchi K., M. Morita, C. R. Fischer, M. Yoichi, Y. Tanji, and H. Unno (2003) *Applied and Environmental Microbiology* **69**, 170–176.

Montag D., S. Hashemolhosseini, and U. Henning (1990) *Journal of Molecular Biology* **216**, 327–334.

Moreno F. and C. Wandersman (1980) *Journal of Bacteriology* **144**, 1182–1185.

Morona R., J. Tommassen, and U. Henning (1985) *European Journal of Biochemistry* **150**, 161–169.

O'Flynn G., R. P. Ross, G. F. Fitzgerald, and A. Coffey (2004) *Applied and Environmental Microbiology* **70**, 3417–3424.

Plasterk R. H., T. A. Ilmer, and P. Van de Putte (1983) *Virology* **127**, 24–36.

Sandmeier H. (1994) *Molecular Microbiology* **12**, 343–350.

Scholl D., S. Adhya, and C. R. Merril (2002) *Journal of Bacteriology* **184**, 2833–2836.

Scholl D., S. Rogers, S. Adhya, and C. R. Merril (2001) *Journal of Virology* **75**, 2509–2515.

Slopek S., I. Durlakova, B. Weber-Dobrowska, A. Kucharewica-Krukowska, M. Dabrowski, and R. Bisikiewic (1981a) *Archivum Immunologiae Et Therapiae Experimentalis* **31**, 267–291.

Slopek S., I. Durlakova, B. Weber-Dobrowska, A. Kucharewica-Krukowska, M. Dabrowski, and R. Bisikiewic (1981b) *Archivum Immunologiae Et Therapiae Experimentalis* **31**, 293–327.

Slopek S., I. Durlakowa, B. Weber-Dobrowska, M. Dabrowski, and A. Kucharewicz-Krukowska (1984) *Archivum Immunologiae et Therapiae Experimentalis* **32**, 317–335.

Smith H. W. and M. B. Huggins (1982) *Journal of General Microbiology* **128**, 307–318.

Stirm S. (1968) *Nature* **10**, 637–639.

Tanji Y., T. Shimada, H. Fukudomi, K. Miyanaga, Y. Nakai, and H. Unno (2005) *Journal of Bioscience and Bioengineering* **100**, 280–287.

Tanji Y., T. Shimada, M. Yoichi, K. Miyanaga, K. Hori, and H. Unno (2004) *Applied Microbiology and Biotechnology* **64**, 270–274.

Tétart F., C. Desplats, and H. M. Krisch (1998) *Journal of Molecular Biology* **282**, 543–556.

Tétart F., C. Desplats, M. Kutateladze, C. Monod, H. W. Ackermann, and H. M. Krisch (2001) *Journal of Bacteriology* **183**, 358–366.

Tétart F., F. Repoila, C. Monod, and H. M. Krisch (1996) *Journal of Molecular Biology* **258**, 726–731.

Toro H., S. B. Price, S. McKee, F. J. Hoerr, J. Krehling, M. Perdue, and L. Bauermeister (2005) *Avian Diseases* **49**, 118–124.

Uhl M. A. and J. F. Miller (1996) *The EMBO Journal* **15**, 1028–1036.

Weber-Dabrowska B., M. Dabrowski, and S. Slopek (1987) *Archivum Immunologiae Et Therapiae Experimentalis* **35**, 363–368.

Wilson J. H., R. B. Luftig, and W. B. Wood (1970) *Journal of Molecular Biology* **51**, 423–434.

Yoichi M., M. Abe, K. Miyanaga, H. Unno, and Y. Tanji (2005) *Journal of Biotechnology* **115**, 101–107.

Yu S. L., H. W. Ding, J. N. Seah, K. M. Wu, Y. C. Chang, M. F. Tam, and W. Syu Jr. (1998) *Journal of Biomedical Science* **5**, 370–382.

Yu S. L., K. L. Ko, C. S. Chen, Y. C. Chang, and W. Syu, Jr. (2000) *Journal of Bacteriology* **182**, 5962–5968.

CHAPTER 10

IDENTIFYING PHAGE LYTIC ENZYMES: PAST, PRESENT, AND FUTURE

JONATHAN E. SCHMITZ, RAYMOND SCHUCH,
and VINCENT A. FISCHETTI
Laboratory of Bacterial Pathogenesis and Immunology, The Rockefeller University,
New York, NY

1. INTRODUCTION

The alarming increases in bacterial resistance to traditional antibiotics have spurred the scientific community to renew its focus on the development of novel therapeutic strategies. In this setting, bacteriophage lytic enzymes have received considerable attention as potential enzybiotic agents. These viral proteins—also referred to as phage lysins, endolysins, or simply lysins—are a class of peptidoglycan hydrolases capable of killing Gram-positive bacteria, whose cell walls are directly accessible from the extracellular space (Fischetti 2005). The lysins of Gram-positive bacteria demonstrate a modular architecture: their N-terminal domains possess either muramidase (i.e., lysozyme), glucosaminidase, endopeptidase (targeting several possible bonds), or alanine-amidase activity, while their C-terminal regions are responsible for binding the target cell envelope. A limited number of bifunctional lysins that possess two distinct catalytic domains have also been reported. Lytic enzymes are encoded by virtually all ds-DNA bacteriophages, consisting mainly of tailed phages of the order Caudovirales (comprising the family Myoviridae, Siphoviridae, and Podoviridae), but also including the much rarer tailless phages of the family Tectiviridae (Verheust et al. 2004).

Enzybiotics: Antibiotic Enzymes as Drugs and Therapeutics. Edited by Tomas G. Villa and Patricia Veiga-Crespo

The appeal of phage lysins lies in their mechanism of action, their potency, and the specificity they demonstrate toward individual species, strains, or serovars (typically whatever types of organisms the encoding phage infects). Between studies on isolated bacteria (planktonic and in biofilms) and those involving animal models of colonization/infection, representative lysins have already demonstrated efficient bacteriolytic activity against a large portion of major Gram-positive pathogens. These findings support obvious therapeutic applications, in addition to the potential value of lysins in the areas of agricultural, veterinary, and food science. A number of recent reviews have focused on phage lysins (Fischetti 2005; Loessner 2005; Borysowski et al. 2006; Fischetti et al. 2006; Chapter 5 of this book), and the reader is referred to these reviews for detailed information on phage lysins' biochemistry and biotechnological applications.

Despite the work dedicated to these molecules to date, the number of lysins that have been recombinantly expressed and functionally tested is still only a small fraction of the total number of lysins encoded by global phage. Bacteriophages, in fact, are thought to be the single greatest source of genetic information on the planet, with an astounding 10^{31} total phage particles estimated in the biosphere (Hendrix 2002; Hatfull 2008). Even among the known phage/prophage of Gram-positive pathogens, only a minority have seen their lytic enzymes cloned, expressed, and examined as anti-infective agents. In itself, the sheer number of lysins with the potential for development as enzybiotics makes this class of proteins extremely attractive pharmacologically. Considering this magnitude, future research will undoubtedly continue to isolate new lysins to complement those already in development. And while it is the activity of these enzymes that will ultimately garner attention, the success of such work is fundamentally dependent on the techniques employed to identify the proteins in the first place. The purpose of the current review, therefore, is to address the issue of lysin identification. Following a summary of lysin isolation prior to modern molecular cloning, we will outline the various techniques currently available for identifying lysin-encoding genes within viral and bacterial genomes. The relative advantages and disadvantages of these approaches will be discussed, especially in the context of ongoing technological advances. Furthermore, we will present recent and nontraditional strategies for identifying phage lysins, as well as future directions and challenges facing this line of research. The basic goal of this review is not to provide detailed screening protocols (although appropriate references are included), but rather to offer general experimental strategies that can be employed when searching for new enzybiotics. Overall, through the efficient identification of novel phage lysins, researchers can only broaden the potential impact of these proteins and hasten their development into true clinical tools.

2. HISTORICAL PERSPECTIVES ON LYSIN IDENTIFICATION

While medical interest in phage lysins is a recent phenomenon, initial observations of these enzymes and attempts to purify them date back much earlier. In 1921, Felix d'Hérelle first proposed the existence of a phage-associated enzyme that was capable of lysing bacteria independently from total phage action. His idea was based on experiments with alcohol-denatured phage; d'Hérelle (1921) was able to extract an active agent that could lyse bacilli but not propagate between cultures. Soon after, Vladimir Sertic (1929) reported the isolation of a "lysine d'une race du bactériophagie" that was responsible for creating altered morphological zones that surrounded *Escherichia coli* plaques proper. The lysin hypothesis was contested at the time, and it is difficult to judge whether these initial observations were truly due to the activity of what we now know as phage lytic enzymes. Nevertheless, other contemporary studies did report the tendency of nonviable Gram-positive bacteria to lyse when in the presence of live bacteria and their corresponding phage (Gratia and Rhodes 1923; Twort 1925; Bronfenbrenner and Muckenfuss 1927). In retrospect, this activity could be attributed to the diffusion of lysin from dead cells.

It was not until several decades later that the existence and activity of lysins were broadly accepted by the scientific community. In 1955, Ralston et al. described a lytic agent that appeared in the supernatant of a *Staphylococcus aureus* culture following infection by bacteriophage P_{14}. This protein, which they termed virolysin, could be separated from intact phage through ultracentrifugation and concentrated by ammonium sulfate precipitation (Ralston et al. 1957). Their work was notable in that the phage enzyme was isolated and characterized alongside an endogenous lytic enzyme produced by the host staphylococci—the distinction between these molecules allowed the phage origin of virolysin to be established. It was subsequently demonstrated that antigenically distinct lytic enzymes could be isolated when the *S. aureus* host was infected with diverse phage (Ralston and McIvor 1964). In later works, two staphylococcal lysins (from phages 80 and 53) were purified beyond ammonium sulfate precipitation by ion exchange chromatography (Doughty and Mann 1967; Sonstein et al. 1971). One should note that, in these initial attempts to isolate *S. aureus* lysins, the final products did not demonstrate the degree of activity currently associated with lytic enzymes. Of the above (semi-pure) enzymes, only the phage-53 lysin could successfully lyse viable staphylococci; the phage-80 lysin was only active against isolated cell walls and the P_{14}-virolysin required the bacteria to be "sensitized" by one of several additional agents. It is unclear whether this diminished activity was due to less-than-ideal purification conditions at the time, or simply reflects the efficacy of these particular enzymes.

During the same period as this work on staphylococcal enzymes, similar research was progressing on other bacteria-bacteriophage combinations. Again with the strategy of isolating the proteins from culture supernatant, lysins were investigated from phages whose hosts included enterobacteria (Maass and Weidel 1963; Inouye and Tsugita 1966; Rao and Burma 1971), bacilli (Murphy 1957; Welker 1967), lactococci (Tourville and Johnstone 1966), and streptococci (Reiter and Oram 1963; Oram and Reiter 1965). Overall, one of the most extensively studied lysins of this time was that of the C_1 phage infecting Lancefield group C streptococci. Two 1957 publications documented its ability to lyse not only live streptococci of the same type, but also live groups A and E streptococci (Krause 1957; Maxted 1957). Over the following years, several increasingly sophisticated attempts were made to purify this lysin using a combination of ammonium sulfate precipitation, gel filtration, and calcium phosphate adsorption (Krause 1958; Doughty and Hayashi 1962). In the process, it became a valuable tool for selectively removing and characterizing antigenic components of the streptococcal cell envelope. The final obstacle to achieving a highly pure protein preparation (the presence of reactive sulfhydyl groups) was overcome by Fischetti et al. (1971), who utilized reversible sodium tetrathionate-protecting groups to stabilize the lysin during chromatography. Interestingly, these nonrecombinant purification schemes for the C_1-phage lysin, currently referred to as PlyC, endured into the modern era of lysins as enzybiotics. In 2001, PlyC was one of the first lysins to effectively decolonize a bacterial pathogen *in vivo* (Nelson et al. 2001), even though the encoding genomic region was not cloned and characterized until several years later (Nelson et al. 2006).

3. INTO THE AGE OF MOLECULAR CLONING

Following the elucidation of the central dogma of molecular biology, attempts to characterize (and ultimately identify) phage lytic enzymes through genetic techniques began in earnest by the late 1960s. The first lysin amino acid sequence was reported by Inouye and Tsugita (1966) for the *E. coli* T4 phage muramidase. It was determined through Edman analysis of culture-purified enzyme, and it suggested an approximate nucleotide sequence by reverse translation. This work on the T4 lysin, in fact, provided important evidence confirming the very nature of the triplet genetic code. By inducing mutations in the T4 genome and observing the corresponding frameshifts in the purified lysin, the authors were able to verify Crick's hypothesis regarding the language of DNA codons (Terzaghi et al. 1966; Okada et al. 1968). The definitive nucleotide sequence for the T4 lysin was not published until 1983, when Owen et al. (1983) successfully cloned the gene by its ability to rescue a lysis-defective phage strain. This sequence, in turn, allowed several additional lysin genes to be recognized through nucleotide homology. In 1985, the *Salmonella* P22-lysin was identified in this manner from a sequenced fragment of the viral genome (Rennell and Poteete 1985). The first nucleotide sequence of a Gram-positive lysin, from *Bacillus subtilis* phage φ29, was likewise reported 1 year later (Garvey et al. 1986). It is important to note that the above work was conducted at the same time as the advent of modern protein expression technology; shortly after their cloning, the T4 and φ29 lysins were the first such enzymes to be expressed and purified recombinantly from *E. coli* (Perry et al. 1985; Saedi et al. 1987).

With molecular techniques at their disposal, researchers could now identify lysins directly from phage DNA, and this ability was soon utilized to search for enzymes against Gram-positive pathogens. One of the first bacterial species targeted in this regard was *Streptococcus pneumoniae*, as the 1980s and 1990s saw the cloning, expression, and functional analysis of a number of pneumococcal phage lysins. The same general strategy was employed in identifying each of these proteins: following the cloning of an initial prototype enzyme, the gene was utilized to identify related lysins through Southern blot analysis. Ironically, the prototype enzyme for these pneumococcal lysins was not actually viral in origin, but rather a genomic peptidoglycan hydrolase encoded by *S. pneumoniae*, LytA. Like the endogenous *S. aureus* lysin from Ralston et al. (1957), LytA is a member of a diverse class of cell wall-degrading enzymes known as autolysins. Encoded by virtually all bacteria, the autolysins are involved in processes such as bacterial growth, division, sporulation, and signaling (Vollmer et al. 2008). Specifically, LytA is an alanine amidase with various roles in pneumococcal physiology, including the release of cytoplasmic virulence factors during pathogenesis (Jedrzejas 2001) and the predation of non-competent cells by competent ones within pneumococcal communities (Guiral et al. 2005). In general, the autolysins are evolutionarily related to phage lytic enzymes, as they often demonstrate sequence homology and similar domain architectures—this relationship was certainly true for LytA.

In 1985, García et al. (1985) cloned LytA through a complementation strategy in which an *S. pneumoniae* genomic fragment was identified by its ability to rescue autolytic activity in a mutant *lytA* strain. The resultant sequence was subsequently used to probe genomic DNA from several *S. pneumoniae* phages in order to isolate their lysin genes. Successful hybridizations were noted in many instances, and the corresponding bands were cloned and sequenced to reveal viral LytA homologs. Specifically, the lysins from the following phage were identified (with corresponding enzyme names): Cp-1 (CPL-1; García et al. 1988); Cp-7 and Cp-9 (CPL-7 and CPL-9; García et al. 1990); HB-3 (HBL; Romero et al. 1990); EJ-1 (EJL; Díaz et al. 1992); and Dp-1 (Pal; Sheehan et al. 1997). These enzymes demonstrated varied overall sequence homology to LytA itself. HBL and EJL are highly homologous to the autolysin throughout their entire sequence, as they share both LytA's N-terminal amidase domain and its C-terminal choline-binding domain. By contrast, CPL-7 and CPL-9 possess muramidase activity and are homologous to LytA only at the C-terminal-binding end. Pal demonstrates only C-terminal homology; although the lysin is an alanine amidase, its enzymatic domain does not align with that of LytA. Of these lysins, CPL-1 has received by far the most attention subsequently as an *in vivo* enzybiotic, as successful trials have been conducted involving rodent models of bacteremia (Loeffler et al. 2003), endocarditis (Entenza et al. 2005), otitis media (McCullers et al. 2007), and meningitis (Grandgirard et al. 2008).

As for the hybridization strategy itself, the technique is no longer commonly used for identifying lytic enzymes. A potential reason for this is logistical: with more recent cloning strategies (described in the following sections), Southern blotting can be relatively cumbersome by comparison. There also exists the possibility of limiting the results due to the sequence of the original DNA probe. With this approach, it is only possible to identify genes with some homology to a prototype sequence. But if a phage is relatively novel—for instance, in terms of its host bacterium or viral morphotype—suitable homologs might not exist for its lysin. This is especially problematic considering that enzymes with highly novel sequences represent some of the most attractive targets for future discovery.

Nevertheless, if one has good reason to suspect that a desired lysin is similar in sequence to one already characterized, techniques based on nucleotide homology can still be quite effective. In this regard, several studies have utilized polymerase chain reaction (PCR)-based approaches as more rapid alternatives to Southern blotting. For example, Morita et al. (2001b) designed primers from the genomic regions surrounding the lysin of a *B. subtilis* phage, which they used to amplify a related enzyme for *Bacillus amyloliquefaciens*. Romero et al. (2004) likewise synthesized various primers based on known LytA-like sequences when attempting to clone the lysins from two *Streptococcus mitis* phages. These authors successfully identified a primer pair that amplified a portion of both enzymes; these partial sequences were subsequently used to characterize the remainder of the genes by genomic primer walking (i.e., chain termination sequencing with the genome as the direct template). We should mention that, although they have not been applied specifically to lytic enzymes, several other techniques are available that could identify a complete lysin gene from only a partial sequence. These include inverse PCR (Ochman et al. 1988) and semi-random PCR (Hermann et al. 2000).

4. PHAGE LYSINS AND FUNCTIONAL SCREENING

To avoid the possibility of sequence-based bias, it is ultimately necessary to identify lysin-encoding genes by the enzymatic activity of their translated proteins. This approach is the foundation of functional genomic screening, and it has become a common tool for lysin identification over the past decade. The experimental specifics of lysin screening can differ slightly, and these variables are reviewed in the proceeding paragraphs. Overall, however, such methods represent variations on the same general theme: (a) phage genomic DNA is isolated and digested into fragments; (b) the fragments are ligated into an expression vector and transformed into a host organism; (c) the transformants are clonally propagated and exposed to an inducing agent to force transcription of the genomic inserts; and (d) the clones are analyzed for the acquisition of a phenotype that indicates the presence of a lysin-encoding gene. For the first three steps, the experimental considerations are fairly general in nature and unrelated to the activity of the targeted enzymes. The original source DNA can be derived from either lytic phage, isolated from the environment or purchased from commercial sources, or lysogenic prophage, induced from host bacteria with an appropriate stressing agent. For the fragmentation step, most lysin screens have utilized a standard shotgun approach to create a random array of genomic fragments. Here, the phage DNA is partially digested with restriction enzyme, usually one with a 4-bp consensus sequence. A final length distribution of 1.5–3 kb is ideal, as it represents 2–3 times the length of typical Gram-positive lysins.

While alternate methods have been used on occasion,1 the second step typically involves ligation of these fragments into an expression plasmid with transformation of an *E. coli* host. One should note that other vectors, in theory, could support longer DNA inserts (i.e., cosmids or bacterial artificial chromosomes). However, these systems suffer from the fact that they would rely upon native promoters for recombinant expression. Given the small size of *Caudoviral* genomes (several dozens to several hundreds kilobases), plasmid-based screens are still capable of identifying lytic clones with high efficiency. In past studies, for example, hits have typically been observed at a frequency of 0.1%– 2% of total colonies (Schuch et al. 2008). When choosing a particular plasmid for lysin screening, the same variables must be taken into consideration as during the recombinant expression of any protein. These include the type of promoter, induction conditions, codon usage, and the host *E. coli* strain. For a given lysin, the expression level and solubility can vary significantly from one system to the next, often in unpredictable ways. Nevertheless, for a detailed protocol that we have found generally reliable, the reader is referred to Schuch et al. (2008). The screen outlined here utilizes an arabinose-inducible pBAD plasmid, which is attractive for its tight transcriptional control and the cost-effective nature of the inducing agent.

For the final steps of a lysin screen, induction and selection of positive clones, the following experimental manipulations are typically involved. Transformed *E. coli* are spread onto agar plates that lack inducing agents, allowing clones to proliferate without transcription of genomic inserts. These master plates are replicated onto screening plates whose agar has been supplemented with inducing agents. After propagating with forced transcription, the clones are exposed to chloroform vapor to permeabilize the *E. coli* and allow free diffusion of the expressed proteins. The clones are overlaid with a soft agar media containing Gram-positive cells. The plates are observed over time for clones over which there develops a zone of diminished bacterial density, indicating the presence of a lysin-encoding gene (see Fig. 10.1). The corresponding clone on the master plate is subsequently identified and expanded for sequencing and large-scale expression. In early functional screens, the process occasionally differed in minor aspects (for instance, in the logistics of replica plating or cell permeabilization). In current studies, however, this procedure has become the norm.

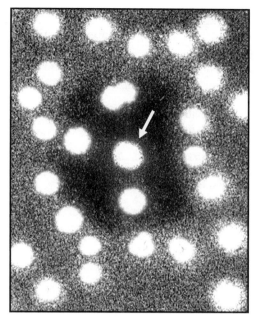

Figure 10.1. Lysin-induced clearing zone. To identify a phage lystic enzyme in a functional screen, Gram-positive bacteria are overlayed on permeabilized *E. coli* clones expressing phage genomic inserts. The desired clone (indicated above with an arrow) is identified by the development of a surrounding halo of Gram-positive lysis. The particular example shown here involves *B. anthracis* (strain 222) cells and the PlyB lysin from the BcpI phage (Porter et al. 2006). One should note that the size and intensity of the halo can vary depending on the particular lysin and the overlaid bacterial species/strain.

The main enduring variable in the process of lysin identification lies at the end of the procedure, as several options are available regarding the type and quantity of Gram-positive bacteria in the soft agar overlay step. Table 10.1 summarizes these techniques and the individual studies in which they have been employed over the years. In a few studies, permeabilized clones were overlaid with concentrated *Micrococcus* cells (Jayaswal et al. 1990; Bon et al. 1997). Decreased micrococcal turbidity is a classic method of quantifying the activity of eukaryotic peptidoglycan hydrolases (Shugar 1952). The majority of screens, however, have utilized the host bacteria of the phage whose genome was being screened. This makes intuitive sense considering the specificity that lytic enzymes demonstrate toward host organisms. At the same time, several variations do exist as to how these cells are applied. The permeabilized clones can be overlaid with either (a) concentrated-viable bacteria, (b) concentrated-nonviable bacteria

TABLE 10.1. Lysins through Functional Genomic Screening

Year	Screening Strategy	Target Bacteria	Reference
1990	Micrococcal overlay	*Staphylococcus aureus*	Jayaswal et al.
1997	Micrococcal overlay	*S. aureus*	Bon et al.
1989	λ-phage expression; Target GPB: Conc. viable	*Lactococcus lactis*	Shearman et al.
1990	Target GPB: Conc. viable	*Lactobacillus bulgaricus*	Boizet et al.
1992	Target GPB: Conc. viable	*L. lactis*	Platteeuw and de Vos
1993	Target GPB: Conc. viable	*L. lactis*	Ward et al.
1995	Target GPB: Conc. viable	*Listeria monocytogenes*	Loessner et al.
1997	Target GPB: Conc. viable	*Bacillus cereus*	Loessner et al.
1998	Target GPB: Conc. viable	*S. aureus*	Loessner et al.
1999	Target GPB: Conc. viable	*S. aureus*	Loessner et al.
2002	Target GPB: Conc. viable	*Bacillus anthracis*	Schuch et al.
2005	Target GPB: Conc. viable	*Streptococcus agalactiae*	Cheng et al.
2004	Target GPB: Conc. nonviable	*Lactobacillus helveticus*	Deutsch et al.
2005	Target GPB: Conc. nonviable	*Lactobacillus gasseri*	Yokoi et al. (b)
2005	Target GPB: Conc. nonviable	*Staphylococcus warneri*	Yokoi et al. (a)
1986	Target GPB: Dilute viable	*Lactobacillus delbrueckii*	Trautwetter et al.
2004	Target GPB: Dilute viable	*Entercoccus faecalis*	Yoong et al.
2006	Target GPB: Dilute viable	*B. anthracis*	Portet et al.
2008	Target GPB: Dilute viable	*B. anthracis*	Schmitz et al. (a)
2001	Target GPB: Not reported	*Lactobacillus plantarum*	Yoon et al.
1995	λ-phage complementation	*L. gasseri*	Henrich et al.

Listed here is (to our knowledge) a complete list of recombinantly expressed phage lytic enzymes against Gram-positive bacteria (GBP) for which the original means of identification was a functional genomic screen. They are organized by screening methodology, and include the corresponding target bacteria and year of identification. The methodologies include overlay by micrococcal cells; overlay by concentrated, viable GPB of the target species; overlay by concentrated, nonviable GPB; and overlay by dilute, viable GPB. For two entries, (Shearman et al. 1989; Henrich et al. 1995), a λ-phage screening strategy was used in place of a plasmid-based approach (see Endnote 1 for more information). Included in the table are a number of entries for lactococci and lactobacilli; for these lysins, the primary interest lies in the area of dairy science. Conc., concentrated.

(typically autoclaved), or (c) dilute-viable bacteria. In the first two cases, plates are observed for clones around which bacterial density decreases, while, in the third, they are observed for clones around which bacteria fail to proliferate. Previous examples of each approach include, respectively: several *Listeria monocytogenes* lysins (Loessner et al. 1995); an enzyme targeting a novel *Staphylococcus* strain (Yokoi et al. 2005a); and the amidase of the γ diagnostic phage of *Bacillus anthracis* (Schuch et al. 2002).

While there has never been a dedicated study comparing the relative efficacy of these three variations, we have utilized each type of overlay within our laboratory and found them all to be generally effective. One notable benefit, however, of employing dilute-viable cells is its particularly high level of sensitivity. Since only enough lysin is required to prevent the growth of a small initial population, clearing zones are often evident even when an enzyme is not well expressed under the screening conditions. Using this method, we have observed clearing zones in instances when a particular clone does not produce sufficient lysin for recombinant purification or even detection by Coomassie-staining (unpublished observations). While this sensitivity can necessitate re-cloning into a different expression vector, it does address the important initial goal of lysin gene identification. Several other factors must likewise be considered when selecting a particular overlay technique. For instance, certain bacteria can react poorly to heat killing (e.g., with aggregation or lysis), rendering the concentrated-nonviable approach ineffective. Other bacterial organisms can interact nonspecifically with the *E. coli* library clones when proliferating in soft agar, leading to widespread *pseudo*-clearing clones for the dilute-viable method. While it is difficult to predict what approach is ideal for a given species/strain, this should not prevent one from successfully cloning a lytic enzyme. All three techniques are ultimately straightforward, and (as replica plating is not a time-consuming step) one can readily conduct multiple types of overlays from each master plate.

5. RECENT ADDITIONS TO FUNCTIONAL SCREENING

Despite the general success of the proceeding techniques, they are not the only functional methods available for cloning phage lysins. Indeed, several additional strategies have been devised recently that either complement or expedite these approaches. One such example is holin-based screening for lytic enzymes; the holins are phage-encoded transmembrane proteins expressed late in the infective process. They insert nonspecifically into the cytoplasmic membrane of the host bacterium, creating pores through which the lysin may diffuse to reach its peptidoglycan target. (For a review of holins, refer to Young 2000; or Chapter 6 of this book.) It is the combined action of lysin and holin that leads to host-cell lysis and the release of progeny viral particles. The necessity of both proteins during phage infection explains why lytic enzymes can be over-expressed recombinantly in *E. coli*: even if a given lysin possesses activity against the *E. coli* peptidoglycan, it cannot exert a toxic effect as long as it is sequestered in the cytoplasm. By contrast, co-expression of both lysin and holin (or even holin by itself) can lead to marked toxicity.

Not only are lysins and holins expressed together during phage infection, but they are often encoded adjacent to one another in phage genomes. As a result, lysin-containing fragments in shotgun libraries commonly encode holins as well, creating the potential for selective toxicity of exactly the clones one hopes to identify. When a lysin-encoding clone is identified in an enzyme-based screen, it is generally one in which either (a) the holin and lysin happen not to be encoded adjacently in the particular genome, (b) the holin (fortuitously) is not sufficiently expressed, or (c) only a limited amount of genomic DNA surrounds the lysin, excluding the complete holin. Due to the small size of phage genomes and the resultant high proportion of lysin-encoding clones, the issue of holin toxicity has not proven to be a tremendous obstacle in past screens. Nevertheless, several studies have looked to avoid the situation altogether by instead selecting for holin-encoding clones. By targeting the holin genes, it is possible to identify adjacent lysins without actually observing lysin activity.

In this regard, Delisle et al. (2006) utilized a "plasmid release" protocol to identify a phage lytic enzyme for the putative dental pathogen *Actinomyces naeslundii*. For this study, mixed *E. coli* transformants were grown in a single-liquid culture. Following induced expression, holin-encoding cells would undergo lysis, releasing their plasmid into the culture media. The plasmids were then purified and used to retransform a new set of competent *E. coli*. Through several rounds of this procedure, the authors were able to enrich for a set of clones encoding the holin-lysin region, ultimately allowing them to subclone and express the lytic enzyme. This method was subsequently adapted for a plate-based screen to eliminate the need for multiple rounds of enrichment (Schmitz et al. 2009). Here, transformed clones were propagated on blood agar lacking an arabinose-inducing agent. Once they had reached macroscopic size, the colonies were exposed to a mist of nebulized arabinose to force recombinant expression. Holin-encoding clones would undergo colony lysis, which could be selected through the development of a hemolytic effect in the surrounding blood agar (see Fig. 10.2). This method was employed not to screen the DNA of a single genome, but rather the mixed phage DNA extracted in bulk from an environmental sample (this *metagenomic* screen is discussed further in Section 8 of this chapter).

One additional technology with potential relevance to lysin identification is that of *whole-genome amplification*. PCR-based methods have

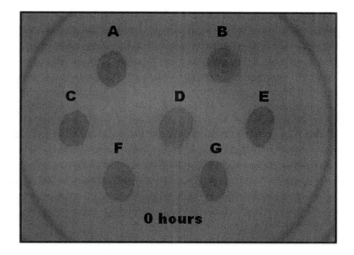

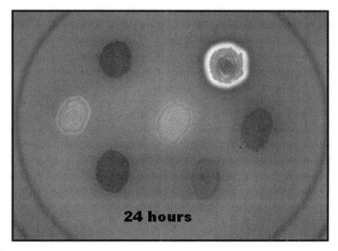

Figure 10.2. Hemolysis-based holin screening. When recombinant protein expression is induced in *E. coli*, holin-encoding clones often undergo lysis. This is detectable by a zone of hemolysis in blood agar, presumably a nonspecific effect due to the release of intracellular bacterial contents. Depicted here are pre- and post-induction examples of two clones encoding lysin/holin genes, along with positive (B) and negative (A, E–G) controls: (A) empty plasmid; (B) aerolysin, a secreted bacterial hemolysin; (C–D) holin + lysin; (E) random phage DNA; (F–G) lysin alone. It should be noted that the technique is not perfectly sensitive or specific for lysin/holin-encoding cassettes, as other toxic proteins encoded by phage could induce a similar effect. Nevertheless, the method provides a potential alternative for cloning lytic enzymes that avoids complications from holin-based toxicity.

recently come into prominence that allow for general amplification of viral (along with bacterial or eukaryotic) DNA. These include the use of high-diversity primers and the ultra-processive φ29 polymerase (for a review that discusses genome amplification specifically in the context of viruses, see Delwart 2007). The significance of these techniques lies in the access they provide to exceedingly small biological samples, as once-undetectable genetic material is now available for analysis. From the perspective of lysin screening, whole-genome amplification can significantly expedite the preparation time for library construction. Purification of phage DNA is a relatively time-intensive process compared with that of cellular organisms. Depending on the phage and the growth properties of the host, obtaining a high-enough viral titer for microgram quantities of DNA (typically required for shotgun cloning) can represent several days to several weeks of work. A similar situation exists when attempting to analyze DNA extracted from uncultured environmental phage, as viruses represent only a small fraction of total environmental biomass (Breitbart et al. 2003; Edwards and Rohwer 2005). However, by amplifying a small initial quantity of phage DNA, one can sidestep this issue and obtain an essentially limitless supply of genomic material.

In this regard, an amplification protocol was recently developed by our laboratory with functional screening of phage genomes in mind. Referred to as expressed linker-amplified shotgun libraries (E-LASL), it combines PCR-amplification with topoisomerase cloning for rapid construction of expressible plasmid libraries (Schmitz et al. 2008). In this approach, genomic DNA is fragmented, ligated to short linker sequences, and amplified with linker-targeted primers (a schematic of the technique is provided in Fig. 10.3). Taq polymerase is utilized in the last step, as it creates nonspecific adenine overhangs at the 3'ends of all amplicons. This modification allows for their direct ligation into linearized vectors through the use of vaccinia-virus topoisomerase, an enzyme that associates with terminal CCCTT-motifs and ligates single-stranded threonine overhangs to complementary adenine overhangs (Shuman 1994). The commercial availability of topoisomerase-cloning kits makes the approach particularly straightforward. As proof-of-principle in our initial study, the E-LASL technique was utilized to clone the lysins from several *B. anthracis* phage, using only a small amount of genomic material from each (~100ng). While E-LASL is by no means applicable only to phage DNA or lysin screening, it has become a preferred tool for identifying lytic enzymes within our group.

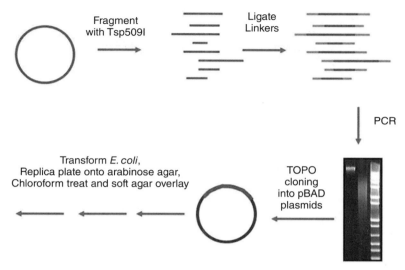

Figure 10.3. Constructing an E-LASL. Sample DNA (represented by the circle at the upper left) is enzymatically fragmented by brief exposure to Tsp509I restriction enzyme. A short segment of linker DNA with a complementary overhang is ligated to both ends of the fragments. PCR is conducted with *Taq* polymerase and a primer targeted against the linker sequence. The resultant amplicons have a range of molecular weights, but all posses 3'-adenine overhangs. The agarose gel at bottom right depicts bacteriophage genomic DNA (left lane) alongside its E-LASL amplification products (middle lane). The amplicons are cloned directly into linearized pBAD plasmids using commercial topoisomerase cloning. *E. coli* are transformed with the plasmids and screened for the acquisition of a desired phenotype.

6. LYTIC ENZYMES AND WHOLE-GENOME SEQUENCING

Despite this growing array of techniques, functional screening still represents only one side of the ongoing effort to clone lytic enzymes. In fact, only ~50% of phage lysins that have been recombinantly expressed to date were first identified in this manner. Many others, at the same time, have been the result of genomic sequencing and nucleotide homology analysis. The field of whole-genome sequencing has expanded rapidly in recent years (for viruses, bacteria, and eukaryotes), driven in large part by the development of high-throughput technologies like 454 pyrosequencing and Illumina/Solexa sequencing-by-synthesis (Strausberg et al. 2008). Publications documenting complete phage genomes have become commonplace, and a growing number of studies now focus on the genomes of numerous, interrelated phage. To date, 500-plus complete phage genomes are present within the National Center for Biotechnology Information database (not counting prophage), and this number is expected to increase substantially in the near future (Hatfull 2008). The impact of genomic sequencing on lysin research is simple: whenever the genome of a new phage is reported, another lysin gene is uncovered. Their enzymatic motifs and overall structure are sufficiently conserved that standard algorithms (e.g., BLAST and PFAM) generally make it possible for one to recognize a lytic enzyme from a sequence alone. Overall, identifying these enzymes through nucleotide homology is not a new concept; it was mentioned before that two of the earliest known lysin genes (from the *Salmonella* P22 and *B. subtilis* φ29 phage) were discovered based on their similarity to the *E. coli* T4 lysin. What distinguishes today's bioinformatic analyses is the large (and ever-growing) size of public databases, which not only allow one to locate a lysin gene itself, but also often predict its enzymatic mechanism, domain phylogeny, or even catalytic residues.2 These genes can then be PCR-cloned and tested for expression and activity.

Table 10.2 provides a chronological list of recombinantly expressed lysins whose genes were initially identified through DNA sequence analysis. With a single exception (O'Flaherty et al. 2005), the genes were amplified and cloned directly from genomic DNA. In the one instance, a staphylococcal lysin was interrupted by an intronic sequence, requiring the investigators to extract mRNA from infected bacteria and prepare cDNA as an amplification template. Aside from the increasing number of lysins in recent years, Table 10.2 reveals several important trends. For instance, whereas earlier lysins were often discovered within partial genomic sequences (i.e., a fortuitous sampling of restriction fragments), recently cloned enzymes have generally been identified from complete phage genomes. Presumably, this reflects both gains in sequencing technology, as well as improvements in functional techniques that make it possible to clone a lysin without resorting to partial sequencing. One should likewise note the four lytic enzymes amplified not from the genomes of individual phage, but rather from the genomes of their bacterial hosts. These include PlyL (Low et al. 2005) and PlyPH (Yoong et al. 2006) from the *Ames* strain of *B. anthracis*, along with the LambdaSa1 and LambdaSa2 lysins from *Streptococcus agalactiae* 2603 V/R (Pritchard et al. 2007). Bacterial genomes in themselves represent attractive sources of lytic enzymes in the form of integrated viral DNA, including both viable prophage and defective prophage remnants. As the genomes of more and more bacterial species (and individual strains) are sequenced and annotated, the number of prophage

TABLE 10.2. Lysins through Genomic Sequencing/Homology Analysis

Year	Sequecning Descriptiom	Target Bacteria	Reference
1986	Partial phage genome	*Bacillus subtilis*	Garvey et al.
1994	Partial phage genome	*Lactococcus lactis*	Arendt et al.
1994	Partial phage genome	*L. lactis*	Birkeland
1996	Partial phage genome	*Lactobacillus sp.*	Oki et al.
1997	Complete phage genome	*L. lactis*	Chandry et al.
1999	Partial phage genome	*Oenococcus oeni*	Parreira et al.
2000	Complete phage genome	*Lactobacillus casei*	Kashige et al.
2002	Complete phage genome	*Clostridium perfringens*	Zimmer et al.
2004	Partial phage genome	*Streptococcus agalactiae*	Pritchard et al.
2004	Complete phage genome	*Bacillus thuringiensis*	Verheust et al.
2005	Complete phage genome	*Staphylococcus aureus*	Takác et al.
2005	Prophage/host genome	*Bacillus anthracis*	Low et al.
2006	Prophage/host genome	*B. anthracis*	Yoong et al.
2006	Complete phage genome	*Listeria monocytogenes*	Korndörfer et al.
2007	Complete phage genome	*S. aureus*	Rashel et al.
2007	Complete phage genome	*S. aureus*	Sass and Bierbaum
2007	Prophage/host genome	*S. agalactiae*	Pritchard et al.
2008	Partial phage genome	*Streptococcus uberis*	Celia et al.
2008	Complete phage genome	*Clostridium difficile*	Mayer et al.

This table lists recombinantly expressed lysins against Gram-positive bacteria that were originally identified through DNA sequencing and bioinformatic comparison to known proteins. Three categories are present within the table: enzymes identified through partial sequencing of a phage genome, those identified through complete sequencing of a phage genome, and prophage lysins identified through complete sequencing of the bacterial host genome.

lysins that could serve as potential enzybiotics will only continue to grow.

Whether viral or proviral, lysin sequences are additionally valuable in that they allow the construction of chimeric proteins that juxtapose the various domains of individual enzymes. The truly modular nature of Gram-positive lysins lends itself to such manipulation, allowing for the expression of enzymes with unique combinations of biochemical properties. Additionally, by fusing the binding domain of one lysin with the enzymatic domain of another, it is possible to construct enzymes with cross-engineered bacterial specificity. The concept of chimeric lysins has actually existed for some time, although relatively few examples have been tested experimentally. Díaz et al. (1990, 1991) first exchanged the modular domains of several of pneumococcal lytic enzymes (including phage enzymes and the LytA autolysin) and demonstrated preserved enzymatic and binding function. Sheehan et al. (1996) later fused the enzymatic domain of a *Lactococcus lactis* phage with the binding domain of an *S. pneumoniae* phage and observed that the latter was sufficient for anti-pneumococcal activity. More recently, Donovan et al. (2006) connected both a partial- and a full-length *S. agalactiae* phage lysin to the complete *Staphylococcus simulans* lysostaphin gene; interestingly, these chimeras demonstrated activity against both *S. agalactiae* and various staphylococci. In a somewhat different approach, Cheng and Fischetti (2007) mutagenized the gene of an *S. agalactiae* lysin and found that a truncated, N-terminal construct possessed increased bacteriolytic activity. While this last finding is not considered typical, the underlying message of all these studies is clear: that laboratory manipulation of known sequences can only increase the vast supply of potential enzybiotics already encoded by global phage and their bacterial hosts.

7. SEQUENCE-BASED VERSUS FUNCTIONAL: PROS AND CONS

At this point, it is worthwhile to consider the relative advantages and disadvantages of identifying lysins through functional screening versus genomic sequencing and PCR-cloning. One of the most attractive features of the latter approach is its straightforward nature. If one has isolated a phage and wishes to identify its lysin, genomic sequencing is a method for which success is highly likely. The primary obstacle here is not cloning the gene itself, but rather identifying a suitable vector for active expression. This challenge is true for all recombinant proteins, however, and is equally applicable to functional lysin screens. A sequence-based approach likewise allows one to mine genomic databases for lysins that have not yet been studied in detail. In the process, one can compare their putative biochemical properties and overall uniqueness through bioinformatic analysis. For example, a lysin with a catalytic domain that is unique for particular host bacteria is far more attractive than a lysin that is highly similar to a previously expressed enzyme. Overall, while it is not always the case that a desired sequence is present within public databases, it is always advantageous to check, as this easily represents the most rapid singular approach to expressing a novel lysin.

Of course, identifying lysins through nucleotide homology does not always present the best option—quite frequently, the reason is purely logistical. Despite ongoing technological improvements, genomic sequencing does not represent yet an insignificant investment of time and resources, and high-throughput access is far from universal.

For the majority of laboratories, functional screening is still the most cost-effective and rapid method for identifying lytic enzymes from individual phage genomes. With sequence analysis alone, moreover, there exists a small chance of overlooking a lytic enzyme due to an atypical sequence. While they are uncommon, lysins whose genes deviate significantly from the norm are occasionally identified. The most prominent example is the streptococcal C_1 lysin (PlyC), described previously in this chapter for its early role in lysin purification. The C_1 genomic sequence, reported in 2003, revealed an abnormal lytic region in which a putative holin (ORF8) lied adjacent to several ambiguous open reading frames (Nelson et al. 2003). While the theoretical translation of ORF9 corresponded to a partial Edman sequence of the purified enzyme, none of the ORFs demonstrated typical lysin features or possessed a molecular mass corresponding to PlyC. Only when the authors functionally screened the C_1 genome and dissected the ORFs of the lytic clone were they able to determine that PlyC represents a unique multimeric phage lysin (resulting from ORFs 9 and 11; Nelson et al. 2006). In fact, PlyC remains without any homologs among sequenced phage, and it is a remarkable coincidence that one of the first lysins ever purified is so atypical.

Just recently, another highly novel lysin was reported for the φIN93 phage of the extremophilic bacterium *Thermus aquaticus* (Matsushita and Yanase 2008). Employing a more classical approach, the authors first purified the enzyme from infected culture supernatant and subjected it to Edman sequencing. The N-terminal amino acids were then cross-referenced to the sequenced genome to locate the lysin-encoding ORF, which until that point had remained unidentified due to its lack of recognized domains or homology to other proteins. While the authors did not utilize a functional screen to identify the enzyme, per se, this case again demonstrates how genomic sequences alone can occasionally be insufficient for lysin identification. Overall, the value of this particular enzyme lies not in its anti-infective potential—*T. aquaticus* is neither Gram-positive nor pathogenic—but in its ability to retain activity at high temperatures, an attractive industrial feature.

Finally, it is important to emphasize that phage can encode unique antibacterial proteins other than traditional lytic enzymes. A growing number of functional screens have identified individual phage proteins that lead to marked toxicity when expressed recombinantly in the host bacterium (Liu et al. 2004; Sau et al. 2008). They are believed to represent anti-host factors that allow the viruses to disrupt internal cellular physiology following infection. From the perspective of drug development, these proteins (or synthetic analogs) face the challenge of having to be present within the bacterial cytoplasm. Nevertheless, limited experimental evidence also suggests that phage can encode proteins, in addition to their lysins, that exert an exogenous antibacterial effect. For instance, functional screening of the *Bacillus cereus* Bcp-1 phage genome revealed two distinct clones that prevented the growth of *Bacillus* cells in soft agar overlays. One clone encoded a typical modular lysin, PlyB, that was subsequently purified, crystallized, and functionally examined (Porter et al. 2006). The other encoded a short, difficult-to-purify protein termed Killer of Anthrax (KOA), which is similar to only a limited number of hypothetical proteins from several other *Bacillus* phage genomes (Schuch, unpublished observations). While it is half the size of standard Gram-positive lysins and does not contain any traditional sequences common to major lysin families, KOA is nonetheless encoded proximal to a holin-like ORF and possesses lytic activity against a range of *B. cereus* organisms, including *B. anthracis*. It remains unclear what roles PlyB and KOA play (independently or in conjunction) during Bcp-1 infection. Overall, it is difficult to predict what sort of antibacterial compounds (enzymatic or otherwise) might be encoded by global phage, mainly because such a relative few have undergone functional analysis. So while genomic sequencing will continue to play a crucial role in identifying lytic enzymes, the ongoing importance of expression-based approaches cannot be discounted.

8. FUTURE STRATEGIES FOR LYSIN IDENTIFICATION

With the methods developed to date, there already exists a diverse complement of techniques for cloning lytic enzymes with biotechnological potential. Looking ahead, however, it is important to consider whether any additional strategies could identify novel enzybiotics. One possibility, for instance, involves the use of *metagenomics*. This term refers to the direct extraction of DNA from environmental samples without prior laboratory cultivation of any individual organisms (reviewed in Daniel 2005; Ferrer et al. 2005; Voget et al. 2005; Green and Keller 2006). Its appeal lies in the access it provides to difficult-to-isolate microbes, as it is estimated that only a small portion of environmental species are culturable under standard laboratory conditions. While the initial focus of metagenomics research was on environmental bacteria, the field has since expanded to include investigations of uncultured viruses (reviewed in Edwards and Rohwer 2005; Delwart 2007). In these studies, viral particles (consisting primarily of phage) are separated from environmental bacteria and other debris, and the extracted DNA is often PCR-amplified to compensate for low yields. A number of increasingly large-scale sequencing projects have been dedicated to environmental phage populations, including ones purified from soil (Fierer et al. 2007), sea water (Breitbart et al. 2004; Angly et al. 2006), and various fecal samples (Breitbart et al. 2003; Finkbeiner et al. 2008). These studies have ranged in scope from several hundreds kilobases to over 180 megabases, the latter facilitated by high-through-out pyrosequencing technology. Measured in base pairs, the amount of information contained here far outpaces that of any individual phage genome. Naturally, these metagenomic sequences have included puta-tive lysins, along with homologs of other typical phage genes and (quite notably) a high proportion of sequences of unknown function.3 Viral metagenomics, in fact, has provided some of the strongest evidence that phage represent the largest source of global genetic diversity.

To complement these impressive sequencing studies, future research in viral metagenomics could likewise involve functional screens for proteins of interest. Novel phage lysins, of course, represent one entic-ing example. In this regard, our laboratory recently developed a pro-tocol for identifying arbitrary lytic clones within mixed phage populations (Schmitz et al. 2009). The procedure involves a toxicity screen (described in Section 5), in which colonies are selected for holin-induced lysis following exposure to nebulized arabinose. In a secondary step, then, the initial hits are overlaid with autoclaved Gram-negative bacteria (specifically *Pseudomonas aeruginosa*) to assay directly for the production of lytic enzyme. Due to their outer lipid membrane, Gram-negative organisms are considered less promising targets for lysin therapy. However, if this layer is compromised—for instance, by auto-claving—these organisms become highly sensitive to cell wall hydro-lases, presumably due to the thinness of the Gram-negative peptidoglycan layer. In our study, we utilized this sensitivity to identify 27 actively expressed lysins from a mixed fecal sample, including Gram-positive enzymes, Gram-negative enzymes, and several lysins demonstrating unusual modular architectures. Unfortunately, when tested against Gram-positive pathogens, these proteins failed to demonstrate high-enough activity to warrant further pursuit as enzybiotics. Nevertheless, this initial metagenomic screen was fairly small in scope: with contin-ued screening of larger populations and the development of further methodologies, uncultured phage could prove to be a valuable source of lytic enzymes or other anti-infective agents.

Overall, the ideal lysin screen would be one that combines the large number of enzymes encoded by metagenomic samples with the specificity that individual lysins demonstrate toward their host bacteria. One possible solution (and an area of current investigation) could involve screens in which genomic DNA from numerous strains of a single pathogen is pooled to form a multigenomic library. Each strain would likely contain integrated prophage, such that a combination of diverse strains should encode a large collection of lysins. A functional screen of this library would generate many hits, like a metagenomic screen, with the exception that all enzymes would target the same bacteria de facto. The only alternatives to accessing numerous lysins, as such, would involve whole-genome sequencing of multiple bacteria or (perhaps) large-scale proviral induction. One caveat to this approach is that, due to the size of bacterial genomes, the lysin-encoding fragments would represent a smaller proportion of the total library. In this regard, there exist a limited number of examples of a prophage lysin cloned through a functional screen of its host's genome (Jayaswal et al. 1990; Zink et al. 1995). In Jayaswal et al. (1990), for example, the authors identified one positive clone out of ~1700 screened, a lower rate than typically observed for isolated phage. In a multigenomic screen, however, this effect could be offset in part by poly-lysogenized strains that contain multiple lysins.

9. ENZYBIOTICS BEYOND PHAGE LYSINS

One additional advantage of screening bacteria-derived DNA (whether genomic, multigenomic, or metagenomic) is that it can access peptidoglycan hydrolases in addition to those provided by phage. Although mentioned here only briefly, bacteria likewise encode autolytic proteins that mediate activities such as growth, division, and sporulation. Some autolysins, like pneumococcal LytA, demonstrate high homology to phage lysins, whereas others are more divergent. A broad review of these proteins and their diverse cellular functions was recently compiled by Vollmer et al. (2008). While phage lysins have received more attention, various reports have documented the ability of recombinant autolysins to act as lytic agents against Gram-positive organisms (for instance, Dhalluin et al. 2005; Fukushima et al. 2008; Yokoi et al. 2008). Their *in vivo* functions might differ from viral proteins, but the process of developing autolysins as antibacterial agents would follow essentially the same principles. In fact, the categories of peptidoglycan hydrolases with enzybiotic potential extends even beyond these proteins. Many bacteria encode proteins/peptides that target other species, often closely related ones—of these molecules, several notable examples function through a cell wall lytic mechanism.4 The most prominent example is lysostaphin, a poly-functional hydrolase originally identified in *S. simulans* that targets rival species/strains of staphylococci (Kumar 2008). In light of the growing resistance to traditional antibiotics, lysostaphin has received considerable attention as a possible weapon against *Methicillin-resistant* S. aureus (MRSA) and non-MRSA isolates of *S. aureus*. Comparable enzymes, namely millericin B and zoocin A, have been isolated from streptococcal strains (Beukes et al. 2000; Akesson et al. 2007). Such proteins serve as an important reminder that the enzybiotics field encompasses more than phage lysins alone, and necessitates that future screening techniques cast a large net in search of novel antibiotics.

Broadening the spectrum further still, phage themselves can encode peptidoglycan hydrolases in addition to lysins proper. Putative enzymatic motifs are often present within proteins of the phage tail assembly (Kanamaru 2002; Kenny et al. 2004; Piuri and Hatfull 2006) or even the head (Moak and Molineux 2004), presumably to facilitate the initial injection of viral DNA. And although it is not their natural purpose, recent evidence has shown that these enzymes can lead to Gram-positive lysis when added exogenously. Rashel et al. (2008) identified a tail-associated protein from the genome of an *S. aureus* phage that contains two putative lytic motifs; when individually expressed and purified, both domains induced staphylococcal death. Despite these findings, no tail enzymes have ever been identified during a functional screen of a phage genome (to our knowledge). Several reasons could underline this discrepancy—for instance, some enzymes might require the presence of the entire macromolecular tail assembly for proper activity. Moreover, structural lysins often represent component domains within much larger polypeptides; in the context of a functional screen, these domains would not be expressed individually and the bulk proteins could prove difficult to obtain due to their size. Nevertheless, if future research could reliably harness the enzymatic activity of tail lysins, it would represent another promising avenue within the enzybiotics field.

In addition to this possibility, phage tail proteins are also of considerable interest for their ability to traverse the other components of the bacterial cell envelope. These include the outer membrane of Gram-negative bacteria and the mycolic acid layer of mycobacteria, structures that act as formidable obstacles to lysin treatment. While it is true that several reports have documented the use of lytic enzymes against viable Gram-negative organisms (Alakomi et al. 2000; Morita et al. 2001a; Kim 2004; Briers et al. 2008), this activity was generally either quantitatively weak or required the addition of a general membrane-disrupting agent. However, co-treatment of Gram-negative bacteria with a membrane-permeabilizing agent and a peptidoglycan hydrolase is already a common practice: it is the basis of commercial genomic extraction techniques and has various shortcomings from a drug-development perspective. On the other hand, selective proteins capable of permeabilizing the outer membrane of only certain bacteria would be of considerable biomedical interest. In this regard, various Gram-negative species encode antibacterial complexes that kill closely related organisms through a cell envelope depolarization effect (Zink et al. 1995; Strauch et al. 2001; Jabrane et al. 2002). These multimeric structures resemble isolated phage-tail assemblies, and are believed to represent defective proviral remnants that have been co-opted by the host. In a recent study, Williams et al. (2008) demonstrated that, by exchanging a component gene of one such complex with the tail fiber genes of other phage, they could engineer complexes with species-targeted activity. While not examined specifically, the study raises interesting questions about potential synergies between these complexes and peptidoglycan hydrolases. Overall, the future of enzybiotic development could involve not only the identification of new enzymes but also complementary molecules that increase lysin potency and broaden their variety of bacterial targets.

10. CONCLUDING REMARKS

In this field of research, one observation is unmistakably clear: between established strategies of lysin identification and novel techniques, there will be no shortage of enzybiotic candidates in the foreseeable future. As methods and technology continue to advance, what is already a dense field is only likely to become more crowded. Perhaps a greater challenge than merely identifying phage lysins, in fact, will be the ability to compare them systematically on a protein level. Enzyme kinetics, thermal and pH tolerance, immunogenicity, *in vivo* half-life, and biodistribution are only a few of the quantities that could vary from one lysin to the next and, ultimately, they are the factors that will determine an enzyme's therapeutic promise. Unfortunately, none of these properties can be measured (as of yet) with a mere shotgun screen or nucleotide sequence, as protein purification and old-fashioned pharmacological analyses are still required. In this regard, the enzybiotic field is in a unique position, as cloning/sequencing technologies have created a pronounced gap between our ability to identify an enzyme-of-interest and the effort it takes to study it in detail and move it through the drug-development process. Nevertheless, if and when the first phage lysin progresses to human use—and an increasing body of experimental evidence suggests that day is coming—there will already be a large reserve of similar agents set to join it in the fight against infectious disease.

ACKNOWLEDGMENTS

This work was supported by NIH/NIAID grants AI057472 and AI11822 to Vincent A. Fischetti. Jonathan E. Schmitz acknowledges the support of the Pharmaceutical Research and Manufacturers of America Foundation and the NIH MSTP program (Weill Cornell/Rockefeller/Sloan-Kettering grant GM 07739). The authors would like to thank the following individuals for their conversations and insights regarding the phage-lysin field: Dr. Daniel Nelson, Dr. Mattias Collin, Dr. Jutta Loeffler, Dr. Qi Cheng, Dr. Anu Daniel, Dr. Pauline Yoong, and Dr. Gregory Resch.

ENDNOTES

1. In the body of the text, we discuss the prevalent strategy for fragmenting and transforming phage genomic DNA for lysin screening (i.e., shotgun cloning into plasmid vectors). Although no longer generally employed, alternative screening techniques were occasionally used in early studies. For example, when cloning the lysin of a lactococcal phage, Shearman et al. (1989) utilized a λ-phage system that relied upon infected *E. coli* for recombinant lysin expression. Henrich et al. (1995) also employed λ-phage when cloning a *Lactobacillus* lysin. This study involved a defective λ-strain that required ligation of a complementary lysis cassette to form plaques on *E. coli*. Furthermore, one should note that *E. coli* is not the only a host that, in theory, could be used for lysin screening. While it has been utilized exclusively to date, there is nothing about this species that makes it particularly well suited for identifying lytic clones (other than the commercial prevalence of *E. coli* expression systems). Lysins have been expressed recombinantly from other bacterial species for purification purposes—and, in certain cases, these alternate species offered superior expression to *E. coli* (Yoong et al. 2004).

2. It is not always possible, however, to determine such information from a nucleotide sequence alone. While analysis of a lysin's catalytic domain typically reveals a conserved motif—which, in turn, implies an enzymatic

class—lytic enzymes with ambiguous domains are still reported (Porter et al. 2006; Summer et al. 2006; Schmitz et al. 2008b). One recent study describes two *S. agalactiae* lysins predicted to be alanine amidases by sequence analysis; following expression and functional testing, they were instead determined to possess endopeptidase and glucosaminidase activity (Pritchard et al. 2007). Overall, while sequence-based identification of lysin genes *in total* is reliable most of the time (see Section 7), bioinformatic analysis of individual lysin domains is more often biased by limited database information.

3. It is important to note that the sequences of intact genes (lysins or otherwise) can only be obtained from metagenomic samples when vector-based cloning is used. Although pyrosequencing has the ability to generate far more sequence data, the reads themselves are significantly shorter (100–200 bp) and the DNA fragments cannot be amplified by clonal propagation. In such cases, while it is still possible to predict protein functionality through nucleotide homology analysis, the individual reads typically encode only portions of genes. In general, recently developed sequencing technologies are not highly compatible with current methods of functional screening.

4. The term bacteriocin is commonly used to describe proteins and peptides that allow one bacterium to exert a toxic effect on another. The classification of peptidoglycan hydrolases as bacteriocins, however, is an issue on which a definitive consensus does not yet exist. Under some schemata, virtually all antibacterial proteins encoded by bacteria themselves should be considered bacteriocins (Heng and Tagg 2006). In this case, enzymes such as lysostaphin and zoosin A are categorized as class IIIa bacteriocins. At the same time, other systems reserve the term *bacteriocin* for distinct classes of peptides that function through a non-hydrolytic mechanism (Cotter et al. 2005). In either case, it should be emphasized that such schemata function only as organizational tools, and do not affect the *in vivo* function or biotechnological potential of these proteins.

REFERENCES

Akesson M., M. Dufour, G. L. Sloan, and R. S. Simmonds (2007) *FEMS Microbiology Letters* **270**, 155–161.

Alakomi H. L., E. Skyttä, M. Saarela, T. Mattila-Sandholm, K. Latva-Kala, and I. M. Helander (2000) *Applied and Environmental Microbiology* **66**, 2001–2005.

Angly F. E., B. Felts, M. Breitbart, P. Salamon, R. A. Edwards, C. Carlson, A. M. Chan, M. Haynes, S. Kelley, H. Liu, J. M. Mahaffy, J. E. Mueller, J. Nulton, R. Olson, R. Parsons, S. Rayhawk, C. A. Suttle, and F. Rohwer (2006) *PLoS Biology* **4**, e368.

Arendt E. K., C. Daly, G. F. Fitzgerald, and M. van de Guchte (1994) *Applied and Environmental Microbiology* **60**, 1875–1883.

Beukes M., G. Bierbaum, H. Sahl, and J. W. Hastings (2000) *Applied and Environmental Microbiology* **66**, 23–28.

Birkeland N. (1994) *Canadian Journal of Microbiology* **40**, 658–665.

Boizet B., Y. Lahbib-Mansais, L. Dupont, P. Ritzenthaler, and M. Mata (1990) *Gene* **94**, 61–67.

Bon J., N. Mani, and R. K. Jayaswal (1997) *Canadian Journal of Microbiology* **43**, 612–616.

Borysowski J., B. Weber-Dabrowska, and A. Górski (2006) *Experimental Biology and Medicine* **231**, 366–377.

Breitbart M., B. Felts, S. Kelley, J. M. Mahaffy, J. Nulton, P. Salamon, and F. Rohwer (2004) *Proceedings of the Royal Society of London B* **271**, 565–574.

Breitbart M., I. Hewson, B. Felts, J. M. Mahaffy, J. Nulton, P. Salamon, and F. Rohwer (2003) *Journal of Bacteriology* **185**, 6220–6223.

Briers Y., A. Cornelissen, A. Aertsen, K. Hertveldt, C. W. Michiels, G. Volckaert, and R. Lavigne (2008) *FEMS Microbiology Letters* **280**, 113–119.

Bronfenbrenner J. and R. Muckenfuss (1927) *Journal of Experimental Medicine* **45**, 887–909.

Celia L. K., D. Nelson, and D. E. Kerr (2008) *Veterinary Microbiology* **130**, 107–117.

Chandry P. S., S. C. Moore, J. D. Boyce, B. E. Davidson, and A. J. Hillier (1997) *Molecular Microbiology* **26**, 49–64.

Cheng Q. and V. A. Fischetti (2007) *Applied Microbiology and Biotechnology* **74**, 1284–1291.

Cheng Q., D. Nelson, S. Zhu, and V. A. Fischetti (2005) *Antimicrobial Agents and Chemotherapy* **49**, 111–117.

Cotter P. D., C. Hill, and R. P. Ross (2005) *Nature Reviews Microbiology* **3**, 777–788.

Daniel R. (2005) *Nature Reviews Microbiology* **3**, 470–478.

Delisle A. L., G. J. Barcak, and M. Guo (2006) *Applied and Environmental Microbiology* **72**, 110–117.

Delwart E. L. (2007) *Reviews in Medical Virology* **17**, 115–131.

Deutsch S., S. Guezenec, M. Piot, S. Foster, and S. Lortal (2004) *Applied and Environmental Microbiology* **70**, 96–103.

Dhalluin A., I. Bourgeois, M. Pestel-Caron, E. Camiade, G. Raux, P. Courtin, M. Chapot-Chartier, and J. Pons (2005) *Microbiology* **151**, 2343–2351.

d'Hérelle F. (1921) *Le bactériophage. Son rôle dans l'immunité, Monographies de l'Institut Pasteur*, Masson et cie, Paris.

Díaz E., R. López, and J. L. García (1990) *Proceedings of the National Academy of Sciences of the United States of America* **87**, 8125–8129.

Díaz E., R. López, and J. L. García (1991) *Journal of Biological Chemistry* **266**, 5464–5471.

Díaz E., R. López, and J. L. García (1992) *Journal of Bacteriology* **174**, 5516–5525.

Donovan D. M., S. Dong, W. Garrett, G. Rousseau, S. Moineau, and D. G. Pritchard (2006) *Applied and Environmental Microbiology* **72**, 2988–2996.

Doughty C. C. and J. A. Hayashi (1962) *Journal of Bacteriology* **83**, 1058–1068.

Doughty C. C. and J. A. Mann (1967) *Journal of Bacteriology* **93**, 1089–1095.

Edwards R. A. and F. Rohwer (2005) *Nature Reviews Microbiology* **3**, 504–510.

Entenza J. M., J. M. Loeffler, D. Grandgirard, V. A. Fischetti, and P. Moreillon (2005) *Antimicrobial Agents and Chemotherapy* **49**, 4789–4792.

Ferrer M., F. Martinez-Abarca, and P. N. Golyshin (2005) *Current Opinion in Biotechnology* **16**, 588–593.

Fierer N., M. Breitbart, J. Nulton, P. Salamon, C. Lozupone, R. Jones, M. Robeson, R. A. Edwards, B. Felts, S. Rayhawk, R. Knight, F. Rohwer, and R. B. Jackson (2007) *Applied and Environmental Microbiology* **73**, 7059–7066.

Finkbeiner S. R., A. F. Allred, P. I. Tarr, E. J. Klein, C. D. Kirkwood, and D. Wang (2008) *PLoS Pathogens* **4**, e10000011.

Fischetti V. A. (2005) *Trends in Microbiology* **13**, 491–496.

Fischetti V. A., E. C. Gotschlich, and A. W. Bernheimer (1971) *Journal of Experimental Medicine* **133**, 1105–1117.

Fischetti V. A., D. Nelson, and R. Schuch (2006) *Nature Biotechnology* **24**, 1508–1511.

Fukushima T., T. Kitajima, H. Yamaguchi, Q. Ouyang, K. Furuhata, H. Yamamoto, T. Shida, and J. Sekiguchi (2008) *Journal of Biological Chemistry* **283**, 11117–11125.

García E., J. L. García, P. García, A. Arrarás, and J. M. Sánchez-Puelles (1988) *Proceedings of the National Academy of Sciences of the United States of America* **85**, 914–918.

García P., J. L. García, E. García, J. M. Sánchez-Puelles, and R. López (1990) *Gene* **86**, 81–88.

García E., J. García, C. Ronda, P. García, and R. López (1985) *Molecular and General Genetics* **201**, 225–230.

Garvey K. J., M. S. Saedi, and J. Ito (1986) *Nucleic Acids Research* **14**, 10001–10008.

Grandgirard D., J. M. Loeffler, V. A. Fischetti, and S. L. Leib (2008) *The Journal of Infectious Diseases* **197**, 1519–1522.

Gratia A. and B. Rhodes (1923) *Comptes Rendus Hebdomadaires des Séances et Mémoires de la Société de Biologie et des ses Filiales* **89**, 1171–1172.

Green B. D. and M. Keller (2006) *Current Opinion in Biotechnology* **17**, 236–240.

Guiral S., T. J. Mitchell, B. Martin, and J. Claverys (2005) *Proceedings of the National Academy of Sciences of the United States of America* **102**, 8710–8715.

Hatfull G. F. (2008) *Current Opinion in Microbiology* **11**, 447–453.

Hendrix R. W. (2002) *Theoretical Population Biology* **61**, 471–480.

Heng N. C. K. and J. R. Tagg (2006) *Nature Reviews Microbiology* **4**, doi: 10.1038/nrmicro1273-c1.

Henrich B., B. Binishofer, and U. Bläsi (1995) *Journal of Bacteriology* **177**, 723–732.

Hermann S. R., J. A. Miller, S. O'Neil, T. T. Tsao, R. M. Harding, and J. L. Dale (2000) *Biotechniques* **29**, 1176–1178.

Inouye, M. and A. Tsugita (1966) *Journal of Molecular Biology* **22**, 193–196.

Jabrane A., A. Sabri, P. Compére, P. Jacques, I. Vandenberghe, J. Van Beeumen, and P. Thenart (2002) *Applied and Environmental Microbiology* **68**, 5704–5710.

Jayaswal R. K., Y. Lee, and B. Wilkerson (1990) *Journal of Bacteriology* **172**, 5783–5788.

Jedrzejas M. J. (2001) *Microbiology and Molecular Biology Reviews* **65**, 187–207.

Kanamaru S., P. G. Leiman, V. A. Kostychenko, P. R. Chipman, V. V. Mesyanzhinov, F. Arisaka, and M. G. Rossman (2002) *Nature* **415**, 553–557.

Kashige N., Y. Nakashima, F. Miake, and K. Watanabe (2000) *Archives of Virology* **145**, 1521–1534.

Kenny J. G., S. McGrath, G. F. Fitzgerald, and D. van Sinderen (2004) *Journal of Bacteriology* **186**, 3480–3491.

Kim W., H. Salm, and K. Geider (2004) *Microbiology* **150**, 2702–2714.

Korndörfer I. P., J. Danzer, M. Schmelcher, M. Zimmer, A. Skerra, and M. J. Loessner (2006) *Journal of Molecular Biology* **364**, 678–689.

Krause R. M. (1957) *Journal of Experimental Medicine* **106**, 365–384.

Krause R. M. (1958) *Journal of Experimental Medicine* **108**, 803–821.

Kumar J. K. (2008) *Applied Microbiology and Biotechnology* **80**, 555–561.

Liu J., M. Dehbi, G. Moeck, F. Arhin, P. Bauda, D. Bergeron, M. Callejo, V. Ferretti, N. Ha, T. Kwan, J. McCarty, R. Srikumar, D. Williams, J. J. Wu, P. Gros, J. Pelletier, and M. DuBow (2004) *Nature Biotechnology* **22**, 185–191.

Loeffler J. M., S. Djurkovic, and V. A. Fischetti (2003) *Infection and Immunity* **71**, 6199–6204.

Loessner M. J. (2005) *Current Opinion in Microbiology* **8**, 480–487.

Loessner M. J., S. K. Maier, H. Daubek-Puza, G. Wendlinger, and S. Scherer (1997) *Journal of Bacteriology* **179**, 2845–2851.

Loessner M. J., S. Gaeng, and S. Scherer (1999) *Journal of Bacteriology* **181**, 4452–4460.

Loessner M. J., S. Gaeng, G. Wendlinger, S. K. Maier, and S. Scherer (1998) *FEMS Microbiology Letters* **162**, 265–274.

Loessner M. J., G. Wendlinger, and S. Scherer (1995) *Molecular Microbiology* **16**, 1231–1241.

Low L. Y., C. Yang, M. Perego, A. Osterman, and R. C. Liddington (2005) *Journal of Biological Chemistry* **280**, 35433–35439.

Maass D. and W. Weidel (1963) *Biochimica et Biophysica ACTA* **78**, 369–70.

Matsushita I. and H. Yanase (2008) *Biochemical and Biophysical Research Communications* **377**, 89–92.

Maxted W. R. (1957) *Journal of General Microbiology* **16**, 584–595.

Mayer M. J., A. Narbad, and M. J. Gasson (2008) *Journal of Bacteriology* **190**, 6734–6740.

McCullers J. A., A. Karlström, A. R. Iverson, J. M. Loeffler, and V. A. Fischetti (2007) *PLoS Pathogens* **3**, e28.

Moak M. and I. J. Molineux (2004) *Molecular Microbiology* **51**, 1169–1183.

Morita M., Y. Tanji, K. Mizoguchi, A. Soejima, Y. Orito, and H. Unno (2001b) *Journal of Bioscience and Bioengineering* **91**, 469–473.

Morita M., Y. Tanji, Y. Orito, K. Mizoguchi, A. Soejima, and H. Unno (2001a) *FEBS Letters* **500**, 56–59.

Murphy J. S. (1957) *Virology* **4**, 563–581.

Nelson D., L. Loomis, and V. A. Fischetti (2001) *Proceedings of the National Academy of Sciences of the United States of America* **98**, 4107–4112.

Nelson D., R. Schuch, P. Chahales, S. Zhu, and V. A. Fischetti (2006) *Proceedings of the National Academy of Sciences of the United States of America* **103**, 10765–10770.

Nelson D., R. Schuch, S. Zhu, D. M. Tscherne, and V. A. Fischetti (2003) *Journal of Bacteriology* **185**, 3325–3332.

Ochman H., A. S. Gerber, and D. L. Hartl (1988) *Genetics* **120**, 621–623.

O'Flaherty S., A. Coffey, W. Meaney, G. F. Fitzgerald, and R. P. Ross (2005) *Journal of Bacteriology* **187**, 7161–7164.

Okada Y., G. Streisinger, A. Tsugita, and M. Inouye (1968) *Science* **162**, 807–808.

Oki M., M. Kakikawa, K. Yamada, A. Taketo, and K. Kodaira (1996) *Gene* **176**, 215–233.

Oram J. D. and B. Reiter (1965) *Journal of General Microbiology* **40**, 57–70.

Owen J. E., D. W. Schultz, A. Taylor, and G. R. Smith (1983) *Journal of Molecular Biology* **165**, 229–248.

Parreira R., C. São-José, A. Isidro, S. Domingues, G. Viera, and M. A. Santos (1999) *Gene* **226**, 83–93.

Perry L. J., H. L. Heyneker, and R. Wetzel (1985) *Gene* **38**, 259–264.

Piuri M. and G. Hatfull (2006) *Molecular Microbiology* **62**, 1569–1585.

Platteeuw C. and W. M. de Vos (1992) *Gene* **118**, 115–120.

Porter C. J., R. Schuch, A. J. Pelzak, A. M. Buckle, S. McGowan, M. C. J. Wilce, J. Rossjohn, R. Russell, D. Nelson, V. A. Fischetti, and J. A. Whisstock (2006) *Journal of Molecular Biology* **366**, 540–550.

Pritchard D. G., S. Dong, J. R. Baker, and J. A. Engler (2004) *Microbiology* **150**, 2079–2087.

Pritchard D. G., S. Dong, M. C. Kirk, R. T. Cartee, and J. R. Baker (2007) *Applied and Environmental Microbiology* **73**, 7150–7154.

Ralston D. J., B. S. Baer, M. Lieberman, and A. P. Krueger (1955) *Proceedings of the Society of Experimental Biology and Medicine* **89**, 502–507.

Ralston D. J., M. Lieberman, B. Baer, and A. P. Krueger (1957) *Journal of General Physiology* **40**, 791–807.

Ralston D. J. and M. McIvor (1964) *Journal of Bacteriology* **88**, 676–681.

Rao G. R. and D. P. Burma (1971) *Journal of Biological Chemistry* **246**, 6474–6479.

Rashel M., J. Uchiyama, I. Takemura, H. Hoshiba, T. Ujihara, H. Takatsuji, K. Honke, and S. Matsuzaki (2008) *FEMS Microbiology Letters* **284**, 9–16.

Rashel M., J. Uchiyama, T. Ujihara, Y. Uehara, S. Kuramoto, S. Sugihara, K. Yagyu, A. Muraoka, M. Sugai, K. Hiramatsu, K. Honke, and S. Matsuzaki (2007) *The Journal of Infectious Diseases* **196**, 1237–1247.

Reiter B. and J. D. Oram (1963) *Journal of General Microbiology* **32**, 29–32.

Rennell D. and A. R. Poteete (1985) *Virology* **143**, 280–289.

Romero A., R. López, and P. García (1990) *Journal of Virology* **64**, 137–142.

Romero A., R López, and P. García (2004) *Journal of Bacteriology* **186**, 8229–8239.

Saedi M. S., K. J. Garvey, and J. Ito (1987) *Proceedings of the National Academy of Sciences of the United States of America* **84**, 955–958.

Sass P. and G. Bierbaum (2007) *Applied and Environmental Microbiology* **73**, 347–352.

Sau S., P. Chattoraj, T. Ganguly, P. K. Chanda, and N. C. Mandal (2008) *Current Protein and Peptide Science* **9**, 284–290.

Schmitz J. E., A. Daniel, M. Collin, R. Schuch, and V. A. Fischetti (2008) *Applied and Environmental Microbiology* **74**, 1649–1652.

Schmitz J. E., R. Schuch, and V. A. Fischetti (2009) (submitted for publication).

Schuch R., D. Nelson, and V. A. Fischetti (2008) *Methods in Molecular Biology* **502**, 307–319.

Schuch R., D. Nelson, and V. A. Fischetti (2002) *Nature* **418**, 884–889.

Sertic V. (1929) *Comptes Rendus Hebdomadaires des Séances et Mémoires de la Société de Biologie et des ses Filiales* **100**, 477–479.

Shearman C., H. Underwood, K. Jury, and M. Gasson (1989) *Molecular and General Genetics* **218**, 214–221.

Sheehan M. M., J. L. García, and R. López (1997) *Molecular Microbiology* **25**, 717–725.

Sheehan M. M., J. L. García, R. López, and P. García (1996) *FEMS Microbiology Letters* **140**, 23–28.

Shugar D. (1952) *Biochimica et Biophysica ACTA* **8**, 302–309.

Shuman S. (1994) *Journal of Biological Chemistry* **269**, 32678–32684.

Sonstein S. A., J. M. Hammel, and A. Bondi (1971) *Journal of Bacteriology* **107**, 499–514.

Strauch E., H. Kaspar, C. Schaudinn, P. Dersch, K. Madela, C. Gewinner, S. Hertwig, J. Wecke, and B. Appel (2001) *Applied and Environmental Microbiology* **67**, 5634–5642.

Strausberg R. L., S. Levy, and Y. Rogers (2008) *Drug Discovery Today* **13**, 569–577.

Summer E. J., C. F. Gonzalez, M. Bomer, T. Carlile, A. Embry, A. M. Kucherka, J. Lee, L. Mebane, W. C. Morrison, L. Mark, M. D. King, J. J. LiPuma, A. K. Vidaver, and R. Young (2006) *Journal of Bacteriology* **188**, 255–268.

Takác M., A. Witte, and U. Bläsi (2005) *Microbiology* **151**, 2331–2342.

Terzaghi E., Y. Okada, G. Streisinger, J. Emrich, M. Inouye, and A. Tsugita (1966) *Proceedings of the National Academy of Sciences of the United States of America* **56**, 500–507.

Tourville D. R. and D. B. Johnstone (1966) *Journal of Diary Science* **49**, 158–162.

Trautwetter A., P. Ritzenthaler, T. Alatossava, and M. Mata-Gilsinger (1986) *Journal of Virology* **59**, 551–555.

Twort F. W. (1925) *Lancet* **206**, 642–644.

Verheust C., N. Fornelos, and J. Mahillon (2004) *FEMS Microbiology Letters* **237**, 289–295.

Voget S., H. Steele, and W. R. Streit (2005) *Minerva Biotecnologica* **17**, 47–53.

Vollmer W., B. Joris, P. Charlier, and S. Foster (2008) *FEMS Microbiology Reviews* **32**, 259–286.

Ward L. J. H., T. P. J. Beresford, M. W. Lubbers, B. D. W. Jarvis, and A. W. Jarvis (1993) *Canadian Journal of Microbiology* **39**, 767–774.

Welker N. E. (1967) *Journal of Virology* **1**, 617–625.

Williams S. R., D. Gebhart, D. W. Martin, and D. Scholl (2008) *Applied and Environmental Microbiology* **74**, 3868–3876.

Yokoi K., N. Kawahigashi, M. Uchida, K. Sugahara, M. Shinohara, K. Kawasaki, S. Nakamura, A. Taketo, and K. Kodaira (2005a) *Gene* **351**, 97–108.

Yokoi K., M. Shinohara, N. Kawahigashi, K. Nakagawa, K. Kawasaki, S. Nakamura, A. Taketo, and K. Kodaira (2005b) *International Journal of Food Microbiology* **99**, 297–308.

Yokoi K., K. Sugahara, A. Iguchi, G. Nishitani, M. Ikeda, T. Shimada, N. Inagaki, A. Yamakawa, A. Taketo, and K. Kodaira (2008) *Gene* **416**, 66–76.

Yoon S., J. Kim, F. Breidt, and H. Fleming (2001) *International Journal of Food Microbiology* **65**, 63–74.

Yoong P., R. Schuch, D. Nelson, and V. A. Fischetti (2004) *Journal of Bacteriology* **186**, 4808–4812.

Yoong P., R. Schuch, D. Nelson, and V. A. Fischetti (2006) *Journal of Bacteriology* **188**, 2711–2714.

Young I., I. Wang, and W. D. Roof (2000) *Trends in Microbiology* **8**, 120–128.

Zimmer M., N. Vukov, S. Scherer, and M. J. Loessner (2002) *Applied and Environmental Microbiology* **68**, 5311–5317.

Zink R., M. J. Loessner, and S. Scherer (1995) *Microbiology* **141**, 2577–2584.

CHAPTER 11

USE OF GENETICALLY MODIFIED PHAGES TO DELIVER SUICIDAL GENES TO TARGET BACTERIA

LAWRENCE D. GOODRIDGE

Department of Animal Sciences, Colorado State University, Fort Collins, CO

1. INTRODUCTION

The renaissance of phage therapy, combined with the age of molecular biology, has led to the development of newer tools for use in the fight against infectious bacterial disease. While cocktails of natural phages continue to be developed as therapeutics, gene products of the phages with the capability of neutralizing various bacterial proteins, or lysing the bacterial cell, have been exploited. In addition, genetically modified phages are increasingly being developed based on modification of phage properties such as host range and lytic ability, and to solve some of the problems associated with therapeutic use of natural phages. One of the newest phage therapies involves the creation of recombinant phages that have been altered to encode lethal genes, which are delivered to the host bacteria. The expression of the lethal gene produces a gene product that inactivates the target cell. Alternatively, phages may be physically labeled with toxic molecules, which destroy the bacterial cell when brought into close proximity (due to binding of the phage). Finally, a major issue of phage therapy is the interaction of phages with the host immune system. Several approaches have been developed that allow for genetically altered phages to remain in circulation for longer periods than wild-type phages. All of these approaches combine to contribute to the continued evolution of phage therapy as a realistic and practical antibacterial treatment.

Enzybiotics: Antibiotic Enzymes as Drugs and Therapeutics. Edited by Tomas G. Villa and Patricia Veiga-Crespo

2. LETHAL AGENT DELIVERY SYSTEMS (LADS)

LADS utilize a phage-based *in vivo* packaging system to create a targeted phage head capable of delivering naturally occurring molecules with bacteriocidal activity to drug-resistant bacteria (Norris et al. 2000). The system consists of a transfer plasmid carrying the genes encoding the antimicrobial agents, a plasmid origin of replication, the P1 lytic origin of replication, and a minimal packaging (PAC) site. The plasmid is maintained in a phage P1 lysogen unable to package its own DNA. The defective lysogen provides all of the replication factors required for activation of the P1 origin of replication on the transfer plasmid and all of the structural components necessary to form mature virions. The lysogen also carries the c1.100 temperature-sensitive repressor mutation, which represses functions leading to vegetative phage production. Induction of the lysogen by a temperature shift results in multiplication of DNA, packaging of the transfer plasmid into P1 phage heads, and lysis of the production strain. Virions are harvested and used to deliver the transfer plasmid to the pathogen. The phage head contains multiple copies of transfer vector DNA, and following delivery to the target bacterial cell, the plasmid DNA recircularizes and expression of the lethal agent (which is under the control of environmental, virulence-regulated, or species-specific promoters) results in rapid cell death. Lethal agents delivered by LADS are naturally occurring lethal genes associated with plasmids, phage, or bacterial chromosomes such as *doc*, *chpBK*, and *gef*. A plethora of such genes exists (as reviewed by Holcík and Iyer [1997]). Norris et al. (2000) tested a number of such lethality systems in *Escherichia coli*, and the doc toxic protein has been shown to be lethal in *E. coli* and is either bacteriocidal or bacteriostatic in other Gram-negative and Gram-positive bacteria including *Pseudomonas aeruginosa*, *Staphylococcus aureus*, and *Enterococcus faecalis*.

Shaak (2004) described the development of toxin–phage bacteriocides (TPBs) (defined as phages whose genomes are modified to contain toxin genes that facilitate bacterial death). The peptide toxins are lethal to the bacterial cell only within the cytoplasm, with little toxic activity observed outside of the cell. Importantly, the TPB retains its activity as a phage, and is therefore capable of completing the lytic phase of its life cycle. Completion of the lytic phase results in both the production of additional TPB and host cell lysis. To create the TPB, the nucleic acid sequence of the peptide toxin A or a derivative is introduced to the phage genome using standard homologous recombinant techniques, which can be achieved *in vivo* and *in vitro*. Alternatively, the recombinant phage genome is packaged into phage particles.

An infected bacterial cell may be killed by several mechanisms as a result of toxin–phage infection, including lysis of the bacterial cell, expression of the toxic peptide, or by a combination of the intracellular expression of the toxic molecule and lytic growth (Shaak 2004).

Other approaches based on the general principle of LADS have been developed. Westwater et al. (2003) produced a recombinant phage that carried a post-segregational killing system and demonstrated the efficacy of the phage to kill target bacteria. Post-segregational killing systems consist of two components including a stable toxin and an unstable antitoxin. One of the best known systems is the host killing/ suppressor of killing (hok/sok) system of the plasmid R1 of *E. coli* (Kuowei and Wood 1994). Other examples include the *pemI-pemK* genes of plasmid R100, the *phd-doc* genes of phage P1, and the *ccdA-ccdB* genes of plasmid F (Jensen and Gerdes 1995; Gerdes et al. 1997; Couturier et al. 1998; Engelberg-Kulka and Glaser 1999). Analogous post-segregational killing systems, such as *chpAI-chpAk*, *sof-gef*, *kicA-kicB*, *relB-relE*, and *chpBI-chpBk*, have been identified in *E. coli* K12 (Poulsen et al. 1989; Masuda et al. 1993; Feng et al. 1994; Aizenman et al. 1996; Gotfredsen and Gerdes 1998). All of these systems work in a similar fashion. The system is controlled by two genes that code for a toxin with a long half-life, and an antitoxin with a shorter half-life, respectively. After cell division, daughter cells that do not contain a copy of the plasmid die because the toxin from the parent cell has a longer half-life than that of the antitoxin. Therefore, only cells that carry a plasmid can produce more antitoxin and survive. This principle is responsible for the name of these systems, since cell death occurs following segregation of the plasmid. As such, post-segregational killing systems are believed to increase the stability of extrachromosomal elements by selectively killing plasmid-free cells, resulting in the proliferation of plasmid-bearing cells within the population (Holcík and Iyer 1997).

In the Westwater et al. (2003) study, the principle of using phage to deliver components of a post-segregational killing system was evaluated by producing recombinant phages based on the M13 phagemid system and the post-segregational killing toxins Gef and ChpBk. The phages were tested for efficacy using *in vitro* and *in vivo* experiments. For the *in vitro* studies, cells (at a concentration of 10^6 CFU/ml) and isopropyl β-D-1-thiogalactopyranoside (IPTG) (to induce expression of the toxin genes from the phagemid) were added to an equal volume of phage lysate (10^9 PFU/ml) and incubated. Bacterial survival in phage-infected cultures was determined by plate count and any surviving *E. coli* cells were enumerated and compared with the number of bacteria in a phage-free *E. coli* control culture. The *in vivo* experiment consisted of infecting mice (via intraperitoneal injection) with 200 µl of an appropriate *E. coli* strain at 10^8 CFU/ml, followed by immediate injection of 200 µl of phage lysate (10^{10} phagemid-containing particles/ml), and 100 µl of 250 mM IPTG. Peripheral blood from the tip of the tail was collected at 1, 3, and 5 h following injections and analyzed by plate count to determine the concentrations of *E. coli*.

The results indicated that M13 delivery of the lethal-agent phagemids reduced target bacterial numbers by several orders of magnitude *in vitro* and in the *in vitro* bacteremic mouse model of infection. For example, in the *in vitro* experiments, bacterial death following phage delivery and expression of the Gef and ChpBK toxin genes was apparent, and delivery of either toxin to host cells reduced the bacterial concentration by more than 2 logs. Using a nonlethal model of infection, the authors showed that phage delivery of the lethal-agent phagemids pGef and pChpBK resulted in a significant reduction in circulating bacteria compared with controls. Phage titers in the blood were also determined at 3 h postinjection and ranged from 10^7 to 10^8 phagemid-containing particles per milliliter, showing that both bacteria and phage migrated away from the injection site. After 5 h, the mice receiving the Gef and ChpBK phages showed a 98% and 94% reduction in blood bacterial titers, respectively, as compared with the control.

Collectively, these results indicate that LADS have great potential for the control of bacterial pathogens in medical settings as well as in veterinary settings and agricultural environments (preharvest control of foodborne pathogens) (Westwater et al. 2003).

Interestingly, while the toxin component of post-segregational killing systems has been utilized to produce recombinant phages with the ability to deliver toxic molecules to bacteria, Pecota and Wood (1996) showed that such mechanisms probably evolved as anti-phage defense systems. To investigate the possibility of phage exclusion by the hok/sok locus, *E. coli* cells that contained hok/sok on a plasmid were challenged with phages T1, T4, T5, T7, and λ. While the hok/sok system did not seem to affect T1, T5, T7, and λ, upon infection with T4, the optical density of cells containing hok/sok on a high-copy-number plasmid continued to increase while the optical density of cells lacking hok/sok rapidly declined. The presence of hok/sok reduced the efficiency of plating of T4 by 42% and decreased the plaque size by approximately 85%. In addition, single-step growth experiments demonstrated that hok/sok decreased the T4 burst size by 40%, increased the eclipse time from 22 to 30 min, and increased the latent period from 30 to 60 min. In this scenario, it seems that the Hok toxin disrupts the cell at the later stages of phage development by delaying cell lysis, since examination of the time required for T4 infection and killing by hok/ sok indicates that there is sufficient time for the killer locus to disrupt cell metabolism before T4 lysis of the cell. The authors concluded that based on the results, post-segregational killing systems are most likely evolutionary important in terms of effecting phage exclusion during bacterial infections (Pecota and Wood 1996).

Several other toxin gene systems have been utilized to produce recombinant phages for therapeutic purposes. For example, Fairhead (2004a, b) developed an antimicrobial system that utilizes phage-based delivery of the small, acid-soluble spore protein (SASP) genes into pathogenic bacteria. Endospores of *Bacillus* spp. are very resistant to physical and chemical stresses, which is due in part to the fact that the spore DNA is protected from damage by saturation with α/β-type SASPs (Setlow 1995; Setlow and Setlow 1995a, b). There are three types of SASPs, including α, β, and γ, and it has been shown that the α/β-type SASPs act as DNA-binding proteins to protect the DNA from damage, while the γ-type SASP is used to supply amino acids for outgrowth (Hackett and Setlow 1987). Setlow et al. (1991) showed that a gene encoding α/β-type SASPs could be inserted into a plasmid under the control of an inducible promoter, and this caused a vegetative cell to assume spore-like characteristics. In Fairhead's (2004a, b) work, the lysis genes of phages were replaced with a gene encoding the α/β-type SASPs. Upon bacterial infection, all viral genes including the SASP were expressed. The α/β-type SASPs are toxic to the bacterial cell since they bind bacterial DNA irreversibly, stopping all cellular activity. The phage does not multiply within the target bacteria but has been shown to produce rapid killing. The mechanism of action of this system provides limited opportunities for the emergence of resistance, and it is currently being developed to target the Gram-positive pathogens methicillin-resistant *S. aureus* (MRSA) and *Clostridium difficile* (Hanlon 2007).

2.1. Production of Non-lytic Phage Mutants

One of the side effects of many antibiotics and lytic phage-based phage therapy is the release of the lipopolysaccharide of Gram-negative bacteria, which mediate the general pathological aspects of septicemia. To address this issue, Hagens and Bläsi (2003) evaluated the ability of a filamentous phage encoding lethal proteins to kill bacteria without host cell lysis. In this work, bacterial survival was determined after infection of a growing *E. coli* culture with phage M13 encoding either the restriction endonuclease *Bgl*II gene (phage M13R) or two modified phage k S holin genes (phages S105 and the VIIIS105). The results indicated that all phages were very efficient (exerted a high killing efficiency) at destroying the host cells while leaving them structurally intact. For example, infection of *E. coli* strain MC4100F' with either of the modified holin phages (S105 and VIII105) caused a rapid decline in cell viability within 2 h. More than 99% of the bacteria were killed within the 2-h period and variations in the multiplicity of infection (MOI) between 1 and 100 had little effect on the killing efficiency. The researchers observed that growth resumed between 120 and 180 min after infection due to the fact that phage-resistant mutants had emerged. Infection of a growing culture of *E. coli* MC4100F' with M13R at an MOI of 10 led to a rapid decline in the cell concentration, and when compared with phages M13S105 or M13VIIIS105, the killing efficiency was increased, such that 99.9% of the bacteria were killed after 6 h.

When compared with a lytic phage, the release of endotoxin was minimized after infection with the genetically modified phages. For example, the effects of phages M13S105 and M13R on endotoxin release were compared with that of λcI⁻ (a lytic λ phage), and the data indicated that when compared with the initial amount present in the culture supernatant at time 0, the endotoxin levels increased with the lytic λcI⁻ phage 18-fold after 1 h of infection, while only a twofold increase was observed after 1 h of infection with either phage M13S105 or M13R. After 4 h, a 27-fold increase in endotoxin concentration was observed in the λcI⁻-infected culture, in contrast to the six- and seven-fold increases in the supernatant of cultures infected with M13S105 and M13R, respectively.

In a related study, Hagens et al. (2004) developed a genetically engineered, non-replicating, non-lytic phage based on phage Pf3 to combat an experimental *P. aeruginosa* infection. An export protein gene of the *P. aeruginosa* filamentous phage Pf3 was replaced with *Bgl*II. This rendered the Pf3 variant (Pf3R) non-replicative and prevented the release of the therapeutic agent from the target cell. The Pf3R phage efficiently killed a wild-type host *in vitro*, while endotoxin release was kept to a minimum. Treatment of *P. aeruginosa* infections of mice with Pf3R or with a replicating lytic phage resulted in comparable survival rates upon challenge. However, the survival rate following phage therapy with Pf3R increased (compared with the lytic phage) when the mice were challenged with a higher concentration of *P. aeruginosa*. This higher survival rate correlated with a reduced inflammatory response elicited by Pf3R treatment relative to that with the lytic phage. Therefore, this study suggests that the increased survival rate of Pf3R-treated mice could result from reduced endotoxin release. The use of genetically engineered non-lytic phages to combat bacterial genes would also be useful to prevent other components of the bacterial cell from being released. For example, antibiotic treatment of patients with Shiga toxin-producing *E. coli* (STEC) infections is controversial due to the fact that certain antibiotics (e.g., quinolones) induce Shiga toxin (Stx)-encoding phages, leading to increased toxin production and release from the bacteria, thereby increasing morbidity and mortality in patients (Zhang et al. 2000). In treatment of STEC disease, the use of a recombinant non-lytic phage would have the advantage of destroying the bacterial cells without inducing the STEC prophages, which would decrease toxin production (induction of Stx prophages is correlated with a large increase in Stx production [Zhang et al. 2000]). The construction of non-lytic or lysis-deficient phage and their therapeutic uses have been the subjects of several patents (Ramachandran et al. 2002a, b).

The use of natural temperate phages in phage therapeutic applications should be avoided because of lysogenic conversion and possible genetic exchange of virulence genes with the target bacterial cell (Krylov 2001). Temperate phages typically induce a state of lysogeny within the bacterial host when the phage DNA becomes integrated within the host DNA chromosome. Still, the isolation of lytic phages that infect some species of bacteria can be time-consuming because lytic phages of some pathogens are relatively rare. Since prophages are more common in bacteria, they are more easily isolated. As such, one approach to the isolation of phages from bacteria, which do not seem to support the growth of lytic phages, is to induce prophages from the DNA chromosome, and then genetically modify the temperate phages to become permanently lytic. In this scenario, the *vir* gene of the induced phage is mutated, giving rise to a permanently lytic phage that is capable of infecting the target bacterium (Rapson et al. 2003). The *vir* mutants usually contain an alteration in the operator region of the phage DNA that prevents the repressor protein from binding to the operator. This alteration leads to derepression of lysogeny, allowing transcription and translation of the phage DNA to follow with subsequent lysis of the bacterial host.

In a novel application, genetically altered phages have been developed for the treatment of eukaryotic diseases. Hoeprich et al. (2003) utilized the DNA-packaging RNA (pRNA) of the *Bacillus subtilis* phage φ29 to deliver hammerhead ribozymes to mammalian cells. Hammerhead ribozymes specifically cleave the polyA signal of hepatitis B viral mRNA, which inhibits replication. The DNA-packaging

RNA (pRNA) of phage φ29 contains two independent tightly self-folded domains. Circularly permuted pRNAs were constructed without affecting pRNA folding. Connecting the pRNA 5′ and 3′ ends with variable sequences did not disrupt its folding and function. These unique features, which help prevent two common problems including exonuclease degradation and misfolding in the cell, make pRNA an ideal vector to carry therapeutic RNAs. A pRNA-based vector was designed to carry hammerhead ribozymes that cleave the hepatitis B virus (HBV) polyA signal. The chimeric HBV-targeting ribozyme was attached to the pRNA 5′ and 3′ ends as circularly permuted pRNA, and two *cis*-cleaving ribozymes were used to flank and process the chimeric ribozyme. The hammerhead ribozyme was able to fold correctly while escorted by the pRNA. During *in vitro* experiments, the chimeric ribozyme was observed to cleave the polyA signal of HBV mRNA almost completely. Cell culture studies showed that the chimeric ribozyme was able to enhance the inhibition of HBV replication when compared with the ribozyme that was not escorted by pRNA. These findings suggest that pRNA can be used as a vector for imparting stability to ribozymes, antisense, and other therapeutic RNA molecules *in vivo*.

3. LABELED PHAGES AS THERAPEUTIC AGENTS

The phage capsid has been decorated with bactericidal proteins as an alternative to the genetic modification of phage chromosomes to carry toxic genes. One novel study evaluated the use of targeted photosensitizers as bactericidal agents. Light-activated antimicrobial agents (photosensitizers) represent promising alternatives to antibiotics for the treatment of topical infections. To improve efficacy and avoid possible damage to host tissues, targeting of the photosensitizer to the infecting organism is desirable, and this has previously been achieved using antibodies and chemical modification of the agent. Embleton et al. (2005) investigated the possibility of using a phage to deliver the photosensitizer tin(IV) chlorin e6 (SnCe6) to *S. aureus*. SnCe6 was covalently linked to *S. aureus* phage 75, and the ability of the conjugate to kill various strains of *S. aureus* when exposed to red light was determined. Substantial reductions in the concentrations of methicillin- and vancomycin-intermediate strains of *S. aureus* were achieved using low concentrations of the conjugate (containing 1.5 µg/ml SnCe6) and low light doses (21 J/cm^2). Under these conditions, the viability of human epithelial cells (in the absence of bacteria) was largely unaffected. On a molar equivalent basis, the conjugate was more effective than the unconjugated SnCe6, and the destruction of the bacteria was not growth phase dependent. The conjugate was effective against vancomycin-intermediate strains of *S. aureus* even after growth in vancomycin. The results of this study have demonstrated that a phage can be used to deliver a photosensitizer to a target organism, resulting in enhanced and selective killing of the organism. Such attributes make the use of labeled phage a promising approach to be used in the photodynamic therapy of infectious diseases.

Yacoby et al. (2006) evaluated the use of filamentous phages as targeted drug carriers and their ability to destroy pathogenic bacteria. The phages were genetically modified to display a targeting moiety on their surface and were used to deliver a large payload of a cytotoxic drug to the target bacteria. The drug was linked to the phages by means of chemical conjugation through a labile linker and subjected to controlled release. This was achieved by mixing a constant molar ratio of 10^5 chloramphenicol prodrug molecules/phage. The reaction was mixed for 1 h, and phage precipitates were purified by centrifugation. In the conjugated state, the drug is devoid of cytotoxic activity and is only activated following its dissociation from the phage at the target site (the surface of a bacterial cell). The study demonstrated the potential of using filamentous phages as universal drug carriers to control the growth of *S. aureus* in a model system. The authors concluded that selectively targeting drugs to the bacterial cell surface may allow the introduction of nonspecific drugs that have been excluded from antibacterial use because of toxicity or low selectivity, and that this approach may help to combat emerging bacterial antibiotic resistance. However, the labeled phage was limited in its capacity to inhibit bacterial growth due to a low number of prodrug (less than 3000) molecules per phage. In a later study, Yacoby et al. (2007) addressed this issue by designing a novel drug conjugation chemistry, which comprised the use of hydrophilic aminoglycoside antibiotics as branched, solubility-enhancing linkers. Changing the arming chemistry and introducing the use of the amino sugar-based aminoglycosides as branched, hydrophilic linkers enabled the solvation of hydrophobic materials such as chloramphenicol, fluorescein isothiocyanate (FITC), or Z-Phe. This enabled the conjugation of a large concentration of hydrophobic molecules to each phage. Over 40,000 chloramphenicol molecules per phage could be conjugated without compromising the integrity of the phage. The researchers also improved the ability of the phage to deliver the conjugated drug to the surface of the target cell by conjugating polyclonal antibodies to the phage. A conjugation level of 10,000 molecules per phage was observed to be sufficient enough to cause growth inhibition of MRSA, a clinical isolate of *Streptococcus pyogenes*, and *E. coli* O78. This study demonstrated an improvement factor in drug potency of 20,000 in comparison to the free drug (Yacoby et al. 2007). The use of phage to target antimicrobials to the bacterial surface is an intriguing idea, and these results should encourage further work in this area.

Cao et al. (2000) produced *Helicobacter pylori*-antigen-binding single-chain variable fragments (ScFv) and expressed them as a g3p-fusion protein on the surface of M13 phage. To create the recombinant phage, the Recombinant Phage Antibody System (Pharmacia Biotech, Piscataway, NJ) was first used to produce recombinant phage antibodies. Messenger RNA was extracted from hybridoma cells, reverse transcribed and amplified using standard molecular biology techniques. The genes of the heavy and light chains were then assembled as a single-chain fragment variable (ScFv) gene with a linker, and the ScFv gene was then digested and subsequently inserted in a suitable phagemid vector that had been predigested with the same enzymes. The ligation product was transformed into a suitable strain of *E. coli* followed by infection with M13KO7 helper phage to yield recombinant phage that display antibody ScFv fragments. Initial studies confirmed that the phage-displayed antibodies were capable of binding *H. pylori* antigen. For example, the recombinant ScFv phage reacted specifically with a 30-kDa monomeric protein from an *H. pylori* surface antigen preparation and was shown to bind to both the spiral and coccoid forms of the bacterium. *In vitro*, the recombinant phage exhibited a bacteriocidal effect and specifically inhibited the growth of all the six strains of *H. pylori* tested. When *H. pylori* was pretreated with the phage 10 min before oral inoculation of mice, the colonization of the mice by the bacterium was significantly reduced ($P < 0.01$). The results of the study suggest that genetic engineering may be used to generate phages that could be used to treat *H. pylori* infection.

4. AVOIDING THE HOST IMMUNE RESPONSE

Initial experiments that involved the injection of phages into animals led to the observation that phages were rapidly cleared by the immune system (Stent 1963). Two such experiments showed rapid clearance of the phages from the blood and organs of rabbits, but long-term survival in the spleen (Appelmans 1921; Evans 1933). Additional experiments in rodents also showed rapid loss from the circulation. For example, after Nungester and Watrous (1934) intravenously injected 10^9 PFU of a staphylococcal phage into albino rats, a blood concentration of only 10^5 PFU/ml was observed after 5 min, and the phage concentration decreased to 40 PFU/ml after 2 h.

To address the rapid clearing of phages by the recticuloendothelial system (RES), Merril et al. (1996) developed a serial-passage technique in mice to select for phage mutants able to remain in the circulatory system for longer periods of time. The method entailed the initial intraperitoneal injection into each mouse of 10^{11} PFU of phage λW60 grown on either wild-type *E. coli* (strain CRM1) or a mutator *E. coli* strain (CRM2), followed by collection of blood samples from the mice after 7 h. The use of the mutator *E. coli* CRM2 strain allowed for an increased rate of mutation in phage λW60, which increased the chances that one or more of the phage offspring would have properties that permitted evasion of the RES. Phage titers at 7 h postinjection were 10^9 and 10^8 PFU from the mutagenized and non-mutagenized phage, respectively. The phage titers in the circulatory system decreased to 10^2 after 48 h and to undetectable levels after 120 h. The phages that were present after 7 h were isolated and grown to high titers in bacteria (*E. coli* strains CRM1 and CRM2). The high-titer phages were purified and the process was repeated. This serial cycling of phage, by injection into animals, isolation, and regrowth in bacteria was repeated nine more times. Approximately 10^6 PFU remained in the circulatory system after 18 h during the second cycle, and in subsequent cycles, the titers gradually rose, so that 10^9 PFU remained in the circulatory system after 18 h during the fourth cycle.

The final six selection cycles did not provide a significant increase in the number of phage remaining in the circulatory system after 18 h. Following the 10th cycle of the selection process, a single plaque from each of the two experiments was isolated, purified, and grown to high titers on the CRM1 host and designated as Argo1 (cycled on strain CRM1) and Argo2 (cycled on strain CRM2). Both Argo1 and Argo2 displayed similar enhanced capacity to avoid RES clearing. For example, the 18-h survival following intraperitoneal injection of Argo1 was 16,000-fold higher and that of Argo2 was 13,000-fold higher than that of the parental λW60 strain. A similar selection process was used to isolate long-circulating variants of the *Salmonella* phage R34. Following eight selection cycles, long-circulating single-phage plaques were similarly isolated and purified. Two such isolates, designated Argo3 and Argo4, were compared with the parental phage R34 for their rates of clearance from the mouse circulatory system. After intraperitoneal injection of 10^7 PFU, no detectable R34 phages were observed at 24 h. In contrast, 10^2 and 10^3 PFU of Argo3 and Argo4 were detected after 24 h. An *in vivo* study of the long-circulating phage and the wild-type coliphages was conducted using a mouse model. Four groups of mice were injected with 10^8 cfu of *E. coli* CRM1. The first group was a control, with no phage treatment. Within 5 h, these mice showed symptoms of illness including ruffled fur, lethargy, and hunchback posture. After 24 h, they were moribund, and they died within 48 h. Three groups of phage were treated with 10^{10} PFU of phage. All of the mice that were treated with phage survived. However, those treated with W60 (group 2) had severe illness before finally recovering, while those treated with Argo1 (group 3) and Argo2 (group 4) exhibited only minor signs of illness before complete recovery. Analysis of the capsid proteins of the wild-type (W60) and mutant phages (Argo1 and 2) revealed an alkaline shift in the 38-kDa major viral protein in Argo1 compared with W60. The same electrophoretic protein shift was observed in Argo2. The protein was subsequently identified as the major λ capsid head protein E. Sequence analysis of the protein E gene in Argo1 and Argo2 revealed a G → A transition mutation at nt 6606 in both phages. This transition mutation resulted in the substitution of the basic amino acid lysine for the acidic amino acid glutamic acid at position 158 of the λ capsid E protein in both Argo strains. Argo2 protein profiles displayed the presence of an additional altered protein, which also had an alkaline shift. Analysis and characterization showed that the mutation was located within the major capsid head protein D of λ. Based on their results, the authors concluded that the use of toxin-free, bacteria-specific phage strains, combined with the serial-passage technique, may provide a viable method for developing phage into therapeutically effective antibacterial agents (Merril et al. 1996).

Alternative approaches have been employed in an attempt to decrease immunogenicity of phages and decrease clearance by the RES. Kim et al. (2008) chemically modified phages by conjugation of the polymer monomethoxy-polyethylene glycol (mPEG) to phage proteins. This polymer is non-immunogenic, so the rationale for the study was based on the fact that conjugating mPEG to the phages would mask them, thereby increasing their blood circulation time. Two myophages (Felix-O1 [infects *Salmonella* spp.] and A511 [infects *Listeria* spp.]) were used in the experiments. Loss of phage infectivity following PEGylation was found to be proportional to the degree of modification, and could be controlled by adjusting the PEG concentration. When injected into naive mice, PEGylated phages showed a strong increase in circulation half-life, whereas challenge of immunized mice did not reveal a significant difference. The results of the study suggest that the prolonged half-life is due to decreased susceptibility to innate immunity as well as avoidance of cellular defense mechanisms. PEGylated phages elicited significantly reduced levels of T-helper type 1-associated cytokine release (IFN-γ and IL-6), in both naive and immunized mice. This study was the first to demonstrate that PEGylation can increase survival of phage *in vivo* by delaying immune responses, and indicates that this approach can increase efficacy of phage therapy (Kim et al. 2008).

5. CONCLUSION

The advent of molecular biology, originally developed through the study of phage (Cairns et al. 2007), has led to the development of recombinant phages that have been demonstrated to efficiently kill target bacteria while eliminating many of the problems associated with the use of natural phages in phage therapeutic applications. As the field of phage therapy progresses, as more phage genomic sequences are published, and as the function of gene products is elucidated, it is likely that phage therapeutic applications will move away from using the whole phage as an antimicrobial and focus instead on a component of the phage. In fact, such approaches have already begun. One such technology is based on the activity of natural and recombinant phage holins and lysins (reviewed in Chapters 5–7). Other approaches have been based on the use of proteins produced shortly after phage infection of a bacterial host. The early proteins that are produced immediately after the entry of the phage genome into a bacterial cell are responsible for the inhibition of host macromolecular biosynthesis, the initiation of phage-specific replication, and the synthesis of late proteins. Inhibition of synthesis of host proteins that eventually leads to cell death is generally caused by the physical and chemical modification of indispensable host proteins by early proteins (Sau et al. 2008). Finally, the possibility exists to use phages to identify novel chemical antimicrobials. Such is the approach of Liu et al. (2004) who applied the concept of phage-mediated bacterial growth inhibition to antibiotic discovery. In that work, the sequences of 26 *S. aureus* phages were used to identify 31 novel polypeptide families that inhibited growth upon expression in *S. aureus*. The cellular targets for several of the polypeptides were identified and some were shown to be essential components of the host DNA replication and transcription machineries. The interaction between one prototypic pair, ORF104 of phage 77 and DnaI, the putative helicase loader of *S. aureus*, was then used to screen for small molecule inhibitors. Several compounds were subsequently found to inhibit both bacterial growth and DNA synthesis. The results of this work suggest that mimicking the growth-inhibitory effect of phage polypeptides by a chemical compound, coupled with the plethora of phages on earth, will combine to yield new antibiotics to combat infectious diseases (Liu et al. 2004). The development of such exciting and innovative approaches to the discovery of phage-based antimicrobials bodes well for the continued evolution of this still emerging field.

REFERENCES

Aizenman E., H. Engelberg-Kulka, and G. Glaser (1996) *Proceedings of the National Academy of Sciences of the USA* **93**, 6059–6063.

Appelmans R. (1921) *Comptes rendus des séances de la Société de biologie et de ses filiales* **85**, 722–724.

Cairns J., G. S. Stent, and J. D. Watson (2007) *Phage and the Origins of Molecular Biology. The Centennial Edition*. Cold Spring Harbor, NY: Cold Spring Harbor Laboratory Press.

Cao J., Y. Sun, T. Berglindh, B. Mellgård, Z. Li, B. Mårdh, and S. Mårdh (2000) *Biochimica et Biophysica Acta* **1474**, 107–113.

Couturier M., E. M. Bahassi, and L. Van Melderen (1998) *Trends in Microbiology* **6**, 269–275.

Embleton M. L., S. P. Nair, W. Heywood, D. C. Menon, B. D. Cookson, and M. Wilson (2005) *Antimicrobial Agents and Chemotherapy* **49**, 3690–3696.

Engelberg-Kulka H. and G. Glaser (1999) *Annual Review of Microbiology* **53**, 43–70.

Evans A. C. (1933) *Public Health Reports* **48**, 411–446.

Fairhead H. (2004a) Antimicrobial compositions and uses thereof. Patent number WO 2004 113375.

Fairhead H. (2004b) Small acid-soluble proteins and uses thereof. U.S. Patent 20040097705.

Feng J., K. Yamanaka, H. Niki, T. Ogura, and S. Hiraga (1994) *Molecular & General Genetics* **243**, 136–147.

Gerdes K., A. P. Gultyaev, T. Franch, K. Pedersen, and N. D. Mikkelsen (1997) *Annual Review of Genetics* **31**, 1–31.

Gotfredsen M. and K. Gerdes (1998) *Molecular Microbiology* **29**, 1065–1076.

Hackett R. H. and P. Setlow (1987) *Journal of Bacteriology* **169**, 1985–1992.

Hagens S. and U. Bläsi (2003) *Letters in Applied Microbiology* **37**, 318–323.

Hagens S., A. Habel, U. von Ahsen, A. von Gabain, and U. Bläsi (2004) *Antimicrobial Agents and Chemotherapy* **48**, 3817–3822.

Hanlon G. W. (2007) *International Journal of Antimicrobial Agents* **30**, 118–128.

Hoeprich S., Q. Zhou, S. Guo, D. Shu, G. Qi, Y. Wang, and P. Guo (2003) *Gene Therapy* **10**, 1258–1267.

Holcík M. and V. N. Iyer (1997) *Microbiology* **143**, 3403–3416.

Jensen R. B. and K. Gerdes (1995) *Molecular Microbiology* **17**, 205–210.

Kim K. P., J. D. Cha, E. H. Jang, J. Klumpp, S. Hagens, W. D. Hardt, K. Y. Lee, and M. J. Loessner (2008) *Microbial Biotechnology* **1**, 1247–1257.

Krylov V. N. (2001) *Russian Journal of Genetics* **37**, 715–730.

Kuowei W. and T. K. Wood (1994) *Biotechnology and Bioengineering* **44**, 912–921.

Liu J., M. Dehbi, G. Moeck, F. Arhin, P. Bauda, D. Bergeron, M. Callejo, V. Ferretti, N. Ha, T. Kwan, J. Macarty, R. Srikumar, D. Williams, J. J. Wu, P. Gros, J. Pelletier, and M. DuBow (2004) *Nature Biotechnology* **22**, 185–191.

Masuda Y., K. Miyakawa, Y. Nishimura, and E. Ohtsubo (1993) *Journal of Bacteriology* **175**, 6850–6856.

Merril C. R., B. Biswas, R. Carlton, N. C. Jensen, G. J. Creed, S. Zullo, and S. Adhya (1996) *Proceedings of the National Academy of Sciences of the USA* **93**, 3188–3192.

Norris J. S., C. Westwater, and D. Schofield (2000) *Gene Therapy* **7**, 723–725.

Nungester W. J. and R. M. Watrous (1934) *Proceedings of the Society for Experimental Biology and Medicine* **31**, 901–905.

Pecota D. C. and T. K. Wood (1996) *Journal of Bacteriology* **178**, 2044–2050.

Poulsen L. K., N. W. Larsen, S. Molin, and P. Andersson (1989) *Molecular Microbiology* **3**, 1463–1472.

Ramachandran J., S. Padmanabhan, and B. Siriam (2002a) U.S. Patent 6913753 B2.

Ramachandran J., S. Padmanabhan, and B. Siriam (2002b) U.S. Patent 6896882 B2.

Rapson M. E., F. A. Burden, L. P. Glancy, D.A. Hodgson, and N. H. Mann (2003) Patent number WO03080823.

Sau S., P. Chattoraj, T. Ganguly, P. K. Chanda, and N. C. Mandal (2008) *Current Protein & Peptide Science* **9**, 284–290.

Shaak D. L. (2004) U.S. Patent 6759229.

Setlow P. (1995) *Annual Reviews in Microbiology* **49**, 29–54.

Setlow B., A. R. Hand, and P. Setlow (1991) *Journal of Bacteriology* **173**, 1642–1653.

Setlow B. and P. Setlow (1995a) *Applied and Environmental Microbiology* **61**, 2787–2790.

Setlow B. and P. Setlow (1995b) *Journal of Bacteriology* **177**, 4149–4151.

Stent G. (1963) *Molecular Biology of Bacterial Viruses*. San Francisco: WH Freeman.

Westwater C., L. M. Kasman, D. A. Schofield, P. A. Werner, J. W. Dolan, M. G. Schmidt, and J. S. Norris (2003) *Antimicrobial Agents and Chemotherapy* **47**, 1301–1307.

Yacoby I., H. Bar, and I. Benhar (2007) *Antimicrobial Agents and Chemotherapy* **51**, 2156–2163.

Yacoby I., M. Shamis, H. Bar, D. Shabat, and I. Benhar (2006) *Antimicrobial Agents and Chemotherapy* **50**, 2087–2097.

Zhang X., A. D. McDaniel, L. E. Wolf, G. T. Keusch, M. K. Waldor, and D. W. K. Acheson (2000) *The Journal of Infectious Diseases* **181**, 664–670.

CONCLUDING REMARKS: THE FUTURE OF ENZYBIOTICS

PATRICIA VEIGA-CRESPO[1] and TOMAS G. VILLA[1,2]
[1]Departament of Microbiology, Faculty of Pharmacy, University of Santiago de Compostela, Spain
[2]School of Biotechnology, University of Santiago de Compostela, Spain

1. ENZYBIOTICS AND ANTIBIOTIC-RESISTANT MICROBIAL STRAINS

In the present book we have tried to update some old and well-known ideas, one of them concerning an ever-present race between science (by developing new antibiotics) and microorganisms (by increasing the mutation rate, thus favoring the appearance of resistant strains). It is quite evident that for the last 60 years bacteria have been developing new and more effective ways to resist antibiotics. Enzybiotics' solution to this problem is to apply natural antimicrobial enzymes (lysins) or even whole bacteriophages to prevent the growth of pathogenic bacteria or fungi. Lysins are therefore proteins, normally elaborated by viruses or even bacteria that destroy bacteria or fungi by degrading their cell walls.

It is evident that enzybiotics are highly targeted and may rapidly kill microbial pathogens without inducing resistance phenotypes. The first two targets for enzybiotics were group B streptococci and pneumococci that can cause serious, potentially life-threatening disease, and if one takes into account the demonstrated efficiency of these enzymes, it is foreseeable that soon we shall have a large variety of usable enzybiotics. Pharmaceutical application of enzybiotics may be either in the form of bacteriophages or lysins (for bacteria) or 1,3-β-D-glucanases or

Enzybiotics: Antibiotic Enzymes as Drugs and Therapeutics. Edited by Tomas G. Villa and Patricia Veiga-Crespo
Copyright © 2010 John Wiley & Sons, Inc.

chitinases (for fungi). In the latter case the use of mycovirus to fight back fungal infections is yet difficult to visualize because of their immunizing abilities. As lysin therapy is indeed different from phage therapy, a whole new pharmacological approach must be developed in order to safely supply the enzybiotics to the sick, in terms of diminishing their immunological response.

It is a different situation when one applies bacteriophage therapy directly. When a bacteriophage infects a bacterial cell, it replicates inside to form abundant viral progeny, and by doing so it kills the host cell by disrupting the peptidoglycan layer of the cell wall. But bacteria have adapted to such invasions by accelerating their mutational rate, thus becoming resistant to bacteriophage infections. Therefore, it is common that phage therapy requires combinations of three or more different phages to assure 90% efficacy. We also know that at the end of the bacteriophage life cycle, expression of holin and lysin genes occurs. Holins are required to provide access to the cell wall through the cell membrane, whereas lysins supplied externally are able to attack the cell wall directly (at least in Gram-positive bacteria) without the requirement of holins. It follows then that possibly holins will not be included in the arsenal of enzybiotics, at least in the near future. As lysins (considering lysins in their broadest possible way, i.e., muramidases, lysozymes, proteases, 1,3-β-D-glucanases, or chitinases) are the enzymes that both bacteria and fungi have to use, in order to provide a physiological environment for the cell wall to grow and expand normally, microbial resistance to lysins through mutations does not occur, probably because they would be lethal, as the bacterial or fungal cells would not be able to expand their cell walls in an appropriate way.

Many experiments have shown different lysins to be effective against a variety of antibiotic-resistant bacteria, and most important, that there can be synergy between antibiotics and lysins, thus suggesting that the use of lysins will not have a negative effect on the genesis of new antibiotics. As indicated by McCullers et al. (2007), the primary focus of enzybiotics is to develop lysins as decolonizing agents. It is known that most of us carry potential pathogenic bacteria on our oral, nasal, urogenital, and intestinal mucosal surfaces, and they live there in perfect comensal relationship. The invasion by other pathogenic microorganisms such as viruses, including flu-causing orthomyxovirus, may induce migration of pathogenic bacteria that in turn may induce otitis, pneumonia, and other bacterial infections. It is believed then that decolonizing at-risk individuals with low doses of topically administered lysins will reduce or even prevent these secondary bacterial infections. So, as pointed out by Fischetti, the strategy falls between the traditional antibiotics that are given only to treat an invasive infection, and vaccines, which are widely administered to prevent diseases in the future. Since it can take weeks or months to stimulate an immune response to vaccines, they are not usually useful in the acute prevention of infection, and at the same time bacterial or fungal pathogens are becoming more resistant to antibiotics. Therefore, enzybiotics are interestingly located between antibiotics and vaccines, with possibilities of pivoting to either side.

Another point to address here concerns the wide range of antibiotic-resistant bacteria or "superbugs" that colonize hospitals throughout the world. According to the Centers for Disease Control and Prevention, more than 70% of the bacteria causing infections in hospitals are resistant to at least one antibiotic. One of these bacteria are the Methicillin-resistant *Staphylococcus aureus* (MRSA) strains, or the multiple drug-resistant *Mycobacterium tuberculosis*, many cases of which are associated with AIDS patients. At this point we should be able to apply enzybiotic therapy to lower the amount of antibiotics used. This is an appealing idea because such a decrease will automatically lower the number of bacterial or fungal resistant strains. We do not foresee total supplantation of antibiotics by enzybiotics; instead and as proposed in this book, we hope that they would be implemented in synergism with antibiotics.

In the light of the foregoing it is clear that the pharmacological fight against microbial pathogens based on "all antibiotics" has a real weakness: old pathogens that had been almost eradicated are reemerging. One has to bear in mind that we are exploiting the fact that the best agents to kill bacteria and fungi come from other bacteria or from other fungi, but microorganisms have mastered, through hundreds of millions of years of evolution, to mutate rapidly so as to create new resistant strains by inactivating the active principle or simply by growing faster than the antibiotic-producing strains. The development of genomics and the knowledge of entire genomes of microbial pathogens will be of invaluable aid for the next generation of scientists working in the discovery and creation of new antibiotics with the final goal of staying one step ahead of the ever-mutating microorganisms. One way to overcome this is by challenging the microbial pathogens with two antibiotics at the same time, each of them disrupting a particular and important biochemical pathway; chances to mutate at both points and at the same time are extremely low. The situation may be even better if we are able to create new drugs exhibiting both antibiotic activities in the same molecule (see later). Another strategy is to ambush the bacteria with their current enemies: viruses. As indicated in previous chapters this idea was put into play soon after the bacteriophages were discovered by Twort and D'Herelle at the beginning of the 20th century (see Chapter 2). The idea was forgotten in Western countries because of the ever-hungry pharmaceutical industry, but not in Eastern Europe countries, which continued for decades to develop new phage combinations to fight microbial pathogens. In recent years, however, the situation is changing, so new companies are being created to produce either bacteriophages or their respective lysins, which are successfully used in therapeutic treatments. In particular, Vincent Fischetti's group at Rockefeller University is enlisting the help of bacteriophages to treat a variety of bacterial infections. Treating humans with live viruses— even bacteriophages—is always risky, so instead Fischetti isolated the lysins to attack bacteria from the outside; so far, he has been able to develop enzymes against pneumococci, streptococci, as well as anthrax.

Until all these new drugs are pharmaceutically ready and generally available, we must dig into the mechanistic abilities that bacteria and fungi have to mutate and evade the drugs we use to control the diseases they produce. No doubt that sequencing whole genomes of as many pathogens as possible will help us in the task of going a bit ahead of pathogens; if we are able to succeed in this, our staying around will be assured.

2. ENZYBIOTICS AND EMERGING PATHOGENS

Emerging pathogens are considered as such by public-health services throughout the world as those pathogens that increase the incidence of an epidemic outbreak as, for example, *Cryptosporidium*, *Escherichia coli* O157:H7, Hantavirus, multidrug-resistant pneumococci, and vancomycin-resistant enterococci. Besides the old and well-known bacterial and fungal pathogens, we have to keep our eyes open when it comes to the description of new microbial taxons in both the prokaryotic and eukaryotic worlds. It is a fact to all microbiologists that we know only a small portion of the microbial population on Earth, so it is quite possible that new pathogens are in our future, as we keep discovering new species in rare and untouched ecosystems. We believe that all new bacterial and fungal descriptions should include their ability to cause disease in both plant and animal kingdoms. And in the case of new bacteria, their respective bacteriophages should be described.

Current emerging and reemerging pathogens include enterohemorrhagic *Escherichia coli*, Legionellosis, a disease transmitted by technical vectors, *Burkholderia/Stenotrophomonas*, *Helicobacter pylori*, *Chlamydiae*, *Borrelia burgdorferi*, *Streptococcus pyogenes*, *S. aureus* and *Staphylococcus epidermidis*, Enterrococci, *Bordetella pertussis* and related species *parapertussis and bronchiseptica*, *Tropheryma whippelii* (perhaps the best paradigm of the re-emergence of an old disease), *Campylobacter jejuni*, Mycoplasmas (another paradigm of reemerging pathogens in both humans and animals), and Gram-negative plant pathogenic bacteria. Besides these one should also include all types of animal and plant viruses, particularly those that are able to replicate in animals as well as in plants; unfortunately, everyday we learn of new viral strains that share this ability. Most recently, the bacterial genus *Dietzia* has been established. The Gram morphology as well as colony appearance of species belonging to this genus is very similar to *Rhodococcus* sp., and in fact, lacking accurate methods for the identification of this bacterial genus it may be simply misclassified as *Rhodococcus* or even *Gordonia*.

Foodborne pathogens are the leading causes of approximately 1.8 million deaths annually in less developed countries, whereas in developed countries foodborne pathogens are responsible for millions of gastrointestinal diseases each year, costing huge amounts of money in terms of lost productivity and medical care. Again, new foodborne pathogens are likely to emerge forced by the use of antibiotics and the evolutionary drive of bacteria or fungi. *Clostridium*, *Shigella*, *Salmonella*, *Listeria monocytogenes*, *Campylobacter*, *S. aureus*, *Vibrio* spp., and *Yersinia enterocolitica* are among the most important ones, although many others cannot be excluded as human habits or manufacture practices change.

Studies on emerging diseases have been a real concern in many scientific journals through different reviews and have focused mainly on the putative spectrum of different emerging pathogens as well as the epidemiologic reasons for their emergence. The experts who have addressed this problem from only an epidemiologic point of view, but forgetting that microorganisms are subjected to evolutionary avatars, often disagree about the feasibility of predicting and preventing the emergence of new pathogens. As many enzybiotics are based on the use of entire bacteriophages that are also subjected to general biological evolution, it follows that one has to take into account these aspects in predicting not only what, but also where or even when a new pathogen will emerge. If we can use modern enzybiotics to either facilitate the identification and the blocking of pathogens that represent great threats to human kind, or to provide methods for inhibiting the emergence of microbial pathogens, then their use would represent a representative advance for science and for the well being of humans. It must be remembered, however, that if we assume that the evolution of virulence of a given pathogen is subjected to forces related to competitive benefits with the host, in such a way that it would tend to lower virulence toward mutualism, the application of too much stress on the pathogen, such as a mixture of antibiotics and enzybiotics, would unbalance the evolution rate of the pathogen, giving rise therefore to a more and more unstable relationship between pathogen and host. In the end, the balance would probably fall on the side of the microorganisms as their duplication time is far higher than that of plants or animals.

Nosocomial infections are always present. If we compare the infections caused by resistant nosocomial microorganisms those produced sensitive ones, the severity of the first are higher so we tend to naturally link this with the antibiotic-resistant phenotype and also to suggest that the increased severity is not simply a result of the antibiotic to exert its action on the bacterial cell, but that "something else" has happened to the microorganisms. We call this evolution power toward the genesis of new bacterial resistant strains. In terms of years of antibiotic usage by our society and duplication of bacterial times, it is possible that this has really happened.

Another point that should be addressed here is that many bacteriophages are able to lysogenize bacterial cells, and if they carry *tox* genes, the bacterial host will now be a toxigenic or even hypertoxigenic strain. As bacteriophages do recombine among themselves, in order to lower this possibility, and if the therapy is finally based on the use of bacteriophages, these must follow unequivocally a lytic cycle without the possibility of entering a lysogenic status. Also, the use of enzybiotics to artificially lower the load of microorganisms present in the animals' bodies is not necessarily good, since the acquired natural immunity would also be lowered, so small variations in the amount of microorganisms or in their virulence would cause infection outbreaks. The positive aspect of universally using enzybiotics would be to control emerging disease at the beginning of the spreading or even before the spreading starts. Without a doubt, enzybiotics could contribute to that and, as argued before, without the risk of generating resistant strains as antibiotics do.

3. GENETIC CROSS-TALK AMONG MICROORGANISMS

Bacteria may exhibit several ways of genetic cross-talk, including mating, transformation, or transduction; fungi belonging to *Ascomycotes* or *Basidiomycotes* show perfect sexual mating and in some cases may have processes resembling bacterial transformation and transduction. These genetic aspects must be taken into account when the generation of new resistant strains is contemplated. Transformation and possibly transduction among Gram-positive bacteria takes place constantly in all systems, probably since the very beginning of times, whereas in Gram-negative bacteria it probably takes place at lower rate compared with the former. In any case, because bacteria are one of the first motors of evolution in the microbial world, we must be ready for the early blocking of such new strains. In the case of fungi this cross-talk is probably not as important as in bacteria and very much depends on their sexual activities. Pre-Triassic fungi must have had more cross-talk than today's as that period saw the genetic stabilization of fungi.

The use of enzybiotics in combination with antibiotics, as indicated before, would overcome the eventual appearance of monogenic resistant strains originated by horizontal gene transmission.

Perhaps the most paradigmatic niche where the above ideas may apply is what Dinalo and Relman (2009) call "cross-talk in the gut," referring in particular to the human gut. The bulk of the bacterial load in the human gut (and in all mammals) is by far the highest as compared with the rest of the body, and the density of bacterial cells in the colon has been estimated up to 10^{11}–10^{12} cells/mL. Obviously, the relationship is mostly mutualistic. The biodiversity in the gut seems to be species-dependent, that is, all humans from any part of the globe share basically the same type of bacteria; the same generalization would apply, for instance, for elephants. With such a huge bacterial population in such a close environment genetic cross-talk between bacteria must occur at a high level and, what is more important, between bacteria and humans, not genetically but biochemically. The genesis of new mutant strains resistant to antibiotics or even to phages and therefore to enzybiotics in their most classical sense may have dramatic projections for the success of enzybiotic-based antibacterial therapy. In this sense application of entire and viable phages in the treatment of gut infections would probably not be the best idea; it would be better to use encapsulated lysins or perhaps "intelligent lysins" that would recognize the pathogen specifically. Application of live-based bacteriophages in other pathologies would not be subjected to this pressure because the pathogen would be the only bacterium.

4. TAYLOR-MADE AND HYBRID ENZYBIOTICS

Application of bacteriophage-based therapies must rely first on accurate bacterial diagnosis and correct identification. We obviously cannot supply a bacteriophage able to replicate in *S. aureus* when the pathogen is for instance Gram-negative bacteria. After correct bacterial diagnosis we must choose the type of bacteriophages; as far as we are concerned only lytic bacteriophages should be chosen, and within them those exhibiting the highest affinity constant between the virus envelope and bacterial receptor, and finally those virus with the shortest cycle possible. Probably—although we do not have enough data as yet—the type of nucleic acid of the virus may be important. In this sense single-stranded DNA or RNA viruses would be better candidates than more complicated ones such as T4 and related ones.

Hybrid enzybiotics, such as bacteriophages having their genomes coming from different phages or lysins formed by the fusion of different open reading frames (ORFs) so that they not only exhibit the ability to disrupt the bacterial peptidoglycan layer but also show protease activity or even act like an antibody to accelerate the complement fixation on the pathogen, will probably be a reality in the near future. Often bacterial or most currently phage endolysins show poor solubility so that their respective genes must be worked on in order to render a more appropriate lysin. Recently, Manoharadas et al. (2009) have overcome this situation for a P16 endolysin from *S. aureus* phage P68; they constructed a chimeric endolysin (P16–17) comprised of the inferred N-terminal d-alanyl-glycyl endopeptidase domain and the C-terminal cell wall targeting domain of the *S. aureus* phage P16 endolysin and the P17 minor coat protein, respectively. This resulted in soluble P16–17 protein, which exhibited antimicrobial activity toward *S. aureus*. In addition, P16–17 augmented the antimicrobial efficacy of the antibiotic gentamicin.

5. ANTIVIRAL ENZYBIOTICS

Virus-caused disease in mammals is a rather different situation, since we cannot use viruses to infect and destroy the pathogenic virus. Instead, the use of enzybiotics in these particular cases must focus only in the use of enzymes (proteases) able to digest the capsid proteins before the virus gets into the target cell. Alternatively, one could stimulate the cell's antiviral state through the use of interferons or inducers; these aspects, as of today, are not truly what we understand included in the term "enzybiotics." The use of enzybiotics in combination with antiviral drugs seems to have a better prediction of success. It has been known for many years now that viral infections such as influenza virus may facilitate the invasion of secondary pathogens such as *Streptococcus pneumoniae* or *Haemophilus influenzae*, and the same may apply for other type of viruses. It may be proposed that even in the absence of an appropriate antiviral drug a general mix of enzybiotics be applied in order to block the secondary bacterial colonizers.

A potential candidate to be included in the term "enzybiotics against animal virus" could be the ribotoxins, which are a large family of ribonucleolytic proteins secreted by fungi, mostly *Aspergillus* and *Penicillium* species (Lacadena et al. 2007). Recently, Varga and Samson (2008) have found that all species assigned to *Aspergillus* section Clavati carry ribotoxin genes, including *Aspergillus* longivesica, *Aspergillus* clavatonanicus, and *Neocarpenteles* acanthosporus. Comparative analysis of the amino acid sequences of these gene fragments indicates that ribotoxins could be produced by these species. All ribotoxins exert their toxic action by first entering the cells since they have the ability to interact with acid phospholipid-containing membranes and then cleave a unique phosphodiester bond located in the large rRNA gene, known as the sarcin-ricin loop. This cleavage leads to inhibition of protein biosynthesis, followed by cellular death by apoptosis. Ribotoxins (RNase T1 is probably the best candidate to be used as an enzybiotic) do not have, as of today, any known cellular receptor; instead, they tend to kill those cells that show increased membrane permeability. Also, ribotoxins have been used for the construction of immunotoxins because of their cytotoxicity. Ribotoxins have several advantages for use in the design of immunotoxins against cancer cells or eventually viral-infected cells. However, their cytotoxic effects hamper their use as therapeutic agents. Genetically engineered immunotoxins with increased stability and affinity are currently being developed (Lacadena et al. 2007). In addition, identifying new types of ribotoxins could facilitate their application as agents against viral infections. The determination of high-resolution structures of ribotoxins, the characterization of a number of mutants, and finally the use of liposomes suggest that ribotoxins or their variants might be used in the near future in human therapies, including viral-caused diseases (Carreras-Sangrà et al. 2008). Recently, a ribotoxin-like active principle from the mite fungal pathogen *Hirsutella thompsonii* has been isolated and studied (Herrero-Galán 2008). This ribotoxin, named hirsutellin A (HtA), is formed of a single polypeptide chain (130 amino acid residues) that has interesting insecticidal properties. It exhibits activity against tumor cells and probably also on virus-infected cells; it interacts with phospholipid bilayers and exhibits a typical sarcin/ricin-like mode of action on the ribosomes, as typical ribotoxins do. In any case, the activity shown by this type of compound is something that warrants further study in the field of enzybiotics before we can recruit them as a potential tool to fight viral infections. An interesting possibility in order to drastically produce ribotoxins with reduced IgE-binding affinity is the heterologous cloning of these genes in *Lacotoccus lactis*; this bacterium seems to secrete the respective wild-type as well as mutant variants that are quite innocuous when supplied to mice.

REFERENCES

Carreras-Sangrà N., E. Alvarez-García, E. Herrero-Galán, J. Tomé, J. Lacadena, J. Alegre-Cebollada, M. Oñaderra, J. G. Gavilanes, and A. Martínez del Pozo (2008) *Current Pharmaceutical Biotechnology* **9**, 153–160.

Herrero-Galán E., J. Lacadena, A. Martínez del Pozo, D. G. Boucias, N. Olmo, M. Oñaderra, and J. G. Gavilanes. (2008) *T. Proteins* **72**, 217–228.

Lacadena J., E. Alvarez-García, N. Carreras-Sangrà, E. Herrero-Galán, J. Alegre-Cebollada, L. García-Ortega, M. Oñaderra, J. G. Gavilanes, and A. Martínez del Pozo. *FEMS Microbiology Reviews* **31**, 212–237.

McCullers J. A., A. Karlström, A. R. Iverson, J. M. Loeffler, and V. A. Fischetti (2007) *PLoS Pathogens* **3** (3), e28.

Varga J. and R. A. Samson (2008) *Antonie Van Leeuwenhoek* **94**, 481–485.

INDEX

Enzybiotics: Antibiotic Enzymes as Drugs and Therapeutics. Edited by Tomas G. Villa
and Patricia Veiga-Crespo
Copyright © 2010 John Wiley & Sons, Inc.